AF356653

In Vitro Regeneration, Micropropagation and Germplasm Conservation of Horticultural Plants

In Vitro Regeneration, Micropropagation and Germplasm Conservation of Horticultural Plants

Editors

Jean Carlos Bettoni
Min-Rui Wang
Qiao-Chun Wang

Basel • Beijing • Wuhan • Barcelona • Belgrade • Novi Sad • Cluj • Manchester

Editors

Jean Carlos Bettoni
Independent Researcher
Videira
Brazil

Min-Rui Wang
Tropical Crops Genetic
Resources Institute, Chinese
Academy of Tropical
Agricultural Sciences
Danzhou
China

Qiao-Chun Wang
College of Horticulture,
Northwest A&F University
Yangling
China

Editorial Office
MDPI
St. Alban-Anlage 66
4052 Basel, Switzerland

This is a reprint of articles from the Special Issue published online in the open access journal *Horticulturae* (ISSN 2311-7524) (available at: https://www.mdpi.com/journal/horticulturae/special_issues/regeneration_micropropagation_conservation).

For citation purposes, cite each article independently as indicated on the article page online and as indicated below:

Lastname, A.A.; Lastname, B.B. Article Title. *Journal Name* **Year**, *Volume Number*, Page Range.

ISBN 978-3-7258-0065-0 (Hbk)
ISBN 978-3-7258-0066-7 (PDF)
doi.org/10.3390/books978-3-7258-0066-7

Cover image courtesy of Min-Rui Wang

Contents

About the Editors

Jean Carlos Bettoni

Dr. Jean Carlos Bettoni holds a bachelor's degree in Agronomy, a master's degree, and a Ph.D. in Plant Production from Santa Catarina State University—UDESC, Brazil, and a post-doctoral qualification from the USDA ARS-National Laboratory for Genetic Resources Preservation (NLGRP), Fort Collins, Colorado, USA, working on plant cryopreservation and virus and viroid eradication. The success of these projects has allowed the first efforts for the implementation of cryopreserved *Vitis* genebank collections to be initiated at USDA ARS-NLGRP in Fort Collins, Colorado, and resulted in the use of cryotherapy techniques in the apple rootstock breeding program in Geneva, New York, and the USDA-APHIS Quarantine program in Beltsville, Maryland, for pathogen eradication. His areas of expertise are tissue culture, conservation of plant genetic resources (medium- and long-term preservation methods), and virus/viroid eradication in horticultural species. He has been working in this field since 2013. From 2020-2023, he worked as a scientist at The New Zealand Institute for Plant and Food Research Limited in New Zealand. His research focused on the optimization of in vitro conservation for the management and conservation of plant genetic resources of various crop species (potato, apple, kiwifruit, hops, etc.) and native species using cryopreservation techniques and the development of solutions for delivering high-health germplasm. Dr. Jean Carlos Bettoni is interested in germplasm conservation using cryopreservation procedures and pathogen eradication in horticultural species.

Min-Rui Wang

Dr. Min-Rui Wang is currently working at the Tropical Crops Genetic Resources Institute (TCGRI), Chinese Academy of Tropical Agricultural Sciences (CATAS). He is now focusing on the establishment of cryopreservation/cryobanks for valuable tropical crops based on the National Genebank of Tropical Crops platform. He studied at Northwest A&F University from 2009 to 2019 and obtained his PhD through research on cryobiotechnology in apples. He subsequently held a post-doctoral position at NWAFU from 2020 to 2022. While at NWAFU, he also engaged in potato cryotherapy for pathogen eradication and contributed to a shallot research program during his time as a visiting Ph.D. student (2017-2019) at the Norwegian Institute of Bioecomomy Research (NIBIO), Norway. He has published two Springer book chapters and more than 20 articles in international journals and served as a guest editor for a special issue in Horticulturae. He is now particularly interested in overcoming challenges in the cryopreservation of cassavas and some endangered tropical species.

Qiao-Chun Wang

Dr. Qiao-Chun Wang is a full-time professor at the College of Horticulture, Northwest A&F University of China. He obtained his Ph.D. from the Department of Plant Protection of Hebrew University in Jerusalem, Israel, in 2003. From 2003 to 2008, he worked as a visiting professor in the Department of Applied Biology of the University of Helsinki of Finland. Since then, he returned to work in China. His research interests focus on plant cryopreservation, cryotherapy for pathogen eradication, responses of in vitro plants to abiotic and biotic stress, and in vitro plant regeneration. He has published more than 140 scientific articles and book chapters since 1988. He has served as associate editor and editorial board member in several SCI journals and now is managing editor in *Stress Biology*.

Editorial

In Vitro Regeneration, Micropropagation and Germplasm Conservation of Horticultural Plants

Jean Carlos Bettoni [1,*], Min-Rui Wang [2] and Qiao-Chun Wang [3]

[1]   Independent Researcher, 35 Brasil Correia Street, Videira 89560510, Santa Catarina, Brazil
[2]   National Gene Bank of Tropical Crops, Tropical Crops Genetic Resources Institute, Chinese Academy of Tropical Agricultural Sciences, Danzhou 571737, China
[3]   State Key Laboratory of Crop Stress Biology for Arid Region, College of Horticulture, Northwest A&F University, Xianyang 712100, China
*   Correspondence: jcbettoni@gmail.com

Citation: Bettoni, J.C.; Wang, M.-R.; Wang, Q.-C. In Vitro Regeneration, Micropropagation and Germplasm Conservation of Horticultural Plants. *Horticulturae* **2024**, *10*, 45. https://doi.org/10.3390/horticulturae10010045

Received: 19 December 2023
Accepted: 21 December 2023
Published: 2 January 2024

In vitro tissue culture technologies provide novel tools for improving plant production. Organogenesis and somatic embryogenesis are the two pathways for plant regeneration, and have been widely used for in vitro micropropagation and germplasm conservation of horticultural crops [1–5]. This Special Issue collects eleven publications, including nine research articles, one review and one essay article that address in vitro plant regeneration, micropropagation and germplasm conservation of horticultural crops. They have already attracted great interest as about 28,000 reads and more than 35 citations have been recorded.

In vitro tissue culture provides important approaches for both propagation and conservation of cultivated and wild species [6–10]. Working on a perennial herbaceous wild species, Basiri et al. [11] established an efficient micropropagation system for indirect shoot regeneration from root explants of Foxtail lily (*Eremurus spectabilis*). In vitro callus induction and indirect shoot regeneration were induced from root explants cultured on suitable culture media. The shoot development from callus was highly dependent on the saline formulation of the basal medium and the concentration of cytokinin. Regenerated plantlets were successfully rooted in vitro and re-established following the acclimatization process. The results of the present study are expected to contribute to in vitro propagation and ex situ conservation of this species. Furthermore, due to the medicinal properties of this wild species, this protocol has potential applications in the large-scale production of secondary metabolites under laboratory conditions.

In a similar study of indirect organogenesis, Tang et al. [12] successfully established a protocol for the propagation of *Agapanthus praecox* subsp. *orientalis* 'Big Blue'. A callus induction rate of 100% was achieved in root tips collected from tissue-cultured plants grown in a medium containing picloram, kinetin and naphthalene acetic acid (NAA). Adventitious shoots formed in the callus, and further developed into plantlets with roots in 90 days. About 93% of the plants were re-established after acclimatization. The authors found that the concentrations and types of plant growth regulators were crucial to enhance the process of callus and shoot regeneration. This study provided an effective tissue culture system for micropropagation of *A. praecox*, and would facilitate further practical applications for germplasm conservation and genetic improvement of this species.

Using flower buds as explants for micropropagation of *Rhododendron decorum*, Wu et al. [13] described a simple and efficient protocol for in vitro regeneration via indirect organogenesis. Effects of basal medium and plant growth regulators on the formation and proliferation of adventitious shoots and rooting were studied. The highest callus induction (95%) and shoot differentiation (91%) rates were achieved from explants grown on Wood Plant Medium (WPM) supplemented with thidiazuron and NAA. Shoots were successfully rooted in an auxin-enriched medium and more than 90% of plants survived acclimatization. The in vitro regeneration protocol optimized in this study has potential applications in the genetic improvements of this species.

Lee and Chang [14] reported an efficient micropropagation procedure for red-fleshed 'Da Hong' pitaya (*Hylocereus polyrhizus*). To efficiently reduce bacterial and fungal infections, a common problem in traditional vegetative propagation, in vitro cultures were initiated from disinfected pitaya seeds. Robust and healthy plantlets were produced within eight weeks and successfully transplanted into the field without any signs of pathogenic infection. 'Da Hong' pitaya plantlets grown on media supplemented with activated charcoal (AC) exhibited increased growth and development of plantlets. Shoots were efficiently micropropagated when cultured on a culture medium supplemented with 200 mg/L AC and 0.10 mg/L NAA. Although spontaneous rooting occurred during shoot development, root density was almost two-fold higher in plantlets cultured in a medium supplemented with 0.20 mg/L NAA. This protocol supported the production of healthy seedlings in self-pollinating pitaya and should be tested for additional accessions.

The use of liquid media in bioreactors has potential applications to maximize micropropagation and facilitate automation [15–17]. The study of Gago et al. [18] used commercial RITA© bioreactors for the micropropagation of two local plum varieties of *Prunus domestica* from the northwest of Spain. The authors investigated the effect of light intensity, supplementation of CO_2-enriched air and sucrose on the proliferation, rooting and acclimation of the shoots produced in bioreactors. They found that plum shoots cultured in bioreactors under high light intensity and CO_2 enrichment grew and proliferated with 1 and 3% sucrose, but shoot growth was poor when cultured on the medium without sucrose. Successful rooting and acclimation were achieved regardless of sucrose presence in the culture medium in bioreactors, but a lower proportion of rootable shoots occurred when shoots were multiplied without sucrose. Comparing micropropagation produced in bioreactors with that in jars containing semisolid medium, the authors demonstrated that shoot multiplication was much more efficient in bioreactors than in jars. Shoots of both plums cultured in jars or bioreactors with 3% sucrose were successfully rooted irrespective of the culture system. The results of this research provided a novel approach for the massive propagation of plum trees and may provide new perspectives for the propagation of other related plant species.

Also working on propagating plant material in liquid media by using temporal immersion systems, Pérez-Caselles et al. [19] developed an effective micropropagation protocol for the apricot cultivars 'Canino' and 'Mirlo Rojo'. The authors also investigated the effect of the application of silver nanoparticles (AgNPs) on the development of in vitro cultures and their penetration into plant tissue. The addition of AgNPs enhanced the overall plant growth of apricot cultivars. Moreover, the elimination of calcium chloride in the culture medium increased (23-fold) the AgNPs' penetration into the plant tissue without any detrimental effect on the micropropagation of apricot cultivars. Their focus on the increased silver intake of plant tissues will facilitate further investigation on the virucidal activity of AgNPs in plant disease management. Therefore, this study provided a basis for further applications of AgNPs against pathogens in tissue-cultured apricot plants in bioreactors.

In addition to mass propagation and the production of healthy plant materials, tissue culture technology facilitates the safe exchange of plant material within and across countries [20–24]. Li et al. [25] proposed an interesting in vitro incubation system for the long-distance shipping and exchange of plant germplasm based on slow growth in a vacuum-sealed microplate. Potato and ginger were used as model crops to optimize the protocol, which was later applied to sweet potato. The effects of light regime, temperature, iron concentration, plant growth retardants and package types on plant viability were assessed. Cultures were safely transported across thousands of kilometers within China without package or sample damage. Plantlets were recovered and genetic fidelity was confirmed. This protocol is valuable for the safe movement and distribution of tissue-cultured plant germplasm.

Genetic improvements in plant breeding are dependent upon the availability of and easy access to plant genetic resources [6,21,26,27]. In vitro culture technologies have been widely used to establish medium-term (in vitro conservation) and long-term (cryopreservation) preservation methods for the germplasms of horticultural plants [28–31]. Effective cryopreservation procedures have been identified for cryopreserving seeds, pollen, cell cultures, dormant buds and shoot tips [32–36]. Đorđević et al. [37] reported a successful conservation of plum pollen, in which pollen was harvested from flowers in the late balloon stage, adjusted to a moisture content between 6.1 and 6.8% (dry weight basis), and stored in darkness at 4, -20, -80 and $-196\,^{\circ}$C). Pollen that was stored in sub-zero temperatures continued to have stable viability after 12 months, while a temperature of $4\,^{\circ}$C was only suitable for short-term storage of up to 3 months in all tested genotypes. This study provided an easy and practical method to conserve plum pollen for up to one year but also a cheap alternative for short-term storage.

Aiming at the establishment of an efficient procedure for maintaining specific gene combinations of citrus and pineapple cultivars, Ozkaya et al. [38] and Villalobos-Olivera et al. [39] developed shoot tip cryopreservation methods using droplet-vitrification. Ozkaya et al. [38] focused on critical points pre- (pretreatment of donor plants, preculture and dehydration conditions) and post-freeze (recovery medium) in cryogenic procedures for enhancing shoot tip recovery of four citrus cultivars. They evaluated different strategies for improving the cryotolerance of shoot tips to vitrification solutions and investigated recovery media formulations to further increase post-cryopreservation recovery. Villalobos-Olivera et al. [39] investigated the morpho-anatomical and physiological characteristics of cryo-derived pineapple plants after acclimatization. Their study showed that acclimatized pineapple plantlets obtained from cryopreservation had comparable development to conventionally micropropagated and non-cryopreserved plants after 45 days of growth in the greenhouse. These efficient procedures provide valuable information on the use of droplet-vitrification cryopreservation for setting up cryobanks of citrus and pineapple plants.

Micrografting has been widely used to produce virus-free plants and for the formation of whole plants, particularly when shoots (scions) have difficulty forming adventitious roots [40–42]. In a comprehensive review, Wang et al. [43] addressed the application of micrografting to improved micropropagation of horticultural species in the 21st century and discussed factors affecting the success of micrografting. The practical aspects and applications of in vitro micrografting discussed in this review paper should attract the attention of readers and support basic and applied research, as well as the implementation of in vitro micrografting within tissue culture laboratories.

In conclusion, the papers collected in this Special Issue provide a representative and valuable collection of the applications of the in vitro tissue culture technologies used for horticultural species. The Special Issue also prospects for future studies on the application of developed technologies. We hope that the information described in this Special Issue will promote further research and practical implementation of biotechnologies for crop improvement and germplasm conservation.

Author Contributions: All authors listed have made a substantial, direct and intellectual contribution to the work, and approved it for publication. All authors have read and agreed to the published version of the manuscript.

Acknowledgments: The editors hereby acknowledge the invaluable contributions made by all the authors. The editors appreciate the contribution of all reviewers for their significant contributions to this Special Issue. They also wish to thank the managing editor, Allen Dou, for her active participation and valuable efforts.

Conflicts of Interest: The authors declare no conflicts of interest.

References

1. Teixeira da Silva, J.A.; Gulyás, A.; Magyar-Tábori, K.; Wang, M.-R.; Wang, Q.-C.; Dobránszki, J. In vitro tissue culture of apple and other *Malus* species: Recent advances and applications. *Planta* **2019**, *249*, 975–1006. [CrossRef] [PubMed]
2. Hasnain, A.; Naqvi, S.A.H.; Ayesha, S.I.; Khalid, F.; Ellahi, M.; Iqbal, S.; Hassan, M.Z.; Abbas, A.; Adamski, R.; Markowska, D.; et al. Plants in vitro propagation with its applications in food, pharmaceuticals and cosmetic industries; current scenario and future approaches. *Front. Plant Sci.* **2022**, *13*, 1009395. [CrossRef] [PubMed]

3. Pathirana, R.; Carimi, F. Studies on Improving the Efficiency of Somatic Embryogenesis in Grapevine (*Vitis vinifera* L.) and Optimising Ethyl Methanesulfonate Treatment for Mutation Induction. *Plants* **2023**, *12*, 4126. [CrossRef] [PubMed]

4. Xiong, Y.; Chen, S.; Wu, T.; Wu, K.; Li, Y.; Zhang, X.; Teixeira da Silva, J.; Zeng, S.; Ma, G. Shoot organogenesis and somatic embryogenesis from leaf and petiole explants of endangered *Euryodendron excelsum*. *Sci. Rep.* **2022**, *12*, 20506. [CrossRef] [PubMed]

5. Romadanova, N.V.; Mishustina, S.A.; Matakova, G.N.; Kushnarenko, S.V.; Rakhimbaev, I.R.; Reed, B.M. In vitro collection methods for *Malus* shoot cultures used for developing a cryogenic bank in Kazakhstan. *Acta Hortic.* **2016**, *1113*, 271–277. [CrossRef]

6. Panis, B.; Nagel, M.; Van den Houwe, I. Challenges and prospects for the conservation of crop genetic resources in field genebanks, in in vitro collections and/or in liquid nitrogen. *Plants* **2020**, *9*, 1634. [CrossRef] [PubMed]

7. Engelmann, F. Use of biotechnologies for the conservation of plant biodiversity. *In Vitro Cell. Dev. Biol.-Plant* **2011**, *47*, 5–16. [CrossRef]

8. Kulak, V.; Longboat, S.; Brunet, N.D.; Shukla, M.; Saxena, P. In Vitro Technology in Plant Conservation: Relevance to Biocultural Diversity. *Plants* **2022**, *11*, 503. [CrossRef]

9. Pence, V.C. Evaluating costs for the in vitro propagation and preservation of endangered plants. *In Vitro Cell. Dev. Biol.-Plant* **2011**, *47*, 176–187. [CrossRef]

10. Volk, G.M.; Carver, D.P.; Irish, B.M.; Marek, L.; Frances, A.; Greene, S.; Khoury, C.K.; Bamberg, J.; del Rio, A.; Warburton, M.L.; et al. Safeguarding plant genetic resources in the United States during global climate change. *Crop Sci.* **2023**, *63*, 2274–2296. [CrossRef]

11. Basiri, Y.; Etemadi, N.; Alizadeh, M.; Alizargar, J. In Vitro Culture of *Eremurus spectabilis* (Liliaceae), a Rare Ornamental and Medicinal Plant, through Root Explants. *Horticulturae* **2022**, *8*, 202. [CrossRef]

12. Tang, Q.; Guo, X.; Zhang, Y.; Li, Q.; Chen, G.; Sun, H.; Wang, W.; Shen, X. An Optimized Protocol for Indirect Organogenesis from Root Explants of *Agapanthus praecox* subsp. *orientalis* 'Big Blue'. *Horticulturae* **2022**, *8*, 715. [CrossRef]

13. Wu, H.; Ao, Q.; Li, H.; Long, F. Rapid and Efficient Regeneration of Rhododendron decorum from Flower Buds. *Horticulturae* **2023**, *9*, 264. [CrossRef]

14. Lee, Y.-C.; Chang, J.-C. Development of an Improved Micropropagation Protocol for Red-Fleshed Pitaya 'Da Hong' with and without Activated Charcoal and Plant Growth Regulator Combinations. *Horticulturae* **2022**, *8*, 104. [CrossRef]

15. Etienne, H.; Berthouly, M. Temporary immersion systems in plant micropropagation. *Plant Cell Tissue Organ Cult.* **2002**, *69*, 215–231. [CrossRef]

16. Ozyigit, I.I.; Dogan, I.; Hocaoglu-Ozyigit, A.; Yalcin, B.; Erdogan, A.; Yalcin, I.E.; Cabi, E.; Kaya, Y. Production of secondary metabolites using tissue culture-based biotechnological applications. *Front. Plant Sci.* **2023**, *14*, 1132555. [CrossRef]

17. Vidal, N.; Sánchez, C. Use of bioreactor systems in the propagation of forest trees. *Eng. Life Sci.* **2019**, *19*, 896–915. [CrossRef]

18. Gago, D.; Sánchez, C.; Aldrey, A.; Christie, C.B.; Bernal, M.Á.; Vidal, N. Micropropagation of Plum (*Prunus domestica* L.) in Bioreactors Using Photomixotrophic and Photoautotrophic Conditions. *Horticulturae* **2022**, *8*, 286. [CrossRef]

19. Pérez-Caselles, C.; Alburquerque, N.; Faize, L.; Bogdanchikova, N.; García-Ramos, J.C.; Rodríguez-Hernández, A.G.; Pestryakov, A.; Burgos, L. How to Get More Silver? Culture Media Adjustment Targeting Surge of Silver Nanoparticle Penetration in Apricot Tissue during in Vitro Micropropagation. *Horticulturae* **2022**, *8*, 855. [CrossRef]

20. Wang, M.-R.; Cui, Z.-H.; Li, J.-W.; Hao, X.-Y.; Zhao, L.; Wang, Q.-C. In vitro thermotherapy-based methods for plant virus eradication. *Plant Methods* **2018**, *14*, 87. [CrossRef]

21. Bettoni, J.C.; Marković, Z.; Bi, W.; Volk, G.M.; Matsumoto, T.; Wang, Q.-C. Grapevine shoot tip cryopreservation and cryotherapy: Secure storage of disease-free plants. *Plants* **2021**, *10*, 2190. [CrossRef]

22. Wang, M.-R.; Bi, W.-L.; Bettoni, J.C.; Zhang, D.; Volk, G.M.; Wang, Q.-C. Shoot tip cryotherapy for plant pathogen eradication. *Plant Pathol.* **2022**, *71*, 1241–1254. [CrossRef]

23. Kumar, P.L.; Cuervo, M.; Kreuze, J.F.; Muller, G.; Kulkarni, G.; Kumari, S.G.; Massart, S.; Mezzalama, M.; Alakonya, A.; Muchugi, A.; et al. Phytosanitary Interventions for Safe Global Germplasm Exchange and the Prevention of Transboundary Pest Spread: The Role of CGIAR Germplasm Health Units. *Plants* **2021**, *10*, 328. [CrossRef]

24. Reed, B.M.; Engelmann, F.; Dulloo, M.E.; Engels, J.M.M. *Technical Guidelines for the Management of Field and In Vitro Germplasm Collections*; IPGRI handbooks for genebanks no. 7; International Plant Genetic Resources Institute: Rome, Italy, 2004; 106p.

25. Li, J.; He, M.; Xu, X.; Huang, T.; Tian, H.; Zhang, W. In Vitro Techniques for Shipping of Micropropagated Plant Materials. *Horticulturae* **2022**, *8*, 609. [CrossRef]

26. Byrne, P.F.; Volk, G.M.; Gardner, C.; Gore, M.A.; Simon, P.W.; Smith, S. Sustaining the Future of Plant Breeding: The Critical Role of the USDA-ARS National Plant Germplasm System. *Crop Sci.* **2018**, *58*, 451–468. [CrossRef]

27. Swarup, S.; Cargill, E.J.; Crosby, K.; Flagel, L.; Kniskern, J.; Glenn, K.C. Genetic diversity is indispensable for plant breeding to improve crops. *Crop Sci.* **2020**, *61*, 839–852. [CrossRef]

28. Bettoni, J.C.; Bonnart, R.; Volk, G.M. Challenges in implementing plant shoot tip cryopreservation technologies. *Plant Cell Tissue Organ Cult.* **2021**, *144*, 21–34. [CrossRef]

29. Panis, B. Sixty years of plant cryopreservation: From freezing hardy mulberry twigs to establishing reference crop collections for future generations. *Acta Hortic.* **2019**, *1234*, 1–8. [CrossRef]

30. Matsumoto, T. Cryopreservation of plant genetic resources: Conventional and new methods. *Rev. Agric. Sci.* **2017**, *5*, 13–20. [CrossRef]
31. Zhang, A.-L.; Wang, M.-R.; Li, Z.; Panis, B.; Bettoni, J.C.; Vollmer, R.; Xu, L.; Wang, Q.-C. Overcoming Challenges for Shoot Tip Cryopreservation of Root and Tuber Crops. *Agronomy* **2023**, *13*, 219. [CrossRef]
32. Wang, M.-R.; Chen, L.; Teixeira da Silva, J.A.; Volk, G.M.; Wang, Q.-C. Cryobiotechnology of apples (*Malus* spp.): Recent and future prospects. *Plant Cell Rep.* **2018**, *37*, 689–709. [CrossRef]
33. Panis, B.; Piette, B.; Swennen, R. Droplet vitrification of apical meristems: A cryopreservation protocol applicable to all Musaceae. *Plant Sci.* **2005**, *166*, 45–55. [CrossRef]
34. Bettoni, J.C.; Kretzschmar, A.A.; Bonnart, R.; Shepherd, A.; Volk, G.M. Cryopreservation of 12 Vitis species using apical shoot tips derived from plants grown in vitro. *HortScience* **2019**, *54*, 976–981. [CrossRef]
35. Vollmer, R.; Villagaray, R.; Castro, M.; Cárdenas, J.; Pineda, S.; Espirilla, J.; Anglin, N.; Ellis, D.; Azevedo, V. The world's largest potato cryobank at the International Potato Center (CIP)—Status quo, protocol improvement through large-scale experiments and long-term viability monitoring. *Front. Plant Sci.* **2022**, *13*, 1059817. [CrossRef] [PubMed]
36. Volk, G.M.; Bonnart, R.; Shepherd, A.; Yin, Z.; Lee, R.; Polek, M.; Krueger, R. Citrus cryopreservation: Viability of diverse taxa and histological observations. *Plant Cell Tissue Organ Cult.* **2017**, *128*, 327–334. [CrossRef]
37. Đorđević, M.; Vujović, T.; Cerović, R.; Glišić, I.; Milošević, N.; Marić, S.; Radičević, S.; Fotirić Akšić, M.; Meland, M. In Vitro and In Vivo Performance of Plum (*Prunus domestica* L.) Pollen from the Anthers Stored at Distinct Temperatures for Different Periods. *Horticulturae* **2022**, *8*, 616. [CrossRef]
38. Ozkaya, D.E.; Souza, F.V.D.; Kaya, E. Evaluation of Critical Points for Effective Cryopreservation of Four Different *Citrus* spp. Germplasm. *Horticulturae* **2022**, *8*, 995. [CrossRef]
39. Villalobos-Olivera, A.; Lorenzo-Feijoo, J.C.; QuintanaBernabé, N.; Leiva-Mora, M.; Bettoni, J.C.; Martínez-Montero, M.E. Morpho-Anatomical and Physiological Assessments of Cryo-Derived Pineapple Plants (*Ananas comosus* var. *comosus*) after Acclimatization. *Horticulturae* **2023**, *9*, 841. [CrossRef]
40. Chilukamarri, L.; Ashrafzadeh, S.; Leung, D.W.M. In-vitro grafting—Current applications and future prospects. *Sci. Hortic.* **2021**, *280*, 109899. [CrossRef]
41. Wang, M.-R.; Bi, W.-L.; Ren, L.; Zhang, A.-L.; Ma, X.-Y.; Zhang, D.; Volk, G.M.; Wang, Q.-C. Micrografting: An Old Dog Plays New Tricks in Obligate Plant Pathogens. *Plant Dis.* **2022**, *106*, 2545–2557. [CrossRef]
42. Faustino, A.; Pires, R.C.; Caeiro, S.; Rosa, A.; Marreiros, A.; Canhoto, J.; Correia, S.; Marum, L. Micrografting in almond (*Prunus dulcis*) Portuguese cultivars for production of disease-free plants. *Acta Hortic.* **2023**, *1359*, 165–172. [CrossRef]
43. Wang, M.-R.; Bettoni, J.C.; Zhang, A.-L.; Lu, X.; Zhang, D.; Wang, Q.-C. In Vitro Micrografting of Horticultural Plants: Method Development and the Use for Micropropagation. *Horticulturae* **2022**, *8*, 576. [CrossRef]

Article

In Vitro Culture of *Eremurus spectabilis* (Liliaceae), a Rare Ornamental and Medicinal Plant, through Root Explants

Yeganeh Basiri [1], Nematollah Etemadi [1,*], Mahdi Alizadeh [2] and Javad Alizargar [3,4,*]

[1] Department of Horticulture, College of Agriculture, Isfahan University of Technology, Isfahan 8415683111, Iran; hybasiri@gmail.com

[2] Department of Horticulture, Faculty of Plant Production, Gorgan University of Agricultural Sciences and Natural Resources, Gorgan 4918943464, Iran; mahdializadeh@gau.ac.ir

[3] Research Center for Healthcare Industry Innovation, National Taipei University of Nursing and Health Sciences, Taipei 112, Taiwan

[4] School of Nursing, National Taipei University of Nursing and Health Sciences, Taipei 112, Taiwan

* Correspondence: etemadin@cc.iut.ac.ir (N.E.); 8javad@ntunhs.edu.tw (J.A.)

Abstract: *Eremurus spectabilis* M. Bieb, a perennial herbaceous wild species, is commonly used in the horticultural, ornamental, and pharmaceutical markets. Studies on the tissue culture systems for this species would be beneficial for mass multiplication as well as for future breeding programs. An in vitro propagation technique was established here using tuberous root explants as unique and responsive starting materials for culture initiation. The proliferated calli were sub-cultured on shoot proliferation media and regenerated microshoots were assessed. The shoot proliferation rate, leaf number, leaf length, and chlorophyll and carotenoid contents were recorded. The highest callus induction per explant (76.67%), callus dry weight (10.25 mg), callus firmness ratio (3.97), and callus color intensity ratio (2.83) were observed in explants inoculated on Murashige and Skoog (MS) medium supplemented with 10.0 mgL^{-1} 6-benzylaminopurine (BAP). The highest shoot proliferation rates were obtained when calli were sub-cultured on MS or Schenk and Hildebrandt (SH) basal media supplemented with 2.0 mgL^{-1} BAP. The half-strength MS medium fortified with 4.0% sucrose + 2.0 mgL^{-1} indole butyric acid (IBA) + 200 mgL^{-1} activated charcoal was a superior combination for root emergence and rooting parameters. Regenerated plantlets were then successfully adapted to ex vitro conditions. The reported protocol can be exploited at a commercial scale following minor modification, or could be beneficial in the production of secondary metabolites in bioreactors where callus is required as fresh plant material.

Keywords: callus induction; foxtail lily; micropropagation; proliferation rate; acclimatization

Citation: Basiri, Y.; Etemadi, N.; Alizadeh, M.; Alizargar, J. In Vitro Culture of *Eremurus spectabilis* (Liliaceae), a Rare Ornamental and Medicinal Plant, through Root Explants. *Horticulturae* **2022**, *8*, 202. https://doi.org/10.3390/horticulturae8030202

Academic Editors: Jean Carlos Bettoni, Min-Rui Wang, Qiao-Chun Wang and Barbara Ruffoni

Received: 23 January 2022
Accepted: 22 February 2022
Published: 25 February 2022

Publisher's Note: MDPI stays neutral with regard to jurisdictional claims in published maps and institutional affiliations.

1. Introduction

The foxtail lily (*Eremurus spectabilis* M. Bieb.) is a perennial and herbaceous wild plant species of the Liliaceae family (Figure 1). This rare species grows well on dry and rocky slopes and is most widespread in Southern and Central Asia, including Iran, western Pakistan, Turkey, Palestine, Lebanon, Syria, and Caucasus [1]. This plant is rich source of antioxidants, phenolic compounds, and minerals that is consumed as a vegetable in some countries, such as Turkey. Foxtail lily also has medicinal importance and is utilized to prepare a special type of plant-derived adhesive [2,3]. The leaves of this plant are frequently used against eczema, fungal infections, and diabetes [4], and recently the phytochemical composition, antioxidant, and antimicrobial capacities of *E. spectabilis* were examined in vitro [5]. Antioxidant, antimicrobial, and anticancer effects of different extracts from wild edible plant *E. spectabilis* leaves and roots have also already been reported by some Turkish scientists [3], where this plant is a wild vegetable, growing in spring in the east of Turkey.

Figure 1. Foxtail lily (*Eremurus spectabilis*) plant species under its natural habitat—Golestan National Park, Golestan province, Northern Iran (**right**); a close up view of inflorescence during full bloom ((**left**); Bar = 10 cm; Photo: Y. Basiri: April 2019).

Different parts of the plants are used to treat fungal diseases, diabetes, hepatitis, liver and stomach disorders, and some cancer types [5–8]. After drying and grinding, the roots of *E. spectabilis* have a highly useful application in adhesives [9]. In addition to all these beneficial nutritional, medicinal, and industrial properties, this plant species has a high ornamental value due to its long spike-type inflorescences and application as a cut flower [10,11].

In vitro mass micropropagation techniques have been considered a promising procedure for the multiplication of valuable, hard-to-propagate and endangered wild species [12]. Although in vitro regeneration studies have been performed on many genera from the Liliaceae family, such as *Aloe, Lilium, Chlorophytum, Feritillaria*, and *Scilla* [13–20], only a few studies have reported on the in vitro regeneration of *E. spectabilis* worldwide [21,22]. In a recent article [21], the leaf and rhizome explants of *E. spectabilis* were cultured on Murashige and Skoog (MS) media [23], supplemented with 6-benzylaminopurine (BAP) and 2, 4-dichlorophenoxyacetic acid (2, 4-D). However, the results showed no in vitro shoot regeneration regardless of explant types. For instance, Tuncer [22] cultured hypocotyls of 35–40-day-old in vitro-germinated seedlings on MS medium containing 2,4-D + kinetin, thidiazuron + NAA, and BAP + 2,4-D to stimulate bulblet and/or shoot regeneration. As such, the optimization of micropropagation methods and the establishment of proper protocols are required for such species. In the literature, it has been recommended that the micropropagation of some medicinal, endangered, wild, and endemic species must be performed through tissue culture techniques, and that consistent protocols must be optimized [21]. Therefore, such protocols may be utilized for mass multiplication as well as the genetic improvement of these valuable crops. According to such a statement, the importance of the in vitro propagation of Foxtail lily is more obvious. Consequently, the present study aimed to examine the effects of the disinfection method, root explant responses, and exogenous growth regulators on the in vitro propagation of Foxtail lily. This study investigates the possibility of the in vitro mass micropropagation of *E. spectabilis* based on the morphological and physiological responses of root explants. The utilization of root explants itself may be considered an innovation in the tissue culture of this species, as it is botanically a monocot plant and regeneration from leaf explants may not be considered a successful strategy. Moreover, the reported protocol would be significant for being the first in vitro optimization of micropropagation of this endemic species in northern Iran.

2. Materials and Methods

2.1. Plant Materials

The fleshy roots of *E. spectabilis* Clade II [24] were collected from its natural habitat (Figure 1), i.e., Golestan National Park, Golestan province, Iran (UTM zone: 37°32′49.8″ (37.5472°) N and 56°23′22.4″ (56.3896°) E), during the start of the dormant season. The Foxtail lily has tuberous roots. Each root clump was individually planted in a 5 L plastic pot. A blend of coco-peat: perlite (1:1) was used as a potting mixture. The pots were regularly irrigated using normal tap water with 7-day intervals. The pots were also irrigated with Hoagland solution (100 mL per pot) with 14-day intervals. The new, fresh leaves emerged in the spring season, three months after planting. The tuberous root segments were then used for in vitro culture initiation.

2.2. Explant Disinfection

In vitro cultures were initiated using root segments (about 1.0 to 1.5 cm long) obtained from root clumps (Figure 2) in March. Initially, the whole root clamp was kept in water for 3 h to soften and remove each and every soil particle; it was then thoroughly washed with tap water. The small root portions were dissected and were thoroughly washed with tap water supplemented with a few drops of Tween-80 for 30 min and were then agitated in a fungicide solution (4 gL^{-1} of carbendazim) for 1 h on a horizontal shaker. The root explants were rinsed with 70% alcohol (Ethanol) for 40 s prior to surface sterilization using either commercial bleach (NaOCl, 40% for 15 min) or HgCl$_2$ (0.1% for 7 min), followed by at least three-times rinsing with autoclaved double distilled water. Explants were then inoculated into 100 mL flasks containing culture initiation medium.

Figure 2. Tuberous root clump of Foxtail lily (*Eremurus spectabilis*) just before explant preparation (**left**); root explants inoculated for callus initiation ((**right**); Bar = 10 mm; Photo: Y. Basiri: May 2019).

2.3. Culture Initiation and Shoot Regeneration

The MS medium [23] and Schenk and Hildebrandt (SH) medium [25] were used as basal media for culture initiation and callus induction. A combination of NAA (Sigma-Aldrich, St. Louis, MO, USA; 0.2 mgL^{-1}) with either 5.0 or 10 mgL^{-1} BAP in both basal media was evaluated for callus induction (Table 1). The ability of the explants to develop calli was recorded two months after inoculation.

The callus' fresh and dry (60 °C for 24 h) weights were recorded. The firmness of the callus tissue was also scored from 1 to 4, where 1 = very soft, 2 = soft, 3 = firm, and 4 = very firm. The callus color intensity was also rated from 1 to 3, where 1 = pale, 2 = medium color intensity, and 3 = dark color.

The in vitro-grown produced plant pigments such as chlorophyll a and b and carotenoids were extracted with dimethyl sulphoxide (DMSO) and their concentrations were estimated using fresh callus samples according to the method described by Barnes et al. [26], which is a spectrophotometric procedure.

Table 1. Media compositions for callus initiation and shoot regeneration of Foxtail lily (*Eremurus spectabilis*).

Media	Basal Medium	Growth Regulators (mgL^{-1})	Other Modifications
Culture initiation (callus induction)	MS	NAA (0.1) + BAP (5.0)	-
		NAA (0.1) + BAP (10)	-
	SH	NAA (0.1) + BAP (5.0)	-
		NAA (0.1) + BAP (10)	-
Shoot regeneration	MS	IBA (0.1) + BAP (2.0)	AC [1] (200 mgL^{-1})
		IBA (0.1) + BAP (5.0)	AC (200 mgL^{-1})
	SH	IBA (0.1) + BAP (2.0)	AC (200 mgL^{-1})
		IBA (0.1) + BAP (5.0)	AC (200 mgL^{-1})
Root regeneration	MS	IBA (2.0) + NAA (1.0)	-
	SH	IBA (2.0) + NAA (1.0)	-
	$\frac{1}{2}$ MS	IBA (2.0)	4.0% sucrose + 200 mgL^{-1} AC
	$\frac{1}{2}$ SH	IBA (2.0)	4.0% sucrose + 200 mgL^{-1} AC

[1] Activated charcoal.

The proliferated calli were sub-cultured on shoot proliferation medium. The MS and SH basal media were supplemented with indole butyric acid (IBA; 0.1 mgL^{-1}) combined with either 2.0 or 5.0 mgL^{-1} BAP (Table 1). Two types of callus mass, i.e., whole uncut (intact) calli and divided callus explants, were evaluated. To obtain a sufficient biomass, the regenerated shoots were sub-cultured for at least three cycles on the same regenerated basal media. A mild concentration of activated charcoal (AC, 200 mgL^{-1}) was supplemented to the shoot proliferation media. Two months after inoculation, the ability of the callus tissues to regenerate shoots was assessed, and the shoot proliferation rate, leaf number, leaf length, and chlorophyll and carotenoid contents were recorded following the same procedure already mentioned for the estimation of pigments in callus tissues [26]. The growth chamber had controlled photoperiod (16/8 h) and temperature (26 ± 1 °C) conditions. It was equipped with cool-white fluorescent lights (227 µmol·m^{-2}·s^{-1}) 40 cm above the vessel cultures.

2.4. In Vitro Rooting

The regenerated shoots from the calli were sub-cultured on rooting media (Table 1). Both full and half-strength basal media were evaluated for root induction. The level of sucrose in the half-strength media was increased to 40 gL^{-1} (4.0%).

All media cultures utilized in the present experiment were solidified with 8.0 gL^{-1} agar and supplemented with 30 gL^{-1} sucrose. The pH was adjusted to 5.8 prior to autoclaving (1.05 kg·cm^{-2} for 15 min). The growth conditions in the rooting stage were exactly the same as the regeneration stage.

2.5. Acclimatization and Ex Vitro Conditions

The rooted plantlets were subjected to hardening and ex vitro transfer. The grapevine hardening strategy already reported by Alizadeh et al. [27] was evaluated for acclimatization of Foxtail lily in vitro-derived plantlets. The rooted plantlets (21-days-old) were transferred to glass jars with a polypropylene (PP) cap containing perlite:coco peat:vermiculite (2:1:1; volume). The hardening media was pre-soaked with either 1/2 MS or 1/2 SH solution (the media was devoid of vitamins and sucrose—only the mineral parts including macro/micro salts, calcium and iron-EDTA were taken). The pre-soaked substrate was subjected to autoclaving prior to plantlet transfer. A total of 10 replicates were used per treatment. The glass jars were then shifted to cool-white fluorescent lights (227 µmol·m^{-2}·s^{-1}) with controlled photoperiod (16/8 h) and temperature (26 ± 1 °C) conditions. After three weeks, the polypropylene caps were loosened gradually, and finally, removed completely. At regular intervals, the growing plantlets were then misted with sterile distilled water containing 0.1% carbendazim (*w/v*) fungicide. Hardened plantlets were transferred to ex vitro glasshouse conditions at the 6 to 7th weeks of acclimation. Finally, at the end of this period, some growth characteristics such as root length, leaf number, leaf length, and biomass volume were recorded.

2.6. Statistical Analysis

All experiments were arranged in a completely randomized design, and the data were subjected to analysis of variance using SAS software (version 9.4, SAS Institute, Cary, NC, USA). Comparisons of means were tested using the least significant difference (LSD) test, at a 0.05 level of probability (*p*).

3. Results

3.1. Callus Induction

The analysis of variance of the in vitro callus induction data is presented in Table 2. The callus proliferation from the fleshy root explants of the Foxtail lily was also depicted in Figure 3A. The highest callus induction frequency per explant (76.67%), callus dry weight (10.25 g), concentration of chlorophyll a (10.23 $\mathrm{mg \cdot g^{-1}}$ FW), chlorophyll b (19.81 $\mathrm{mg \cdot g^{-1}}$ FW), and carotenoids (6.50 $\mathrm{mg \cdot g^{-1}}$ FW), callus firmness ratio (3.97), and callus color intensity ratio (2.83) were found in explants grown in MS medium containing 10 $\mathrm{mgL^{-1}}$ BAP and 0.1 $\mathrm{mgL^{-1}}$ NAA (Table 2).

Table 2. Effects of basal medium, BAP levels and sterilization method on callus induction rate (C, %), callus dry weight (DW, mg), chlorophyll a (Chl a, $\mathrm{mg \cdot g^{-1}}$ FW), chlorophyll b (Chl b $\mathrm{mg \cdot g^{-1}}$ FW), chlorophyll a + b (Chl b $\mathrm{mg \cdot g^{-1}}$ FW), carotenoids (Ctn, $\mathrm{mg \cdot g^{-1}}$ FW), callus firmness (Cf), color intensity rate (CI), and callus color (CC) of Foxtail lily (*Eremurus spectabilis*).

Medium	Growth Regulators (mgL^{-1})	Sterilization Method	C	DW	Chl a	Chl b	Ctn	Cf	CI	CC
MS	NAA (0.1) + BAP (5.0)	NaClO, 40% for 15 min	36.67 [e]	6.21 [bc]	4.06 [c]	6.01 [b]	1.53 [c]	3.33 [bc]	1.25 [bc]	YG
		HgCl2, 0.1% for 7 min	25.00 [f]	5.19 [cd]	0.16 [e]	0.33 [c]	0.55 [d]	2.00 [f]	1.00 [c]	B
	NAA (0.1) + BAP (10)	NaClO, 40% for 15 min	76.67 [a]	10.25 [a]	10.23 [a]	19.81 [a]	6.50 [a]	3.97 [a]	2.83 [a]	G
		HgCl2, 0.1% for 7 min	55.00 [b]	8.16 [ab]	3.18 [d]	10.07 [b]	3.50 [b]	3.00 [cd]	1.25 [bc]	YG
SH	NAA (0.1) + BAP (5.0)	NaClO, 40% for 15 min	56.67 [b]	6.03 [bc]	4.82 [bc]	10.29 [b]	2.84 [b]	3.67 [ab]	1.33 [b]	YG
		HgCl2, 0.1% for 7 min	46.67 [cd]	4.15 [cd]	0.27 [e]	0.60 [c]	0.20 [d]	2.67 [de]	1.17 [bc]	B
	NAA (0.1) + BAP (10)	NaClO, 40% for 15 min	50.00 [bc]	5.04 [cd]	5.35 [b]	9.91 [b]	3.07 [b]	3.33 [bc]	1.33 [b]	YG
		HgCl2, 0.1% for 7 min	40.00 [de]	3.28 [d]	0.36 [e]	0.84 [c]	0.49 [d]	2.33 [ef]	1.00 [c]	B
	Source of variation									
	Medium (M)		25.00 [ns]	95.99 **	35.02 **	159.03 **	22.49 **	0.19 [ns]	1.69 **	
	Growth regulator (G)		2408.33 **	19.81 *	72.28 **	410.96 **	53.30 **	1.02 *	2.08 **	
	Sterilization method (S)		2133.33 **	34.24 **	315.09 **	875.43 **	63.43 **	15.19 **	4.08 **	
	M × G		5208.33 **	58.92 **	55.04 **	420.50 **	41.25 **	4.69 **	3.00 **	
	M × S		133.33 *	0.21 [ns]	1.48 [ns]	8.35 [ns]	1.17 *	0.19 [ns]	1.33 **	
	G × S		75.00 [ns]	0.66 [ns]	9.74 **	8.88 [ns]	2.83 **	0.02 [ns]	1.69 **	
	M × G × S		75.00 [ns]	1.10 [ns]	9.91 **	16.37 [ns]	3.23 **	0.02 [ns]	1.02 **	
	Error		45.83	3.33	0.55	15.49	0.33	0.19	0.06	

Means followed by the same letter are not significantly different by the LSD multiple range test at $p \leq 0.05$. YG, Yellowish green; B, Black; G, Green. * and **: Significant at 5% and 1% probability levels, respectively; ns: Non-significant.

Figure 3. (**A**): Callus regenerated from Foxtail lily (*Eremurus spectabilis*) root explants inoculated on MS medium containing 10 mgL^{-1} BAP. (**B**): Shoot regeneration from callus on MS medium containing 2.0 mgL^{-1} BAP and 0.1 mgL^{-1} IBA, two months after inoculation (Bar = 10 mm; Photo: Y. Basiri: June 2019).

3.2. Shoot Regeneration

The ANOVA table for the shoot regeneration data is shown in Table 3. According to this data, the culture medium and growth regulator had no significant effect on the number of leaves in regenerated shoots. The best proliferation rate was obtained from callus explants inoculated in either MS or SH media containing 2.0 mgL^{-1} BAP and 0.1 mgL^{-1} IBA (Figure 3B). The application of higher concentrations of BAP (5.0 mgL^{-1}) in both basal media significantly reduced the shoot proliferation rate (Table 3). The highest number of leaves was obtained from uncut, intact explants on MS medium containing 5.0 mgL^{-1} BAP and 0.1 mgL^{-1} IBA, while the highest leaf length was observed with the same explant type and medium containing 2.0 mgL^{-1} BAP. The lowest number of leaves and the smallest leaves were observed in the cultures regenerated from calli sub-cultured on the SH medium containing 5.0 mgL^{-1}. The highest concentration of chlorophyll a, b, and carotenoids were recorded with the intact callus tissues sub-cultured on MS medium containing 2.0 mgL^{-1} BAP and 0.1 mgL^{-1} IBA. Conversely, shoots regenerated from divided callus tissues on SH medium containing either 2.0 or 5.0 mgL^{-1} BAP had the lowest levels of pigments (Table 3).

Table 3. Effects of basal medium, BAP levels and explant type on shoot proliferation (SP, %), leaf number (Ln), leaf length (Ll, cm), chlorophyll a (Chl a, mg·g^{-1} FW), chlorophyll b (Chl b mg·g^{-1} FW), and carotenoids (Ctn, mg·g^{-1} FW) of Foxtail lily (*Eremurus spectabilis*).

Medium	Growth Regulator	Callus Type	SP	Ln	Ll	Chl a	Chl b	Ctn
MS	IBA (0.1) + BAP (2.0)	Intact	6.33 [a]	39.17 [abc]	20.45 [a]	9.96 [a]	18.78 [a]	6.12 [a]
		Divided	5.17 [ab]	41.83 [ab]	16.42 [ab]	9.71 [ab]	16.82 [ab]	6.36 [a]
	IBA (0.1) + BAP (5.0)	Intact	3.67 [bcd]	47.17 [a]	5.83 [c]	7.45 [c]	13.64 [bc]	4.2 [b]
		Divided	4.17 [bc]	33.33 [bc]	6.28 [c]	8.33 [bc]	13.32 [bc]	4.53 [b]
SH	IBA (0.1) + BAP (2.0)	Intact	4.33 [ab]	36.67 [abc]	16.45 [ab]	7.47 [c]	14.34 [bc]	4.63 [b]
		Divided	3.17 [bcd]	37.33 [abc]	12.42 [b]	7.09 [cd]	12.75 [dc]	4.87 [b]
	IBA (0.1) + BAP (5.0)	Intact	1.83 [d]	42.17 [ab]	2.17 [c]	4.96 [e]	9.19 [d]	2.71 [c]
		Divided	2.16 [cd]	28.33 [c]	2.78 [c]	5.81 [de]	9.21 [d]	3.04 [c]
Source of variation								
Medium (M)			46.02 **	216.75 [ns]	172.52 **	76.68 **	218.62 **	26.65 **
Growth regulator (G)			38.52 **	12.00 [ns]	1776.33 **	44.29 **	224.98 **	71.53 **
Callus type (C)			1.69 [ns]	444.08 *	36.75 [ns]	0.90 [ns]	11.08 [ns]	0.99 [ns]
M × G			0.02 [ns]	6.75 [ns]	0.52 [ns]	0.008 [ns]	0.003 [ns]	0.00 [ns]
M × C			0.02 [ns]	3.00 [ns]	0.02 [ns]	0.02 [ns]	0.38 [ns]	0.00 [ns]
G × C			7.52 [ns]	720.75 *	62.56 *	4.11 [ns]	7.92 [ns]	0.02 [ns]
M × G × C			0.02 [ns]	3.00 [ns]	0.02 [ns]	0.008 [ns]	0.002 [ns]	0.00 [ns]
Error			3.35	128.87	13.49	1.92	9.42	0.59

Means followed by the same letter are not significantly different, as determined by the LSD multiple range test at $p \leq 0.05$. * and **: Significant at the 5% and 1% probability levels, respectively; ns: Non-Significant.

3.3. Root Regeneration

The ANOVA table for the root parameters regenerated by in vitro microshoots (Table 4) revealed that the effect of the culture medium on all recorded parameters such as root length was statistically significant (Table 4). Figure 4 shows the emergence of roots from the basal part of shoot clumps. The best combination of basal medium and growth regulator for root length, number of leaves, and amount of chlorophyll *a* and *b* was found to be 1/2 MS + 2.0 mgL^{-1} IBA + 4.0% sucrose + 200 mgL^{-1} activated charcoal. MS + 2.0 mgL^{-1} IBA + 1.0 mgL^{-1} NAA also produced the highest chlorophyll *a* and *b* and carotenoid levels (Table 4).

Table 4. Effects of medium on root length (Rl, cm), biomass volume (Bv), leaf number (Ln), leaf length (Ll, cm), chlorophyll a (Chl a, mg·g^{-1} FW), chlorophyll b (Chl b, mg·g^{-1} FW), and carotenoids (Ctn, mg·g^{-1} FW) of Foxtail lily (*Eremurus spectabilis*).

Medium	Rl	Bv	Ln	Ll	Chl a	Chl b	Ctn
MS + 2.0 mgL^{-1} IBA + 1.0 mgL^{-1} NAA	0 [c]	37.30 [a]	16.77 [a]	4.70 [b]	8.92 [a]	15.19 [ab]	5.65 [a]
SH + 2.0 mgL^{-1} IBA + 1.0 mgL^{-1} NAA	0 [c]	37.70 [a]	6.71 [c]	4.30 [b]	7.23 [b]	12.60 [bc]	3.79 [b]
$\frac{1}{2}$ MS + 2.0 mgL^{-1} IBA + 4.0% sucrose + 200 mgL^{-1} AC	1.11 [a]	41.00 [a]	18.95 [a]	6.40 [a]	8.58 [a]	16.53 [a]	3.41 [b]
$\frac{1}{2}$ SH + 2.0 mgL^{-1} IBA + 4.0% sucrose + 200 mgL^{-1} AC	0.39 [b]	41.50 [a]	11.93 [b]	5.10 [ab]	5.85 [c]	11.00 [c]	2.04 [c]

Table 4. *Cont.*

Medium	Rl	Bv	Ln	Ll	Chl a	Chl b	Ctn
Source of variation							
Medium	2.73 **	47.56 ns	296.44 **	8.29 *	19.62 **	62.20 **	22.23 **
Error	0.062	140.35	17.21	2.65	1.60	12.38	1.03

Means followed by the same letter are not significantly different as determined by the LSD multiple range test at $p \leq 0.05$. * and **: Significant at the 5% and 1% probability levels, respectively; ns: Non-significant.

Figure 4. (**A**): Indirect shoot regeneration from callus explants. (**B**) Shoot clumps transferred to rooting medium. (**C**) Emergence of fleshy roots from the basal part of shoot clump three weeks after inoculation (Bar = 10 mm; Photo: Y. Basiri: July 2019).

3.4. Acclimatization and Ex Vitro Transfer

Plantlets comprising both healthy shoots and roots regenerated from different media were subjected to acclimatization. During the acclimation period, the plantlets fertigated with half MS solution had longer roots, bigger leaves, and higher biomass volume than plantlets that received the half SH solution (Table 5).

Table 5. Effects of nutrient solutions on root length (Rl, cm), leaf number (Ln), leaf length (Ll, cm) and biomass volume (Bv, cm^3) of Foxtail lily (*Eremurus spectabilis*) plantlets during acclimation and ex vitro transfer period.

Nutrient Solutions	Rl	Ln	Ll	Bv
$\frac{1}{2}$ MS	1.14 a	32.60 a	13.66 a	5.80 a
$\frac{1}{2}$ SH	0.66 b	27.80 a	6.98 b	3.70 b
Source of variation				
Nutrient solution	1.15 *	115.20 ns	223.11 **	22.05 *
Error	0.18	76.78	20.25	2.98

Means followed by the same letter are not significantly different as determined by the LSD multiple range test at $p \leq 0.05$. * and **: Significant at the 5% and 1% probability levels, respectively; ns: Non- significant.

4. Discussion

The *E. spectabilis* is an endemic wild species of the Iran/Turan region—especially in central Asia [21]. There is insufficient coverage of this species in the literature, especially with respect to its micropropagation. The fleshy roots of this species may be utilized as explant preparations; however, the contamination of explants is a significant and limiting factor when optimizing micropropagation protocols [28]. A surface-sterilization process aimed at eliminating all microorganisms could guarantee the explant's viability and regeneration capacity, which are affected by the concentration and application period [29].

Therefore, explant disinfecting was assumed as an essential step in the culture establishment of *E. spectabilis*. In the present study, the root explants were efficiently sterilized with NaClO (40% for 15 min). When these were cultured in MS medium supplemented with 10 mgL^{-1} BAP, they produced a green callus—but other treatments including media composition, disinfection methods, and BAP levels produced yellowish-green or blackish calli masses (Table 2). The success of the callus induction and the application of an appropriate disinfection method and growth regulators are in agreement with previous investigations [30–32]. Furthermore, Colgecen et al. [33] examined the potential of different basal media such as SH and MS in callus induction and the proliferation of *Arnebia* explants, and their results showed that the best friable callus was obtained with MS medium, which further supported our results regarding the superior effect of MS medium on callus induction in *E. spectabilis* explants. Regeneration of calli with various colors and structures from *E. spectabilis* explants is also in conformity with other results reported on various explant sources from other plant species [34,35]. Furthermore, it is well documented that there are several types of calli—especially in monocot plants—based on their regenerative and morphological characteristics [36]. However, these different types of callus often have significantly different plant regeneration capabilities [36]. Therefore, the details of the distinction and regenerative capacities of calli in *E. espectabilis* require further investigation in future studies.

The present study indicated significant interactions of basal medium, cytokinin concentration, and disinfection method on callus induction, carotenoid concentration, and chlorophyll content. The highest percentage of callus formation from potato explants had already been obtained using MS medium containing 5.0 [37] and 2.0 mgL^{-1} BAP [38]. The results of the present study corroborate the findings of Carsono and Yoshida [39] and Benderradji et al. [40], who also reported that medium composition could be a source of variation affecting callogenesis in different explants.

The degree of success in any technology employing cell, tissue, or organ culture is related to relatively few major factors. A significant factor is the choice of nutritional components and plant growth regulators [41]. The success of plant tissue culture as a means of plant propagation is greatly influenced by the nature of the culture medium used. There are several basal media reported in the literature [41,42]—among them, the MS medium [23] is the most widely used. Furthermore, auxins and cytokinins are by far the most important for regulating growth and morphogenesis in plant tissue and organ cultures [42]. Hence, in the present study, a combination of NAA with BAP in two basal media (MS, SH) was evaluated for callus induction. Auxins are very widely used in plant tissue culture and usually form an integral part of nutrient media. Auxins promote, mainly in combination with cytokinins, the growth of calli, cell suspensions and division, and cell elongation. Since they are capable of initiating cell division, they are involved in the formation of meristems, giving rise to either unorganised tissue or defined organs [42]. BAP, either singly or combined with NAA, has a positive effect on shoot regeneration [22,27,42]. Based on a recent research work [43], it was revealed that the combination of BAP and NAA was a superior treatment for mico-bulb induction in Lily (*Lilium longiflorum*) explants. They also stated that when auxin or cytokines are exogenously added to culture media, they will trigger the further formation of micro tubers more quickly. This can increase the concentration of endogenous growth regulators in cells, helping the growing process in developing tissues.

The in vitro shoot regenerated from callus explants clearly showed a significant influence of medium type and BAP concentration. Hence, shoot proliferation, leaf length, and the concentrations of chlorophyll *a*, chlorophyll *b*, and carotenoids were also significantly affected. Tuncer [22] reported significant differences among MS media containing different combinations of plant growth regulators ($p < 0.01$) on direct shoot regeneration from hypocotyl explants of *E. spectabilis*. To our best of knowledge, the data presented here demonstrate a complete and applicable protocol for indirect shoot regeneration in callus cultures of Foxtail lily (Figures 3 and 4). The plant growth regulators might play a key role in the shoot regeneration capacity of explants [44]. In most of the tissue culture studies, establishing a suitable ratio of auxin to cytokinin has been applied for high-frequency shoot regeneration [45]. Our study has demonstrated that the optimal level of BAP in shoot proliferation medium is 2.0 mgL^{-1}, which is satisfactory for superior shoot regeneration capacity. Furthermore, some physiological attributes, such as the chlorophyll and carotenoid contents of regenerated shoots, could be improved.

While the efforts of Tuncer [21] to regenerate shoots from rhizome and leaf explants of *E. spectabilis* have not been successful, we observed in vitro microshoots just 6 weeks after inoculation of the organogenic callus on media containing 2.0 mgL^{-1} BAP. Tuncer [21] ascribed phenolic compound-induced browning of cultured explants (rhizome and leaf) as a reason for the loss of the regeneration ability of the Foxtail lily. We supposed that one of the reasons that our proliferated callus tissues had satisfactory regenerating abilities may be due to their green appearance, as Iqbal et al. [46] has previously suggested that green callus tissues have good regeneration abilities and can be successfully applied for shoot proliferation. Irvani et al. [47] reported that the highest shoot regeneration from calli of *Dorema ammoniacum* D. Don plants was achieved on MS medium containing 2.0 mgL^{-1} BAP, which is equivalent to our results regarding the positive role of BAP at the same concentration. The results also showed that when the callus tissues were sub-cultured intact on MS medium containing 5.0 mgL^{-1} BAP, a higher number of leaves were recorded as compared to divided tissues. Furthermore, the most extended leaf length was also found with the same intact calli sub-cultured on medium containing 2.0 mgL^{-1} BAP. The number of leaves and their length are considered essential parameters to describe the leaf biomass and physiological functions in regenerated shoots due to photosynthesis and respiration [48]. Besides the Tuncer report [22], there is no other paper regarding the effects of different explant types and growth regulators on shoot regeneration in Foxtail lilies. The present study revealed that the micropropagation of Foxtail lily through callus culture may be practicable for the mass production of microshoots in this plant species.

The application of auxins for in vitro root induction is common in the micropropagation of different plant species [49,50]. In the present research, the effectiveness of IBA for root regeneration in Foxtail lily explants was also confirmed (Table 4). Auxins increase the expression of various genes involved in root regeneration [51]. The recorded root data revealed that the utilization of half-strength basal medium, combined with higher sucrose concentrations (4.0%) and AC (200 mgL^{-1}), are responsive modifications for root induction in Foxtail lily explants. The same may be followed in the commercial mass multiplication of this species.

The in vitro rooted plantlets must finally be transferred to ex vitro conditions. Consequently, a proper strategy for hardening and ex vitro transfer must be followed. In Foxtail lily, the ex vitro adaptation is slow, as has been already observed in many bulbous plants [52–54]. Interestingly, the grapevine hardening strategy already reported by Alizadeh et al. [27] was successfully exploited for the acclimation of in vitro-raised plantlets. However, from a practical point of view, this strategy may not be feasible in large-scale micropropagation, and needs to be standardized further. However, the fertigation with 1/2 MS solution followed in this strategy could efficiently avoid transplantation shock. Two months after hardening, the recorded morphological attributes, such as root length, leaf size, and biomass volume of the hardened and acclimatized Foxtail lily plantlets, demonstrated elongated roots that could easily penetrate the potting substrate (coco peat:perlite:vermiculite). The following result agrees with Ozel et al. [55], who observed similar results while acclimating *Muscari muscarimi* Medikus plants. In another study comparable to our findings, Sharma et al. [56] reported that the irrigation of rooted plantlets with a quarter strength of MS provided essential nutrients to the acclimatizing plant—thus increasing its chances of survival.

5. Conclusions

In the present study, in vitro callus induction and indirect shoot regeneration were developed from root explants of foxtail lily. Hereafter, the disinfection method and appropriate basal medium were reported for callus induction and further shoot regeneration. Surface sterilization of root explants with NaClO (40%, 15 min) was found to be an effective treatment for culture establishment. The inoculated explants had the highest callogenesis efficiency on MS medium supplemented with 10.0 mgL^{-1} BAP. The highest percentage of shoot regeneration from calli was obtained with 2.0 mgL^{-1} BAP. The half-strength MS medium was supplemented with 2.0 mgL^{-1} IBA + 4.0% sucrose + 200 mgL^{-1} AC, proving to be a suitable combination for root regeneration in micro-shoots. The regenerated plantlets were successfully adapted to ex vitro conditions. The reported in vitro regeneration protocol can be exploited at a commercial scale, following just minor modifications. Furthermore, it could be beneficial for producing valuable secondary metabolites at a large scale with the help of bioreactors, where calli would be required as input materials.

Author Contributions: Conceptualization, Y.B., N.E. and M.A.; Formal analysis, Y.B. and M.A.; Project administration, Y.B. and N.E.; Resources, J.A.; Validation, J.A.; Writing—original draft, Y.B. and M.A.; Writing—review and editing, N.E., M.A. and J.A. All authors have read and agreed to the published version of the manuscript.

Funding: This research received no external funding.

Institutional Review Board Statement: Not applicable.

Informed Consent Statement: Not applicable.

Data Availability Statement: The data presented in this study are available on request from the corresponding authors.

Conflicts of Interest: The authors declare no conflict of interest.

References

1. Kamenetsky, R.; Akhmetova, M. Floral development of *Eremurus altaicus* (Liliaceae). *Isr. J. Plant Sci.* **1994**, *42*, 227–233. [CrossRef]
2. Tosun, M.; Ercisli, S.; Ozer, H.; Turan, M.; Polat, T.; Ozturk, E.; Padem, H.; Kilicgun, H. Chemical composition and antioxidant activity of Foxtail lily (*Eremurus spectabilis*). *Acta Sci. Pol-Hortorum Cultus* **2012**, *11*, 145–153.
3. Tuzcu, Z.; Koclar, G.; Agca, C.A.; Aykutoglu, G.; Dervisoglu, G.; Tartik, M.; Sahin, K. Antioxidant, antimicrobial and anticancer effects of different extracts from wild edible plant *Eremurus spectabilis* leaves and roots. *Int. J. Clin. Exp. Med.* **2017**, *10*, 4787–4797.
4. Altundag, E.; Ozturk, M. Ethnomedicinal studies on the plant resources of east Anatolia, Turkey. *Proc. Soc. Behav. Sci.* **2011**, *19*, 756–777. [CrossRef]
5. Bircan, B.; Kirbag, S. Determination of antioxidant and antimicrobial properties of *Eremurus spectabilis* Bieb. *Artvin Çoruh Univ. Orman. Fak. Derg.* **2012**, *16*, 176–186. [CrossRef]
6. Tuzlaci, E.; Dogan, A. Turkish folk medicinal plants, IX: Ovacık (Tunceli). *Marmara. Pharm. J.* **2012**, *14*, 136–143. [CrossRef]
7. Pourfarzad, A.; Najafi, M.B.H.; Khodaparast, M.H.H.; Khayyat, M.H.; Malekpour, A. Fractionation of (*Eremurus spectabilis*) fructans by ethanol: Box–Behnken design and principal component analysis. *Carbohydr. Polym.* **2014**, *106*, 374–383. [CrossRef]
8. Abubaker, S.R.; Hidayat, H.J. Anti-tumor potential of Local Aslerk (*Eremurus spectabilis*) leaf extracts by HPLC and applying on cancer cell lines in vitro. *Iraqi J. Cancer Med. Genet.* **2015**, *8*, 123–128.
9. Eghtedarnejad, N.; Mansouri, H.R. Building wooden panels glued with a combination of natural adhesive of tannin/*Eremurus* root (*syrysh*). *Eur. J. Wood Prod.* **2016**, *74*, 269–272. [CrossRef]
10. Schiappacasse, F.; Szigeti, J.C.; Manzano, E.; Kamenetsky, R. *Eremurus* as a new cut flower crop in Aysen, Chile: Introduction from the Northern hemisphere. *Acta Hortic.* **2013**, *1002*, 115–121. [CrossRef]
11. Ahmad, I.; Dole, J.M.; Schiappacasse, F.; Saleem, M.; Manzano, E. Optimal postharvest handling protocols for cut 'Line Dance' and 'Tap Dance' *Eremurus* inflorescences. *Sci. Hortic.* **2014**, *179*, 212–220. [CrossRef]
12. Hesami, M.; Jones, A.M.P. Application of artificial intelligence models and optimization algorithms in plant cell and tissue culture. *Appl. Microbiol. Biotechnol.* **2020**, *104*, 9449–9485. [CrossRef] [PubMed]
13. Baksha, R.; Jahan, M.A.A.; Khatun, R.; Munshi, J.L. Micropropagation of *Aloe barbadensis* Mill. through in vitro culture of shoot tip explants. *Plant Tissue Cult. Biotechnol.* **2005**, *15*, 121–126.
14. Sun, L.; Jin, L. Studies on the in vitro rapid propagation of *Lilium X Longliflorum*. *Liaoning Agric. Sci.* **2011**, *6*, 9–12.
15. Skoric, M.; Zivkovic, S.; Savic, J.; Siler, B.; Sabvoljevic, A.; Todorovic, S.; Grubisic, D. Efficient one-step tissue culture protocol for propagation of endemic plant, *Lilium martagon* var. Cattaniae Vis. *Afr. J. Biotechnol.* **2012**, *11*, 1862–1867.

16. Monemi, M.B.; Kazemitabar, S.K.; Khaniki, G.B.; Yasari, E.; Sohrevardi, F.; Pourbagher, R. Tissue culture study of the medicinal plant Leek (*Allium ampeloprasum* L). *Int. J. Mol. Cell. Med.* **2014**, *3*, 118–125.
17. Suh, S.Y.; Baskar, T.B.; Kim, H.H.; Al-Dhabi, N.A.; Park, S.U. Ethylene inhibitors promote shoot organogenesis of *Aloe arborescens* Miller. *South Indian J. Biol. Sci.* **2015**, *1*, 43–46. [CrossRef]
18. Cakmak, D.; Karaoglu, C.; Aasim, M.; Sancak, C.; Ozcan, S. Advancement in protocol for in vitro seed germination, regeneration, bulblet maturation, and acclimatization of *Fritillaria persica*. *Turk. J. Biol.* **2016**, *40*, 878–888. [CrossRef]
19. Pandey, A. Research on in-vitro clonal multiplication protocol in various cultivars of *Chlorophytum*. *Tech Chron. Int. J. Emerg. Trends Sci. Eng. Technol.* **2016**, *1*, 32–42.
20. Razib, M.A.; Mamun, A.N.; Kabir, M.H.; Al Din, S.M.; Hossain, M.E.; Firoz Alam, M. High frequency in vitro propagation of *Aloe vera* L. through shoot tip culture. *Int. J. Appl. Biol. Pharm.* **2016**, *7*, 28–32.
21. Tuncer, B. Investigation of the in vitro regeneration of some medical and aromatic wild plant species. *Appl. Ecol. Environ. Res.* **2017**, *15*, 905–914. [CrossRef]
22. Tuncer, B. In vitro germination and bulblet and shoot propagation for wild edible *Eremurus spectabilis* M.Bieb. *Not. Bot. Horti. Agrobot. Cluj-Napoca* **2020**, *48*, 814–825. [CrossRef]
23. Murashige, T.; Skoog, F. A revised medium for rapid growth and bio-assays with tobacco tissue cultures. *Physiol. Plant.* **1962**, *15*, 473–497. [CrossRef]
24. Hadizadeh, H.; Bahri, B.A.; Qi, P.; Wilde, H.D.; Devos, K.M. Intra-and interspecific diversity analyses in the genus *Eremurus* in Iran using genotyping-by-sequencing reveal geographic population structure. *Hortic. Res.* **2020**, *7*, 1–13. [CrossRef] [PubMed]
25. Schenk, R.V.; Hildebrandt, A.C. Medium and techniques for induction and growth of monocotyledonous and dicotyledonous plant cell cultures. *Can. J. Bot.* **1972**, *50*, 199–204. [CrossRef]
26. Barnes, J.D.; Balaguer, L.; Manrique, E.; Elvira, S.; Davison, A.W. A reappraisal of the use of DMSO for the extraction and determination of chlorophyll 'a' and 'b' in lichens and higher plants. *Environ. Exp. Bot.* **1992**, *32*, 85–90. [CrossRef]
27. Alizadeh, M.; Singh, S.K.; Patel, V.B.; Deshmukh, P.S. In vitro clonal multiplication of two grape (*Vitis* spp.) rootstock genotypes. *Plant Tiss. Cult. Biotechnol.* **2018**, *28*, 1–11. [CrossRef]
28. Dalila, Z.D.; Fahmi, A.B.M.; Nurkhalida, A.K.S. Optimization of sterilization method and callus induction of *Cocos nucifera* Linn. var. Matag from inflorescence. In Proceedings of the International Conference on Plant, Marine and Environmental Science, Kuala Lumpur, Malaysia, 1–2 January 2015.
29. Allan, A. Plant cell culture. In *Plant Cell and Tissue Culture*; Stafford, A., Warren, G., Eds.; Open University Press: London, UK, 1991; pp. 1–24.
30. Petersen, K.K.; Hansen, J.; Krogstrup, P. Significance of different carbon sources and sterilization methods on callus induction and plant regeneration of *Miscanthus x ogiformis* Honda 'Giganteus'. *Plant Cell Tissue Organ Cult.* **1999**, *58*, 189–197. [CrossRef]
31. Roy, A.; Ghosh, S.; Chaudhury, M.; Saha, P.K. Effect of different plant hormones on callus induction in *Gymnema sylvestris* R.Br. (Asclepiadaceae). *Afr. J. Biotechnol.* **2008**, *7*, 2209–2211.
32. Osman, N.I.; Sidik, N.J.; Awal, A. Effects of variations in culture media and hormonal treatments upon callus induction potential in endosperm explant of *Barringtonia racemosa* L. *Asian Pac. J. Trop. Biomed.* **2016**, *6*, 143–147. [CrossRef]
33. Colgecen, H.; Koca, U.; Toker, G. Influence of different sterilization methods on callus initiation and production of pigmented callus in *Arnebia densiflora* Ledeb. *Turk. J. Biol.* **2009**, *35*, 513–520.
34. Ehsandar, S.; Majd, A.; Choukan, R. Callus formation and regeneration of the first modified Iranian potato cultivar (Savalan). *Adv. Crop. Sci.* **2013**, *3*, 201–208.
35. Kumlay, A.M.; Ercisli, S. Callus induction, shoot proliferation and root regeneration of potato (*Solanum tuberosum* L.) stem node and leaf explants under long-day conditions. *Biotechnol. Equip.* **2015**, *29*, 1074–1084.
36. Hesami, M.; Jones, A.M.P. Modeling and optimizing callus growth and development in *Cannabis sativa* using random forest and support vector machine in combination with a genetic algorithm. *Appl. Microbiol. Biotechnol.* **2021**, *105*, 5201–5212. [CrossRef]
37. Khatun, N.; Bari, M.A.; Islam, R.; Huda, S.; Siddque, N.A.; Rahman, M.A.; Mullah, M.U. Callus induction and regeneration from nodal segment of potato cultivar Diamant. *J. Bio. Sci.* **2003**, *3*, 1101–1106.
38. Yasmin, S.; Nasiruddin, K.M.; Begum, R. Regeneration and establishment of potato plantlet through callus formation with BAP and NAA. *Asian J. Plant Sci.* **2003**, *2*, 936–940. [CrossRef]
39. Carsono, N.; Yoshida, T. Identification of callus induction potential of 15 Indonesian rice genotypes. *Plant Prod. Sci.* **2006**, *9*, 65–70. [CrossRef]
40. Benderradji, L.; Brini, F.; Kellou, K.; Ykhlef, N.; Djekoun, A.; Masmoudi, K.; Bouzerzour, H. Callus induction, proliferation, and plantlets regeneration of two bread wheat (*Triticum aestivum* L.) genotypes under saline and heat stress conditions. *Int. Sch. Res. Notices* **2011**, *2012*, 1–8.
41. Gamborge, O.L. Media preparation. In *Plant Tissue Culture Manual*; Lindsey, K., Ed.; Springer: New York, NY, USA, 1991; Volume A1, pp. 1–24.
42. George, E.F. *Plant Propagation by Tissue Culture*; Exegetics Ltd.: Edington, UK, 1993; p. 574.
43. Deswiniyanti, N.W.; Dwipayani Lestari, N.K. In vitro propagation of *Lilum longiflorum* using NAA and BAP plant growth regulator treatment. *KnE Life Sci.* **2020**, *5*, 6437. [CrossRef]
44. Ozcan, S.; Yildiz, M.; Sancak, C.; Ozgen, M. Adventitious shoot regeneration in sainfoin (*Onobrychis viciifolia* Scop.). *Turk. J. Bot.* **1996**, *20*, 497–501.

45. Yıldız, M.; Saglik, C.; Telci, C.; Erkilic, E.G. The effect of in vitro competition on shoot regeneration from hypocotyl explants of *Linum usitatissimum*. *Turk. J. Bot.* **2011**, *35*, 211–218.
46. Iqbal, M.; Jaskani, J.M.; Rafique, R.; Hassan, S.Z.; Iqbal, M.S.; Raheed, M.; Mushtaq, S. Effect of plant growth regulators on callus formation in potato. *J. Sci. Food Agric.* **2014**, *2*, 77–81.
47. Irvani, N.; Solouki, M.; Omidi, M.; Zare, A.R.; Shahnazi, S. Callus induction and plant regeneration in *Dorem ammoniacum* D., an endangered medicinal plant. *Plant Cell Tissue Organ Cult.* **2010**, *100*, 293–299. [CrossRef]
48. Huang, W.; Ratkowsky, D.A.; Hui, C.; Wang, P.; Su, J.; Shi, P. Leaf fresh weight versus dry weight: Which is better for describing the scaling relationship between leaf biomass and leaf area for broad-leaved plants? *Forests* **2019**, *10*, 256. [CrossRef]
49. Simon, S.; Petrášek, P. Why plants need more than one type of auxin. *Plant Sci.* **2011**, *180*, 454–460. [CrossRef]
50. Kulus, D. Influence of growth regulators on the development, quality, and physiological state of in vitro-propagated *Lamprocapnos spectabilis* (L.) Fukuhara. *In Vitro Cell. Dev. Biol. Plant* **2020**, *56*, 447–457. [CrossRef]
51. Overvoorde, P.; Fukaki, H.; Beeckman, T. Auxin control of root development. *Cold Spring Harb. Perspect. Biol.* **2010**, *2*, 1–18. [CrossRef] [PubMed]
52. Paek, K.Y.; Murthy, H.N. High frequency of bulblet regeneration from bulb scale sections of *Fritillaria thunbergii*. *Plant Cell Tissue Organ Cult.* **2002**, *68*, 247–252. [CrossRef]
53. Khawar, K.M.; Cocu, S.A.; Parmaksiz, I.; Sarihan, E.O.; Ozcan, S. Mass proliferation of Madona Lilly (*Lilium candidum* L.) under in vitro conditions. *Pak. J. Bot.* **2005**, *37*, 243–248.
54. Priyakumari, I.; Sheela, V.L. Micropropagation of gladiolus cv. 'Peach Blossom' through enhanced release of axillary buds. *J. Trop. Agric.* **2005**, *43*, 47–50.
55. Ozel, C.A.; Khawar, K.M.; Unal, F. Factors affecting efficient in vitro micropropagation of *Muscari muscarimi* Medikus using twin bulb scale. *Saudi J. Biol. Sci.* **2015**, *22*, 132–138. [CrossRef] [PubMed]
56. Sharma, U.; Kataria, V.; Shekhawat, N.S. In vitro propagation, ex vitro rooting and leaf micromorphology of *Bauhinia racemosa* Lam.: A leguminous tree with medicinal values. *Physiol. Mol. Biol. Plant* **2017**, *23*, 969–977. [CrossRef] [PubMed]

 *horticulturae*

Article

An Optimized Protocol for Indirect Organogenesis from Root Explants of *Agapanthus praecox* subsp. *orientalis* 'Big Blue'

Qianwen Tang [1,†], Xiangxin Guo [1,†], Yuanshan Zhang [1], Qingyun Li [1], Guanqun Chen [1,*], Huale Sun [2], Weiming Wang [3] and Xiaohui Shen [1,*]

[1] School of Design, Shanghai Jiao Tong University, Shanghai 200240, China
[2] Shanghai Shangfang Horticulture Co., Minhang, Shanghai 201114, China
[3] ArborGen Inc., Ridgeville, SC 29472, USA
*   Correspondence: chengq@sjtu.edu.cn (G.C.); shenxh62@sjtu.edu.cn (X.S.)
†   These authors contributed equally to this work.

Citation: Tang, Q.; Guo, X.; Zhang, Y.; Li, Q.; Chen, G.; Sun, H.; Wang, W.; Shen, X. An Optimized Protocol for Indirect Organogenesis from Root Explants of *Agapanthus praecox* subsp. *orientalis* 'Big Blue'. *Horticulturae* **2022**, *8*, 715. https://doi.org/10.3390/horticulturae8080715

Academic Editor: Sergio Ruffo Roberto

Received: 20 June 2022
Accepted: 14 July 2022
Published: 9 August 2022

Publisher's Note: MDPI stays neutral with regard to jurisdictional claims in published maps and institutional affiliations.

Abstract: *Agapanthus praecox* has become a burgeoning variety in the flower market due to its high ornamental value with unique large blue-purple inflorescence. For rapid entering into the market, tissue culture technology or organogenesis has an attractive application over the conventional reproduction approach. In this study, a highly efficient protocol based on indirect organogenesis has been successfully established for *A. praecox* subsp. *orientalis* 'Big Blue'. Two types of explants, root tips versus root segments, were compared for callus induction frequency in response to the induction culture media. The induction media contain Murashige and Skoog's (MS) Basal Salt supplemented with various concentrations of picloram (PIC), 2,4-Dichlorophenoxyacetic acid (2,4-D), thidiazuron (TDZ), kinetin (KT) and naphthalene acetic acid (NAA). Of the two types of explants, root tips were found to be more effective for callus induction than root segments. Among the induction media tested, the highest callus induction rate (100.00%) was achieved when cultured on MS supplemented with 2.0 mg/L PIC, 1.5 mg/L KT and 0.1 mg/L NAA, which was probably accredited to higher endogenous phytohormone contents, especially of 3-indoleacetic (IAA). The optimal medium for callus proliferation was MS + 1.0 mg/L PIC + 1.0 mg/L 6-BA + 0.4 mg/L NAA, and the fresh weight increased by 72.74%. After being transferred onto the adventitious bud induction medium for 25 days, shoots were dedifferentiated from the surface of the flourishing callus, which then developed to the plantlet with roots in 90 days. The plantlets were transplanted in a greenhouse with a survival rate of 92.86%. This study innovatively established an indirect organogenesis tissue culture system of *A. praecox* with roots as explants, which provided a practical reference in its application.

Keywords: *Agapanthus praecox*; root system; callus induction; indirect organogenesis; endogenous hormones

1. Introduction

Agapanthus praecox is a perennial herbaceous plant belonging to Amaryllidaceae with sub-leathery, linear long-lanceolate or strip-shaped basal leaves. A large erect umbrella-type inflorescence contains more than 100 funnel-shaped florets, reaching 50~70 cm in height, which is dark blue to white in color [1]. Native to southern Africa, it prefers abundant sunshine, fertile soil and mild climate. Furthermore, the suitable growth temperature is 20~25 °C. There are six species and nearly 600 varieties of *Agapanthus* genus, mostly distributed in developed countries, such as the United States, United Kingdom, Netherlands and New Zealand. With a well-developed root system and strong fleshy rhizomes, *A. praecox* can effectively prevent sandstorms and soil erosion, reduce environmental pollution, and have important ecological functions. *A. praecox* has a prominent waterlogging tolerance and can be utilized to reduce the pollutants of sanitary sewage with a decline of chemical oxygen demand (COD) and biochemical oxygen demand (BOD), while *A. praecox* is cultivated in an artificial wetland environment alone [2] or together with *Canna indica*, *Zantedeschia aethiopica* and *Watsonia borbonica* [3]. It is an excellent plant material used for rural sewage treatment.

First introduced to China from southern Africa in 2002, *A. praecox* has brought attention of floriculture due to its ornamental values as pot plants, cut flowers, and garden plants [4]. To date, there has been more and more *A. praecox* planted in to flower borders, public parks, roof gardens and road greenbelts in China [5,6]. However, owing to the influence of accumulated temperature in Shanghai and the Yangtze River Delta, it is difficult to bear matured fruits and the seed yield is very low, resulting in a contradiction between high demand and low supplies.

As alternatives, in vitro approaches are commonly implemented to propagate Agapanthus. These approaches include (but are not limited to) adventitious shoots or buds and root induction [7], somatic embryogenesis [7,8], and callus-derived protoplasts [9]. Among them, adventitious bud induction can be divided into direct and indirect organogenesis, and the latter usually relies on the dedifferentiation and redifferentiation of calluses to form plantlets. Substantial work has been done by several scientists throughout the world so far by employing an array of explants for inducing calluses followed by organogenesis and plantlet regeneration with varying levels of accomplishment; for instance, embryos [10], stem bases [1,11], rhizomes [12,13], leaves [14–16], flower buds [13], pedicels [17] have been used. Within the majority of these studies, the callus induction rate ranged from approximately 40 to 54%, and it was clear that rhizome, stems and pedicals preceded leaves as explants, whereas the comparison between others remained unknown [12,13,17]. No such attempt has been made using roots as explants.

For organogenesis, explants are often cultivated on a culture media with various concentrations and types of plant growth regulators (PGRs), such as cytokinins (6-benzylaminopurine (6-BA), 2-isoamyl Alkenyl adenine (2iP) or thiadiazolone (TDZ), as well as with various auxins (3-indoleacetic acid (IAA), 1-naphthaleneacetic acid (NAA) or 2,4-D) or other hormones (PIC) in combination to enhance the process of callus and shoot regeneration. A previous study demonstrated that large numbers (47.3 ± 1.96) of adventitious buds were induced per shoot tip on an MS medium supplemented with 22.2 μM 6-BA, 2.9 μM IAA, and 4.5 μM TDZ [18]. Yellow dense calluses from leaf segments could be induced on an MS medium containing 1 mg/L PIC for 2 months, and milky white and brittle embryogenic calluses (EC) formed on the surface followed by subculture [19,20]. Zou [21] and Ren et al. [22] analyzed the effect of picloram (PIC) concentration on the induction of *A. praecox* calluses and embryogenic calluses (EC) and found that 1.5 mg/L PIC significantly promoted the callus induction ability and embryogenicity. Yue et al. [16] established an indirect organogenesis system using *A. praecox* leaves as explants. The results show that the meristematic state of the leaf tissue determined the callus induction rate. Using the base of new leaves as explants, the callus induction rate was the highest, reaching 85.71%, and the optimum media formulation was MS + 2.0 mg/L PIC.

The explants studied above for the indirect organogenesis are mostly small pedicels, stem bases and leaves, while the limitations are that the collection time of the small pedicels is narrow, removing stem bases are harmful to plants, and leaves are prone to pathogen contamination, culture browning and vitrification. These protocols are still not applicable for the rapid propagation of *A. praecox* seedlings.

The authors intend to take advantage of the monocotyledonous plantlets with well-developed root systems to explore the feasibility of using the root system as explants. The main objective of this study is to effectively expand the types and sources of explants for overcoming bottlenecks in material collections, such as the limitation of the flowering period, providing multiple alternatives for high-quality callus and embryogenic callus induction, genetic improvement and cryopreservation. At the same time, the contents of four endogenous hormones in the process of root-induced calluses are investigated, which will offer us the theoretical basis for the optimization of the hormone ratio and formulation in the tissue culture system of *A. praecox*.

2. Materials and Methods

2.1. Plant Material Collection

In late September 2020, mature seeds of evergreen *A. praecox* ssp. *Orientalis* 'Big Blue' grown in the open field of Kunming city, Yunnan province, China, were collected. After natural air drying, one hundred seeds were randomly selected with 5 replications, and the average dry weight was calculated according to Chen [23]. The morphological indicators, for instance, the length and thickness of zygotic embryos and sizes of seeds, were measured to evaluate the seed quality. The selected seeds of good quality were used for subsequent sterilization and germination experiments.

2.2. Sterilization and Germination

Due to the large number of endophytes in the seeds of *A. praecox*, ensuring aseptic operations required complicated disinfection procedures according to the preliminary experiment. The selected *A. praecox* seeds had their seed wings removed first and were placed in 50 mL centrifuge tubes, then washed with sterile water twice, 20 min each to remove impurities, followed by 75% (v/v) alcohol for 1 min and were shaken continuously. These seeds were then rinsed with sterile water 3 times, and the surface was disinfected with 20% NaClO for 20 min and later rinsed with sterile water 6 times. Finally, sterilized filter paper was used to absorb the residual water off the seeds.

Referring to the method of Fan [1], the sterilized seeds were cultured on an MS medium supplemented with 30 g/L sucrose and 10 g/L agar (pH 5.8, the same below) for germination. Each dish (Φ 9.0 cm) was plated with 9~15 seeds and repeated 3 times. After about 30 days, the seeds germinated into seedlings with flourishing root systems and were ready for the explant collection for callus induction experiments.

2.3. Callus Induction from Root Explants

The roots (1.0~2.0 mm in diameter) of 60-day-old seedlings were directly cut into sections (0.8~1.5 cm in length). Two types of explants, with or without root tips (root segments), were picked and inoculated on callus induction media (CIM). CIM is an MS medium supplemented with 30 g/L sucrose, 3.0 g/L phytagel and three kinds of PGRs with different concentrations, shown in Table 1, designed by an orthogonal experiment.

The morphological change, contamination rate, browning rate and callus induction rate were recorded at 45 days and calculated using the following equation:

$$\%\text{Callus induction} = \left(\frac{n}{N}\right) \times 100$$

where "n" is the number of explants capable of inducing calluses and "N" is the total number of explants used in the experiment. The old leaves and residual roots were excised from the seedlings and then transferred onto the rooting medium (RM) with 2.17 g/L MS, 0.15 mg/L IBA, 30 g/L sucrose and 6 g/L agar. The light-green roots regenerated from the seedlings after 7 days, and the explants could be captured repeatedly for callus induction after 14 days.

2.4. Analysis of Endogenous Hormone Content during Callus Induction

The CIM was an MS medium supplemented with 3.0 mg/L PIC, 30 g/L sucrose and 3 g/L phytagel. Repeating the process at Section 2.3, at 0, 5, 10, 15 and 20 days, the root explants (not less than 0.2 g fresh weight (FW), including root tips and root segments respectively) were harvested and wrapped in the sealed foil, and then stored in a $-20\ ^{\circ}$C refrigerator until the sampling at each time point was finished. All samples were taken out at the same time, thawed, and temporarily stored in a 2~8 $^{\circ}$C refrigerator. Then, an appropriate amount of phosphate-buffered saline (PBS, pH 7.4) was added to each sample, fully homogenized, and centrifuged at 2000 rpm for 20 min. The supernatant was collected; the contents of IAA, ABA, GA_4 and ZT were determined by an enzyme-linked immunosorbent assay (ELISA) according to the instructions of the kit (Shanghai Enzyme Link Biotechnology Co., Ltd., Shanghai, China)

Table 1. CIM design for callus induction from root explants of *A. praecox* seedlings.

Explant Types	Concentration (mg/L)				
	PIC [1]	TDZ [2]	2,4-D [3]	KT [4]	NAA [5]
root tips, root segments	1.00	0.00	0.50	/	/
	1.00	0.40	1.00	/	/
	1.00	0.60	2.00	/	/
	2.00	0.00	1.00	/	/
	2.00	0.40	2.00	/	/
	2.00	0.60	0.50	/	/
	3.00	0.00	2.00	/	/
	3.00	0.40	0.50	/	/
	3.00	0.60	1.00	/	/
root tips	1.00	/	/	0.00	0.10
	1.00	/	/	1.00	0.50
	1.00	/	/	1.50	1.00
	2.00	/	/	0.00	0.50
	2.00	/	/	1.00	1.00
	2.00	/	/	1.50	0.10
	3.00	/	/	0.00	1.00
	3.00	/	/	1.00	0.10
	3.00	/	/	1.50	0.50

[1] Picloram, bought from Sigma-Aldrich (suitable for plant cell culture). [2] Thiadiazolone, bought from Sangon Biotech ($\geq$95.0%). [3] 2,4-Dichlorophenoxyacetic acid, bought from Sigma-Aldrich ($\geq$95%, crystalline). [4] Kinetin, bought from Sangon Biotech ($\geq$98.0%). [5] Naphthalene acetic acid, bought from Shanghai Yuhan Biotech ($\geq$99.5%). The same below.

2.5. Callus Proliferation

According to the methods described by Fan [1] and Iraq [24], 0.20 g of a callus cultured for about 45 days was transferred onto the callus proliferation media (CPM) conducted by an orthogonal experimental design, and the PGRs with three concentrations (1.00~2.00 mg/L PIC, 0~1.00 mg/L 6-BA, and 0.10~0.40 mg/L NAA) were added to the MS medium.

At 0, 5, 10, and 15 d cultured on the CPM, fresh samples of the callus were weighed for drawing proliferation curves in proliferation cultures. The proliferation rate was calculated using the equation below:

$$\%\text{Callus proliferation} = \left(\frac{m}{M}\right) \times 100$$

where 'm' is the total mass of the callus after being cultured at a sampling date point and 'M' is the total mass of the callus used at the beginning of the experiment. Meanwhile, using the acetocarmine method described by Yue et al. [25], the cytological observation of the callus at different states in proliferation cultures over 15 days was carried out.

2.6. Adventitious Buds Induction and Regeneration

The callus induced from root explants was cultured on a proliferation medium (MS supplemented with 1.0 mg/L PIC, 1.0 mg/L 6-BA, and 0.4 mg/L NAA) for 5 subcultures. A well-proliferated callus (1.0 g in FW) was picked up and transferred onto the indirect organogenesis medium which contained MS supplemented with 1.5 mg/L 6-BA, 0.3 mg/L NAA, 30 g/L sucrose, and 10 g/L agar. The state and ratio of normal plantlets were observed and recorded after 25 days. The plantlets with well-developed roots were transplanted to the artificial substrates containing peat and vermiculite (*v:v*, 3:1) for acclimatization under natural light in the culture room. Survival rates of the plantlets were investigated after 30 days.

2.7. Culture Conditions

For callus induction and proliferation, an artificial incubator without light at 25 ± 2 °C and a relative humidity of 45 ± 5% was used. For adventitious buds and plantlet regenerations, the cultures were placed in a growth chamber under light for 14 h with an intensity of 2500~3000 Lx at 25 ± 2 °C, and the relative humidity for buds was 75 ± 10% and 35 ± 5% for plantlets.

2.8. Experimental Design and Statistical Analysis

All the treatments were repeated at least three times. Data were recorded using Excel 2019 (Microsoft Co., Redmond, WA, USA) to calculate the mean value, variance and standard deviation. Data processing, variance analysis (Duncan method, $p < 0.05$), multiple comparisons (LSD method, $p < 0.05$) and correlation charts were performed using SPSS 26.0 and GraphPad Prism 7.0.

3. Results

3.1. Effects of Three PGRs on Callus Induction Rate and Morphogenesis

Table 2 shows the effect of adding PIC, 2,4-D and TDZ to the CIM on the callus induction rates of *A. praecox* roots.

Table 2. Effects of combination of PIC, TDZ and 2,4-D on callus induction from *A. praecox* roots.

	PIC Concentration (mg/L)		TDZ Concentration (mg/L)		2,4-D Concentration (mg/L)		Induction Rate (%)	
	RT [1]	RS [2]	RT	RS	RT	RS	RT	RS
	1.00		0.00		0.50		76.60	25.57
	1.00		0.40		1.00		33.33	11.45
	1.00		0.60		2.00		26.19	23.98
	2.00		0.00		1.00		48.53	26.40
	2.00		0.40		2.00		20.83	25.94
	2.00		0.60		0.50		54.02	11.56
	3.00		0.00		2.00		25.49	36.03
	3.00		0.40		0.50		39.29	23.04
	3.00		0.60		1.00		38.10	16.92
k1 [3]	45.38 a *	17.93 b	50.21 a	32.77 a	56.64 a	19.77 a		
k2 [4]	41.13 a	21.82 a	31.15 b	5.96 b	39.99 b	16.48 a		
k3 [5]	34.29 a	6.88 c	39.43 ab	7.89 b	24.17 b	10.37 b		
R [6]	11.09	14.94	19.06	26.81	32.46	9.40		

[1] Root tips. [2] Root segments. [3] The sum of the numerical values of the test indicators corresponding to the lowest level of concentrations. [4] The sum of the numerical values of the test indicators corresponding to the medium level of concentrations. [5] The sum of the numerical values of the test indicators corresponding to the highest level of concentrations. [6] Range, indicating the magnitude of the factor's influence on the result. * Means followed by the same letters are not significantly different according to Duncan's range test at $p < 0.05$, the same below.

Except for the combination of 3.0 mg/L PIC and 2.0 mg/L 2,4-D, the callus induction rate with root tips as explants was higher than that of the root segment, among which the highest was 76.60%. Among the treatments using root segments as explants, the highest induction rate was only 36.03%, indicating that the root tip was more suitable for callus induction. The R-value shows that 2,4-D and TDZ on callus induction from the root tip and root segment were more effective than that of PIC, up to 32.46 and 26.81, respectively, and of which 2,4-D was the most potent hormone. The optimal PGRs combination was 1.0 mg/L PIC and 0.5 mg/L 2,4-D in an MS medium.

Figure 1 shows the morphogenesis of explant RTs and RMs under optimal culture conditions. It was obvious that the root tips began to swell after 7 days, and some root tips ruptured and dedifferentiated to form calluses at 14 days (Figure 1A,C); however, the morphological changes of the root segments were slightly delayed for 3~5 days (Figure 1B,D), demonstrating that root tips as explants can shorten the time of callus induction.

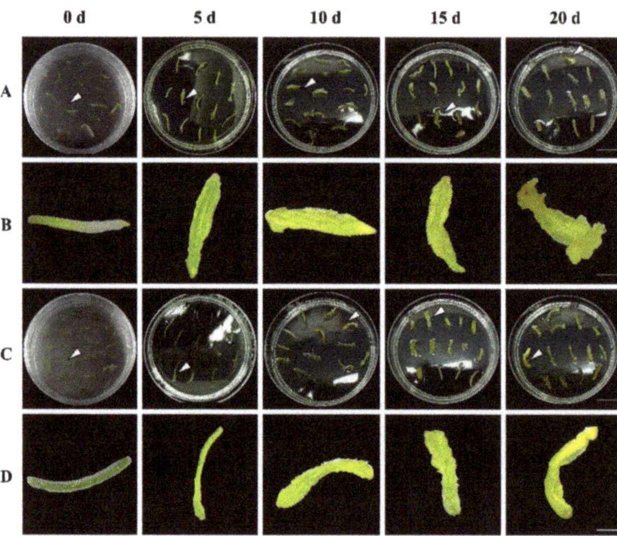

Figure 1. Morphological changes of the root tips and segments of *A. praecox* during callus induction. (**A**) Root tip. Bars = 1.0 cm; (**B**) Explants indicated by arrows in (**A**). Bars = 0.3 cm; (**C**) Root segments. Bars = 1.0 cm; (**D**) Explants indicated by arrows in (**C**). Bars = 0.3 cm.

3.2. Optimization of Three PGRs Combination on Callus Induction from Root Tips

Using the root tip as the best explant material, the effects of different combinations of auxin (PIC, NAA) and cytokinin (KT) on the induction rate of *A. praecox* calluses were further explored, aiming to promote the induction process. The results were shown in Table 3 and Figure 2.

Table 3. Effects of combination of PIC, NAA and KT on callus induction from *A. praecox* root tips.

	PIC Concentration (mg/L)	KT Concentration (mg/L)	NAA Concentration (mg/L)	Induction Rate (%)
	1.00	0.00	0.10	88.89
	1.00	1.00	0.50	61.11
	1.00	1.50	1.00	83.33
	2.00	0.00	0.50	72.22
	2.00	1.00	1.00	66.67
	2.00	1.50	0.10	100.00
	3.00	0.00	1.00	88.89
	3.00	1.00	0.10	88.89
	3.00	1.50	0.50	65.08
k1 [1]	77.78 a	72.22 a	92.59 a	
k2 [2]	79.63 a	82.80 a	66.14 b	
k3 [3]	80.95 a	83.33 a	79.63 ab	
R [4]	3.17	11.11	26.46	

[1] The sum of the numerical values of the test indicators corresponding to the lowest level of concentrations. [2] The sum of the numerical values of the test indicators corresponding to the medium level of concentrations. [3] The sum of the numerical values of the test indicators corresponding to the highest level of concentrations. [4] Range, indicating the magnitude of the factor's influence on the result. Means of k1, k2 and k3 followed by the same letter within a row between different level of harmones are not significantly different at $p \leq 0.05$. The same below.

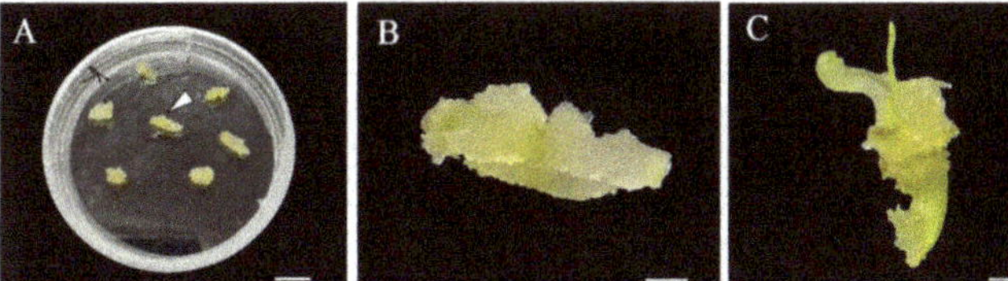

Figure 2. Callus and adventitious buds from root tips of *A. praecox* induced by the combination of hormones PIC, NAA and KT for 20 days. (**A**) The yellowish callus. Bars = 1 cm; (**B**) callus collected, respectively, from the explants indicated by arrows in **A**. Bars = 0.2 cm; (**C**) Adventitious bud. Bars = 0.2 cm.

According to the induction rate, the most optimal PGRs combination was 2.0 mg/L PIC, 1.5 mg/L KT and 0.1 mg/L NAA in MS medium, and the callus induction rate was 100.00%. The induction effects of five treatments were found to be better than the optimized treatment (the highest induction rate was 76.60%) in Table 2. In the range analysis, the R-value of NAA was the largest, followed by KT, and PIC was the smallest, indicating that the effects of the three PGRs on the callus induction rate were in the order of NAA > KT > PIC.

Corresponding to the results in Section 3.1, the root tips began to expand after 7 days of inoculation on the medium, and gradually dedifferentiated into yellowish calluses (Figure 2A,B); when cultured for 20 days, it was obvious that some calluses directly differentiated to produce adventitious buds (Figure 2C).

3.3. Role of Four Endogenous Hormones during Callus Induction

The contents of the four endogenous hormones (IAA, ABA, GA_4 and ZT) in the explants (root tips and root segments, respectively) were investigated to reveal their functions during the primary induction process. Significant differences were found as shown in Figure 3, and the order from high to low was IAA > ABA > GA_4 > ZT, in which the content of IAA was tens to hundreds of times that of ABA and GA4, and tens of thousands of times higher than that of ZT, indicating that IAA may play a crucial role in the process of callus induction.

Furthermore, the IAA content in the root tips of *A. praecox* was significantly higher than that in the root segments at the initial stage, continued to reach the highest level on the 15th day, and then decreased, but the IAA content in the root segments continued to increase in the entire induction process. On the 15th day, the IAA content was the same as that in the root tip and was higher on the 20th day than that in the root tip, which might be the main reason why the initiation of callus induction in the root segment was delayed. The ABA content had a wave-like change that increased and then decreased two times, but no significant difference between different induction times and two types of explants was observed. The contents of GA_4 and ZT in root tips and root segments had an overall upward trend, but there were no significant differences in either.

Figure 4 shows that the ratio of IAA to the other three endogenous hormones contents in *A. praecox* root tips and root segments changed with the prolongation of induction time, and the IAA/ZT ratio in the root tips increased to a maximum value on the fifth day and then gradually decreased (Figure 4A). However, the proportional relationship between IAA to ZT, GA_4 and ABA in the root tips and root segments of plantlets was not obvious (Figure 4B). In addition, IAA/GA_4 had no significant difference in the root tip and root segment, similar to IAA/ABA, and tended to be stable in the root tip, while IAA/GA_4 gradually increased in the root segment and stabilized after 15 days.

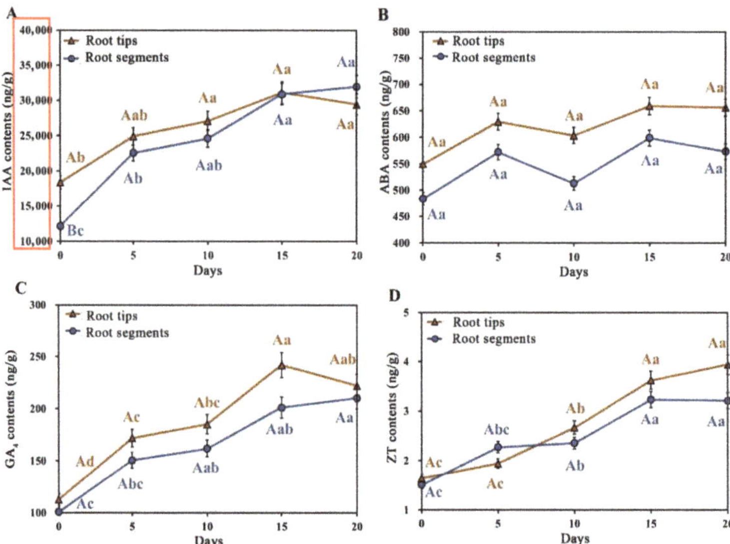

Figure 3. The dynamic changes of endogenous hormones in the root tips and segments of *A. praecox* during callus induction. (**A**) IAA; (**B**) ABA; (**C**) GA$_4$; (**D**) ZT. The different letters indicate significant differences at 5% level by Duncan's range test.

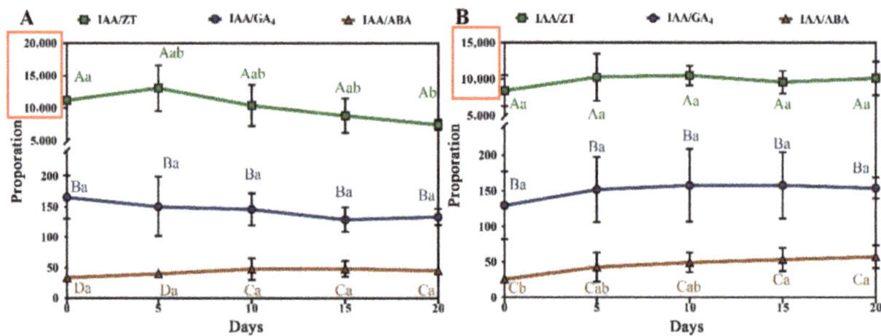

Figure 4. The changes of endogenous hormone proportion in *A. praecox* root tips and segments during callus induction. (**A**): Root tip; (**B**): Root segments. The different letters indicate significant differences at 5% level by Duncan's range test.

3.4. Effects of Three PGRs on Callus Proliferation Rate and State

There were differences between the proliferation effects of calluses among nine treatments containing three kinds of PGRs shown in Table 4. The maximum FW under the addition of 1.0 mg/L PIC, 1.0 mg/L 6-BA, and 0.4 mg/L NAA reached 345.0 mg, the highest proliferation rate of 72.74% after 15 days, proving to be the most stimulating treatment for proliferation. The R-value analysis shows that the impact of three PGRs in the callus proliferation was in the order of PIC > 6-BA > NAA.

There were also differences in the appearance of calluses cultured on proliferation media after 15 days (Figure 5). Milky white and soft callus cells are spherical, uniform in size and regularly arranged, with large and obvious nuclei, dense cytoplasm, and accumulation of starch granules and other nutrients (Figure 5A,B). The callus with a slightly yellowish and slightly compact structure has an obvious difference in cell size, the nucleus is relatively smaller, and the proportion of cytoplasm in the cell volume is increased (Figure 5C,D). The calluses with browning and indistinct differentiation are oblong, have a chaotic arrangement, a large volume difference, indistinct nuclei and a blurred nucleocytoplasmic boundary, and the cell contents can be observed as scattered in the intercellular space under a microscope, indicating cell death (Figure 5E,F). In general, the states of calluses shown in Figure 5A,B were optimal and suitable for subsequent differentiation experiments.

Table 4. Effects of different hormone combinations on proliferation of callus induced from *A. praecox* root explants.

	PIC Concentration (mg/L)	6-BA [1] Concentration (mg/L)	NAA Concentration (mg/L)	Proliferation Rate (%)
	1.00	0.00	0.10	31.69
	1.00	0.40	0.20	43.45
	1.00	1.00	0.40	72.74
	1.50	0.00	0.20	35.43
	1.50	0.40	0.40	23.13
	1.50	1.00	0.10	31.92
	2.00	0.00	0.40	14.87
	2.00	0.40	0.10	43.60
	2.00	1.00	0.20	31.69
k1 [2]	0.493 a	0.273 b	0.357 a	
k2 [3]	0.302 b	0.367 ab	0.363 a	
k3 [4]	0.295 b	0.449 a	0.369 a	
R [5]	0.197	0.176	0.012	

[1] 6-benzylaminopurine, bought from Sigma-Aldrich (suitable for plant cell culture). [2] The sum of the numerical values of the test indicators corresponding to the lowest level of concentrations. [3] The sum of the numerical values of the test indicators corresponding to the medium level of concentrations. [4] The sum of the numerical values of the test indicators corresponding to the highest level of concentrations. [5] Range, indicating the magnitude of the factor's influence on the result.

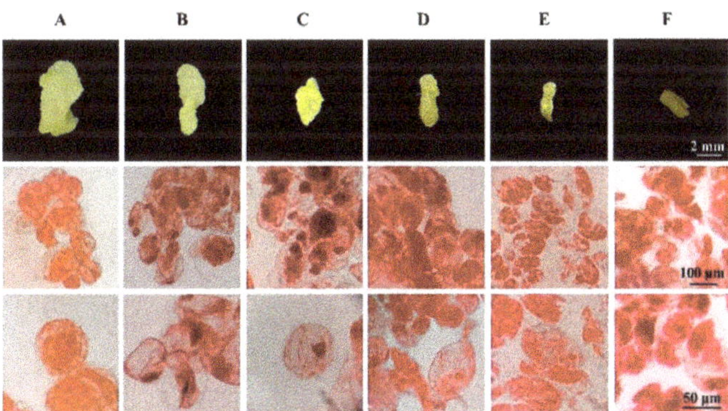

Figure 5. Morphological differences of calluses from *A. praecox* roots after 15 days of culture. (**A,B**) Creamy-white callus; (**C,D**) Yellowish callus; (**E,F**) Yellow-brown callus.

3.5. Hardening and Transplanting of Seedlings Dedifferentiated from Callus

After the callus was transferred to the organogenesis medium for 25 days, part of the callus differentiated into bright green adventitious buds (Figure 6A). These adventitious buds were then cultivated for 90 days, and the color of the upper part of the leaves changed from green to dark green, while the lower part was light green (Figure 6B), from the base of which some roots regenerated (Figure 6C).

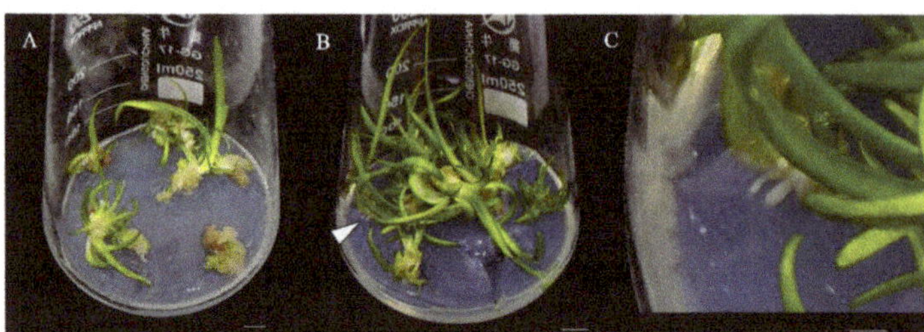

Figure 6. Plantlets regeneration from calluses of *A. praecox* root tips after 25 days and 90 days in organogenesis medium. (**A**) Adventitious bud after 25 days of differentiation. Bars = 1 cm; (**B**) plantlets after 90 days of differentiation. Bars = 1 cm; (**C**) adventitious root indicated by arrows in (**B**). Bars = 0.2 cm.

The culture vessel was partially opened for 3 days for hardening, and the plantlet was transplanted to the substrates containing peat and vermiculite (*v:v*, 3:1) for acclimation. After 30 days, the survival rate of plantlets with a well-developed root system was as high as 100.00% (Figure 7).

Figure 7. Indirect organogenesis tissue culture system of *A. praecox* root explants. (**A**) Germinated seeds after inoculation into MS medium; (**B**,**C**) Seedlings after germination for 20 days (**B**) and

40 days (**C**); (**D**) Root system of 40-day-old seedlings after germination; (**E,F**) 7 days after inoculated on the induction medium, most of the root tips began to swell; (**G,H**) 7 days after inoculated on the induction medium, fewer of root segments began to swell; (**I**) Callus induced from root tips inoculated on the proliferation medium; (**J,K**) Adventitious buds produced through dedifferentiation from the callus and well-developed seedlings on the indirect organogenesis medium for 25 days (**J**) and 90 days (**K**); (**L**) Seedlings after hardening and transplanting on the mixed culture medium for 30 days. (**A–E,G,I–K**) Bars = 1 cm; (**F,H**) Bars = 0.2 cm.

4. Discussion

Agapanthus is a native species to South Africa, and its suitable temperature for growth is below 35 °C. Kunming city is located in the south west region of China, and its climate is similar to that of South Africa, which is more conducive to the growth and development of *A. praecox*.

A. praecox is a perennial herb flower belonging to a monocotyledonous plant with a well-developed root system, which leads us to a new idea of using its roots as explants for the development of tissue culture systems. Chen [26] firstly reported that the induction rates of *Lilium* spp. with different parts of roots (root tips, root segments without tips) as explants were different in an MS medium, and the induction rate with root tips as explants was as high as 96.67%, which was 987.40% higher than that without root tips (only 8.89%). Previous studies have revealed that auxin is biosynthesized in the root tip and maintains a gradient distribution, then is transported to other organs through polarity to stimulate the division, expansion and differentiation of somatic cells, thereby affecting callus formation. The relevant experimental evidence also indicated that the strongest auxin signals appeared in the root tip quiescent center (QC) using the green fluorescent protein (GFP) labeling method [27]. High callus induction of *A. praecox* in this study also proved the importance of the root tip's presence in the root segment as an explant. The callus induction rate on the MS medium containing PIC, KT and NAA reached 100.00%, which is much higher than previous works of Ma (27.71%) [10] and Fan (87.51%) [1]. The expansion of the explant types and the optimization of the induction medium will provide strong technical support for in vitro propagation of Agapanthus plants.

The intrinsic genetics, developmental status of explants and the ratio of plant hormones are the key factors in the process of callus formation that determine the establishment of tissue culture protocol. Existing research data show the types and concentrations and combinations of endogenous PGRs influence the callus induction frequency and callus morphogenesis [28]. In the process of culturing the roots and stem segments of *Centaurium erythraea*, IAA content is higher than that of cytokinins in the roots, and it is also twice the IAA contents in the stem [29]. These PGRs may significantly promote the expansion of dividing cells in explants and callus initiation. In agreement with the conclusions above, our present study found that IAA content was significantly higher than that of ABA, GA_4 and ZT in the root tips of *A. praecox*, and the contents of four endogenous hormones in the root tips were higher than those in the root segments. In addition, the peak of IAA and GA_4 contents in *A. praecox* root explants appeared on the 15th day, coinciding with morphological changes of the explants, which were consistent with Fu et al. [30]. Further observations confirmed that about two weeks after the initiation of induction culture was the critical point for the morphological changes of explants and the formation of calluses. Similar results were reported by Zayed et al., who found that in immature female inflorescence explants of *Phoenix dactylifera*, the contents of IAA and GA_3 reached the highest value at the beginning of callus formation and then gradually decreased; the subsequent peaks of ZT, IAA and ABA contents appeared after an embryogenic callus was initiated and developed [31].

The ratio of endogenous hormone contents also affects callus induction. A higher ratio of auxin to cytokinins was beneficial to improving the callus initiation rate in rice (*Oryza sativa*) anther cultures [32]. Zhang et al. [33] found that the ratio of IAA/ABA in the young leaves of *Sopatholobus suberechtu* increased rapidly on the 14th day of induction, regulating the formation of calluses. In the process of callus induction of *A. praecox* root tips and root segments, the ratio of endogenous hormones also has a similar change pattern with the prolongation of culture time, that is, when the ratio of GA_4/ZT, IAA/ABA is increased, the root tips expanded exceedingly. These results suggest that endogenous hormones play an important role in the dedifferentiation of plant explants to form calluses and exogenous growth regulators should be rationally optimized according to their changing patterns so that the levels of endogenous and exogenous hormones reach a moderate dynamic balance to promote callus formation.

Exogenous additives have a crucial effect on the proliferation of calluses. Previous studies have shown that PIC is a commonly used hormone for callus induction and proliferation culture of *A. praecox* and other monocotyledonous plants, but the optimal concentration of PIC in various types of explants for callus induction was significantly different. The *A. praecox* embryonic callus (EC) cultured on the medium containing 1.5 mg/L PIC had uniform morphology and strong embryonic potential in Zou's research [22]. Yue et al. [25] studied the effect of different concentrations of 6-BA on the proliferation of calluses from *A. praecox* florets and found that the callus under 1.0 mg/L 6-BA was significantly larger than other treatments. Li et al. [34] subcultured the filamentous callus of *Lilium brownie* 'Manissa' and found that the proliferation was faster when the NAA concentration was higher (0.5 mg/L), which was consistent with the optimal PGRs combination for root callus proliferation in this study (1.0 mg/L PIC, 1.0 mg/L 6-BA, and 0.4 mg/L NAA).

5. Conclusions

Current *A. praecox* tissue cultures often use small pedicels and young leaves as explants. However, the collection time of small pedicels is limited, and secondary metabolites in leaves with a high degree of differentiation can easily cause browning of explants and calluses, thus reducing the callus induction rate. This study established an in vitro propagation system of *A. praecox* through indirect organogenesis using its roots as explants. Sterile roots are easy to get and are available without the limitations of collecting seasons, plant growth and development. Abundant calluses and adventitious buds derived from the protocol provide sufficient materials for *A. praecox* micropropagation, somatic embryogenesis research and further genetic improvement in the new varieties.

6. Patents

The results of this paper have applied for a patent (CN 114532227 A).

Author Contributions: Conceptualization, G.C. and X.S.; methodology, X.G. and Q.T.; software, X.G. and Q.L.; formal analysis, X.G.; investigation, X.G. and H.S.; resources, H.S.; data curation, Q.T.; writing—original draft preparation, Q.T.; writing—review and editing, Y.Z., W.W., G.C. and X.S.; visualization, X.G. and Q.T.; supervision, X.S.; project administration, G.C.; funding acquisition, X.S. All authors have read and agreed to the published version of the manuscript.

Funding: This research was funded by National Natural Science Foundation of China (grant number 31901351 and 31370695) and Science and Technology Commission of Shanghai Municipality (grant number 22ZR1428600).

Institutional Review Board Statement: Not applicable.

Informed Consent Statement: Not applicable.

Data Availability Statement: Not applicable.

Acknowledgments: The authors extend their deep appreciation to Yunwei Zi from Yunnan Lan Rong Landscape Engineering Co., Ltd. for providing seeds of *Agapanthus praecox* subsp. *orientalis* 'Big Blue'.

Conflicts of Interest: The authors declare no conflict of interest.

References

1. Fan, X. *Micropropagation In Vitro of Agapanthus praecox ssp. Orientalis 'Big Blue'*; Shanghai Jiao Tong University: Shanghai, China, 2009; pp. 31–45.
2. Zurita, F.; Belmont, M.A.; De Anda, J.; White, J.R. Seeking a way to promote the use of constructed wetlands for domestic wastewater treatment in developing countries. *Water Sci. Technol.* **2011**, *63*, 654–659. [CrossRef] [PubMed]
3. Calheiros, C.; Bessa, V.S.; Mesquita, R.; Brix, H.; Rangel, A.; Castro, P. Constructed wetland with a polyculture of ornamental plants for wastewater treatment at a rural tourism facility. *Ecol. Eng.* **2015**, *79*, 1–7. [CrossRef]
4. Mor, Y.; Halevy, A.H.; Kofranek, A.M.; Reid, M.S. Post-harvest handling of lily of the Nile flowers. *J. Am. Soc. Hortic. Sci.* **1984**, *109*, 494–497.
5. Chen, X.; Lu, L.; Qian, Y.; Fan, Y. Recent advances in germplasm and landscape application of *Agapanthus* spp. *Chin. Landsc. Archit.* **2016**, *32*, 99–105.
6. Snoeijer, W. *Agapanthus: A Revision of the Genus*; Timber Press: Portland, OR, USA, 2004.
7. Supaibulwatana, K.; Mii, M. Organogenesis and somatic embryogenesis from young flower buds of *Agapanthus africanus* Hoffmanns. *Plant Biotechnol.* **1997**, *14*, 23–28. [CrossRef]
8. Suzuki, S.; Nakano, M. Agrobacterium-mediated transformation in Liliaceous ornamental plants. *Jpn. Agric. Res. Q.* **2002**, *36*, 119–127. [CrossRef]
9. Nakano, M.; Tanaka, S.; Oota, M.; Ookawa, E.; Suzuki, S.; Saito, H. Regeneration of diploid and tetraploid plants from calluse-derived protoplast of *Agapanthus praecox* ssp. orientalis (Leighton) Leighton. *Plant Cell Tissue Organ Cult.* **2003**, *72*, 63–69. [CrossRef]
10. Ma, K. *Establishment of Regeneration System Cultured In Vitro of Agapanthus praecox ssp. Orientalis 'Big Blue'*; Northeast Forestry University: Harbin, China, 2007; pp. 24–26.
11. Yaacob, J.S.; Syafawati, J. *Tissue Culture, Cellular Behaviour and Pigment Analysis of Agapanthus praecox ssp. minimus*; University of Malaya: Kuala Lumpur, Malaysia, 2013.
12. Kang, L. Preliminary report on the experiment of regeneration system of *Agpanthus praecox* ssp. minimus 'Storms River'. *Contemp. Hortic.* **2009**, *12*, 7–8 + 25.
13. Hu, Z.; He, Y. Study on tissue culture and plantlet regeneration of *Agapanthus praecox* ssp. orientalis 'Big Blue'. *North. Hortic.* **2011**, *10*, 118–120.
14. Suzuki, S.; Oota, M.; Nakano, M. Embryogenic callus induction from leaf explants of the Liliaceous ornamental plant, *Agapanthus praecox* ssp. orientalis (Leighton) Leighton—Histological study and response to selective agents. *Sci. Hortic.* **2002**, *95*, 123–132. [CrossRef]
15. Suzuki, S.; Supaibulwatana, K.; Mii, M.; Nakano, M. Production of transgenic plants of the Liliaceous ornamental plant *Agapanthus praecox* ssp. orientalis (Leighton) Leighton via Agrobacterium-mediated transformation of embryogenic calli. *Plant Sci.* **2001**, *161*, 89–97. [CrossRef]
16. Yue, J.; Dong, Y.; Wang, X.; Sun, P.; Wang, S.; Zhang, X.; Zhang, Y. A regeneration system for organogenesis and somatic embryogenesis using leaves of *Agapanthus praecox* as explants. *Chin. Bull. Bot.* **2020**, *55*, 588–595.
17. Wang, Y.; Fan, X.; Wang, G.; Zhang, D.; Shen, X. Regeneration of *Agapanthus praecox* ssp. orientalis 'Big Blue' via somatic embryogenesis. *Propag. Ornam. Plants* **2012**, *12*, 148–154.
18. Baskaran, P.; Staden, J.V. Rapid in vitro micropropagation of *Agapanthus praecox*. *S. Afr. J. Bot.* **2013**, *86*, 46–50. [CrossRef]
19. Yaacob, J.S.; Taha, R.M.; Mohajer, S. Organogenesis induction and acclimatization of African blue Lily (*Agapanthus praecox* ssp. minimus). *Int. J. Agric. Biol.* **2014**, *16*, 57–64.
20. Yaacob, J.S.; Saleh, A.; Elias, H.; Abdullah, S.; Mahmad, N.; Mohamed, N. In vitro regeneration and acclimatization protocols of selected ornamental plants (*Agapanthus praecox*, *Justicia betonica* and *Celosia cristata*). *Sains Malays.* **2014**, *43*, 715–722.
21. Zou, M. *Physiological and Biochemical Aspects of Picloram Regulates the Embryonic Potential of Callus in Agapanthus*; Shanghai Jiao Tong University: Shanghai, China, 2015.
22. Ren, N.; Zou, M.; Chen, G.; Shen, X. Difference analysis of redox levels in the callus of *Agapanthus praecox* ssp. orientalis with different embryonic ability. *Mol. Plant Breed.* **2022**, *20*, 1281–1288.
23. Chen, X.; Lv, X.; Yin, L.; Feng, J.; Zhang, D. Comparison with morphology and germination of *Agapanthus praecox* seeds under different cultivation region and environment. *Acta Agric. Shanghai* **2020**, *4*, 43–47.
24. Iraqi, D.; Francine, M. Analysis of carbohydrate metabolism enzymes and cellular contents of sugars and proteins during spruce somatic embryogenesis suggests a regulatory role of exogenous sucrose in embryo development. *J. Exp. Bot.* **2001**, *52*, 2301–2311. [CrossRef]
25. Yue, J.; Wei, Z.; Li, P.; Wang, Z. Influence on callus induction and physiological characters of callus proliferation in *Agapanthus praecox* ssp. orientalis by carbon sources. *Acta Bot. Boreali Occident. Sin.* **2021**, *41*, 439–449.
26. Chen, M.; Zhang, J.; Zhou, Y.; Gu, J.; Shen, Q.; Zhang, P. Study on factors influence somatic embryogenesis fromoots of *Lilium* spp. *J. Nucl. Agric. Sci.* **2015**, *29*, 1494–1501.
27. Tian, H.; Niu, T.; Yu, Q.; Quan, T.; Ding, Z. Auxin gradient is crucial for the maintenance of root distal stem cell identity in Arabidopsis. *Plant Signal. Behav.* **2013**, *8*, e26429. [CrossRef] [PubMed]
28. Zhao, L.; Wang, X.; Zhang, J.; Wang, S. Plant tissue culture and its research advances and application in the forage plants. *Pratacult Sci.* **2011**, *28*, 1140–1148.

29. Trifunović-Momčilov, M.; Motyka, V.; Dragićević, I.Č.; Petrić, M.; Jevremović, S.; Malbeck, J.; Holík, J.; Dobrev, P.I.; Subotić, A. Endogenous phytohormones in spontaneously regenerated *Centaurium erythraea* Rafn. plants grown in vitro. *J. Plant Growth Regul.* **2016**, *35*, 543–552. [CrossRef]
30. Fu, F.; Feng, Z.; Qu, B.; Li, W. Relatianships between induction rate of embryonic callus and endogenous hormones content in Maize. *J. Nucl. Agric. Sci.* **2006**, *1*, 10–14.
31. Zayed, E.M.M.; Abd Elbar, O.H. Morphogenesis of immature female inflorescences of date palm in vitro. *Ann. Agric. Sci.* **2015**, *60*, 113–120. [CrossRef]
32. Wang, X.; Shi, X.; Wu, X.; Wang, X.; Zhou, K. The influence of endogenous hormones on culture capability of different explants in rice. *Sci. Agric. Sin.* **2004**, *12*, 1819–1823.
33. Zhang, Y.; He, g.; Rong, G.; Liu, X.; Ni, S. Changes of endogenous hormones during the culture of callus from *Sopatholobus suberechtus. Biotechnol. Bull.* **2017**, *33*, 66–70.
34. Li, Y. Callus induction and plant regeneration from floral organ of *Lilium* 'Manissa'. *Chin. J. Trop. Crops* **2013**, *34*, 1507–1512.

 *horticulturae*

Essay

Rapid and Efficient Regeneration of *Rhododendron decorum* from Flower Buds

Hairong Wu, Qian Ao, Huie Li * and Fenfang Long

College of Agriculture, Guizhou University, Guiyang 550000, China
* Correspondence: lihuiesh@126.com

Abstract: *Rhododendron decorum* is a woody species with high ornamental and medical value. Herein, we introduce a novel in vitro regeneration method for *R. decorum*. We used flower buds to develop an efficient and rapid plant regeneration protocol. Sterile flower buds of *R. decorum* of a 2 cm size were used as explants to study the effects of the culture medium and plant growth regulators on the callus induction and adventitious shoot differentiation, proliferation, and rooting. According to the results, the optimal medium combination for callus induction was WPM + 1 mg/L TDZ + 0.2 mg/L NAA, and its induction rate reached 95.08%. The optimal medium combination for adventitious shoot differentiation from the callus was WPM + 0.5 mg/L TDZ + 0.1 mg/L NAA, and its differentiation rate reached 91.32%. The optimal medium combination for adventitious shoot proliferation was WPM + 2 mg/L ZT + 0.5 mg/L NAA, for which the proliferation rate reached 95.32% and the proliferation coefficient reached 9.45. The optimal medium combination for rooting from adventitious shoots was WPM + 0.1 mg/L NAA + 1 mg/L IBA, and its rooting rate reached 86.90%. The survival rates of the rooted regenerated plantlets exceeded 90% after acclimatization and transplantation. This regeneration system has the advantages of being simple and highly efficient, and it causes little damage to the shoots of the mother plants, laying a foundation for the plantlet propagation, genetic transformation, and new-variety breeding of *R. decorum*.

Keywords: *Rhododendron decorum*; tissue culture; in vitro culture; flower buds; callus induction; adventitious shoot; regeneration system; plant growth regulators

Citation: Wu, H.; Ao, Q.; Li, H.; Long, F. Rapid and Efficient Regeneration of *Rhododendron decorum* from Flower Buds. *Horticulturae* **2023**, *9*, 264. https://doi.org/10.3390/horticulturae9020264

Academic Editor: Jean Carlos Bettoni

Received: 14 January 2023
Revised: 11 February 2023
Accepted: 12 February 2023
Published: 15 February 2023

1. Introduction

Rhododendron decorum Franch. is a woody ornamental plant species that belongs to the subgenus *Hymenanthes* of *Rhododendron* of Ericaceae. It is distributed in southwest China and northeast Myanmar, and it grows under forests at an altitude of 1000–3700 m. Its inflorescence is huge, white, and graceful, with fragrance and late flowering, which make the plant popular. This plant is suited to a cool and humid climate and humus-rich and slightly acidic soil. It requires sunlight, but it is not resistant to strong sunlight [1,2]. In addition to its high ornamental value, *R. decorum* also has high medicinal value, and its roots, branches, and leaves are all traditional medicinal materials for local people in China [3]. Its corolla is rich in various amino acids, polysaccharides, minor elements, and other substances, and is a good food resource for locals [4–7]. In recent years, given the indiscriminate exploitation of its wild resources, the habitat of *R. decorum* has been seriously damaged, and its resources are being increasingly endangered [8]. Thus, its propagation and conservation are urgently needed.

The species of the subgenus *Hymenanthes* are difficult to propagate by cuttings, and particularly *R. decorum* [9]. The species of this subgenus are mainly propagated by seeds, but the seeds are small, the germination rate is low, the seedling growth cycle is long, and the traits of the seed seedlings are prone to variation [10]. Tissue culture technology can effectively solve these problems. The establishment of a tissue regeneration system not only provides an efficient technique for the propagation of *R. decorum*, but it also serves as an important prerequisite for genetic transformation and gene function verification studies [11–14]. Previous studies on *R. decorum* have mainly focused on genetic polymorphism [15,16], medicinal components [17,18], interspecific hybridization [19,20], and mycorrhizal fungi [21,22]. However, an efficient tissue regeneration system has not been reported.

Therefore, the effects of the flower bud culture medium and plant growth regulators on the callus induction and adventitious shoot differentiation, proliferation, and rooting of *R. decorum* were studied to provide a basis for its efficient propagation, further research on breeding, and related genetic studies.

2. Materials and Methods

2.1. Plant Materials

The flower buds of *R. decorum* were collected from a rhododendron nursery in Guizhou Province, China, on 10 December 2021. The flower buds were soaked in tap water for 2 h. The outer sepals were removed and rinsed five times for 1 min with tap water, and the surfaces were dried with tissue paper. Then, the flower buds were sterilized with 75% alcohol (*v/v*) for 1 min, washed with sterile water five times, sterilized with 5% sodium hypochlorite (*v/v*) for 10 min and/or 0.1% mercuric chloride (*v/v*) for 10–30 min to compare the efficiency of the sterilization, and finally rinsed five times for 1 min with sterile distilled water. All the remaining sepals were finally removed. The white 2 cm sized sterile flower buds in the medium for tissue culture and their contamination and survival rates were observed after 2 weeks and were determined using the following equation: contamination rate (%) = number of flower buds contaminated/total number of flower buds × 100%; survival rate (%) = number of flower buds that survived/total number of flower buds × 100%.

2.2. Culture Medium and Conditions

The uniform flower bud explants were cultured vertically on the media. The culture media included Murashige and Skoog (MS) [23], Woody Plant Medium (WPM) [24], and Driver and Kuniyuki Walnut (DKW) [25]. Agar as the gelling agent (7 g/L) and sucrose (30 g/L) were added to each medium, and 15 g/L of sucrose was added to the rooting culture. The pH values of the media were adjusted to 5.8 with 1 N NaOH or 1 N HCl, and the media were autoclaved at 121 °C for 20 min. The callus induction culture was incubated in artificial climate chambers without light at 25 ± 1 °C, and the other cultures were incubated in artificial climate chambers under a 12 h light cycle with a light intensity of 20–30 $\mu mol \cdot m^{-2} \cdot s^{-1}$ at 25 ± 1 °C. All chemicals and reagents used in this study were purchased from Solarbio Company, Beijing, China.

2.3. Callus Induction

The flower buds were placed on the basal media MS, WPM, and DKW in the dark for the selection of the suitable basal medium for callus induction. Then, after 2 weeks, the flower buds were placed on a basal WPM containing different combinations of thidiazuron (TDZ) (0.1, 0.5, and 1 mg/L) and naphthaleneacetic acid (NAA) (0.1, 0.2, and 0.5 mg/L) in the dark for the screening of the best suitable combination of plant growth regulators for callus induction. All induction rates were determined after 30 days of the culture of the flower buds with the following equation: callus induction rate (%) = induction of the number of flower buds from the callus/total number of flower buds × 100%.

2.4. Shoot Induction

Calluses were transferred to a WPM containing different plant growth regulator combinations of TDZ (0.1, 0.5, and 1 mg/L) and NAA (0.1, 0.2, and 0.5 mg/L) under light conditions for adventitious shoot induction. All induction rates were determined after 30 days of the culture of the calluses with the following equation: induction rate of adventitious shoots (%) = number of callus-induced adventitious shoots/total number of calluses $\times$ 100%.

2.5. Shoot Proliferation

Adventitious shoots up to 1 cm high with two intact leaves were defined as effective shoots. They were transferred to the WPM containing different combinations of zeatin (ZT) (1, 2, and 3 mg/L) and NAA (0.1, 0.2, and 0.5 mg/L) for shoot proliferation. All induction rates were determined after 30 days of culture of the shoots with the following equation: adventitious shoot proliferation rate (%) = number of shoot-induced effective shoots/total number of shoots $\times$ 100%; proliferation coefficient = number of proliferating shoots/total number of shoots.

2.6. Cytological Observation

The cytological changes during callus induction, callus proliferation, and the differentiation of complete adventitious shoots were observed on paraffin sections prepared every 20 days of culture. The shoots were initially fixed in FAA solution (formalin, acetic acid, and 70% ethanol in a 1:1:13 ratio) for 24 h, dehydrated in the tertiary butyl alcohol series, and embedded in liquid paraffin [26]. The transverse sections of a 5–10 μm thickness were sliced using a rotary microtome (Erma, Yoshikawa, Japan). The sections were stained with 1% (w/v) safranin and mounted in glycerin. The specimens were microphotographed with a Leica DM 750 LED microscope, as described.

2.7. Rooting and Acclimatization

After the adventitious shoot proliferation culture, adventitious shoots of approximately 2 cm in height were cultured vertically on a WPM for rooting. Effective shoots were transferred to WPM medium combinations of indole-3-butyric acid (IBA) (0.5 and 1 mg/L) and NAA (0.1 and 0.5 mg/L) for rooting. The rooting status was recorded after 30 days. Regenerated plantlets with six fully developed leaves were transferred from an artificial climate and incubated under natural light for 5–7 days. The tissue culture bottles were opened for 2 days for the refinement of the seedlings. Plantlets were transferred to pots containing a sterilized substrate (vermiculite:peat soil = 3:1) at 121 °C for 1 h, and then covered with polyethylene film. Water was sprayed daily to create saturated conditions of relative humidity. The plantlets remained in a growth room at 24 $\pm$ 1 °C under a 16 h light cycle at a 35 μmol$\cdot$m$^{-2}\cdot$s^{-1} photosynthetic photon flux density, provided by fluorescent lamps. After 4 weeks of acclimatization, the survival rate of the regenerated plantlets was determined using the following equation: rooting rate (%) = number of rooting shoots/numbers of transplanted shoots $\times$ 100%; survival rate (%) = number of surviving regenerated plants/total number of transplanted regenerated plants $\times$ 100%.

2.8. Statistical Analysis of Data

In the regeneration experiments, one flower bud was cultured on the medium of each bottle, and 20 flower buds were cultured for each experiment with three replicates. The results are presented as means $\pm$ standard errors (SEs). The mean and SE values were assessed using Microsoft Excel 2019. IBM SPSS Statistics v26 (Armonk, NY, USA) was used for the variance analyses. The significance of the differences among the mean values was assessed using Duncan's multiple range test at $p \leq 0.05$. The results are presented as the mean $\pm$ SE of three replicates.

3. Results

3.1. Sterilization of Flower Buds

The results of the different combinations of sterilization methods showed that sodium hypochlorite (5%) or mercury chloride (0.1%) resulted in a high rate of contamination and the low survival rate of the explants, whereas the combination of sodium hypochlorite and mercury chloride resulted in a lower rate of contamination than that of sodium hypochlorite and mercury chloride alone (Table 1). The sterilization combination of 5% sodium hypochlorite for 10 min followed by 0.1% mercury chloride for 30 min yielded the best result.

Table 1. Effects of different sterilization methods on explants.

Sodium Hypochlorite (min)	Mercury Chloride (min)	Contamination Rate (%)	Survival Rate (%)
10	0	100.00 ± 0.00 a	0.00 ± 0.00 g
0	10	65.84 ± 2.84 b	34.16 ± 1.54 f
0	20	46.12 ± 3.03 c	53.88 ± 2.08 e
0	30	34.67 ± 4.04 d	65.33 ± 3.48 d
10	10	16.63 ± 1.42 e	83.37 ± 2.02 c
10	20	5.72 ± 1.81 f	94.28 ± 0.69 b
10	30	0.00 ± 0.00 g	100.00 ± 0.00 a

Different letters in same column are significantly different at $p < 0.05$ (DMRT).

3.2. Effects of Different Media on Callus Induction

Calluses could be induced and grew well in the flower buds on the WPM, DKW, and MS media without plant growth regulators, but the induction rate varied significantly (Table 2). Two weeks after the flower buds were cultured in the different basal media, the bases of the flower buds began to swell and produce white calluses. No differences in the sizes or colors of the calluses were observed. The induction rates in the WPM and DKW media were significantly higher than the induction rate in the MS medium, and the WPM medium had the highest induction rate at 93.97%.

Table 2. Effects of basal media on callus induction rate of *R. decorum* flower buds.

Medium	Induction Rate (%)
WPM	93.97 ± 1.51 a
DKW	85.72 ± 3.31 b
MS	65.25 ± 1.93 c

Different letters in same column are significantly different at $p < 0.05$ (DMRT).

3.3. Effect of Plant Growth Regulators on Callus Induction

TDZ and NAA were used to induce the *R. decorum* calluses, and the induction effects are shown in Table 3. After the callus induction in the medium supplemented with plant growth regulators for 2 weeks in the dark, the flower bud bases expanded and produced white and soft calluses. The best callus induction combination was WPM + 1 mg/L TDZ + 0.2 mg/L NAA, and the induction rate was 95.08%. The induction rate increased with the TDZ and NAA concentrations, but the rate decreased when the concentration exceeded the optimal concentration (1 mg/L TDZ and 0.2 mg/L NAA).

Table 3. Effects of TDZ and NAA on callus induction.

TDZ (mg/L)	NAA (mg/L)	Induction Rate (%)
0.1	0.1	36.67 ± 1.53 f
0.1	0.2	51.02 ± 5.29 e
0.1	0.5	54.66 ± 3.51 e
0.5	0.1	65.73 ± 1.52 d
0.5	0.2	72.33 ± 3.79 c
0.5	0.5	78.67 ± 0.58 c
1	0.1	85.66 ± 2.08 b
1	0.2	95.08 ± 1.02 a
1	0.5	78.45 ± 5.13 c

Different letters in same column are significantly different at $p < 0.05$ (DMRT).

3.4. Effects of Different Plant Growth Regulators on Adventitious Shoot Induction

The calluses began to differentiate adventitious shoots by culture on the shoot induction media after 2 weeks (Table 4). The best induction rate was 91.32% on the medium combination WPM + 0.5 mg/L TDZ + 0.1 mg/L NAA. The induction rate initially increased, and it then decreased with the increasing TDZ concentrations. In addition, when the TDZ concentration was low, a high concentration of NAA improved the induction rate of the adventitious shoots, and when the concentration of TDZ was high, a low concentration of NAA was more suitable for the adventitious shoot induction. Although the concentration of the growth regulator exceeded the optimal value, the growth rate of the adventitious shoots decreased with their increasing induction rate, but the adventitious shoot leaves were bright green and healthy. When cultured under a light environment, most of the calluses differentiated into adventitious shoots, and a few calluses did not, which exhibited browning and eventually died (Figure 1).

Table 4. Effects of TDZ and NAA on adventitious shoot induction.

TDZ (mg/L)	NAA (mg/L)	Induction Rate (%)
0.1	0.1	46.05 ± 2.45 f
0.1	0.2	61.21 ± 4.35 d
0.1	0.5	74.00 ± 2.36 c
0.5	0.1	91.32 ± 2.36 a
0.5	0.2	84.02 ± 1.56 b
0.5	0.5	73.07 ± 4.36 c
1	0.1	58.33 ± 0.57 d
1	0.2	55.53 ± 0.88 d
1	0.5	51.45 ± 2.03 e

Different letters in same column are significantly different at $p < 0.05$ (DMRT).

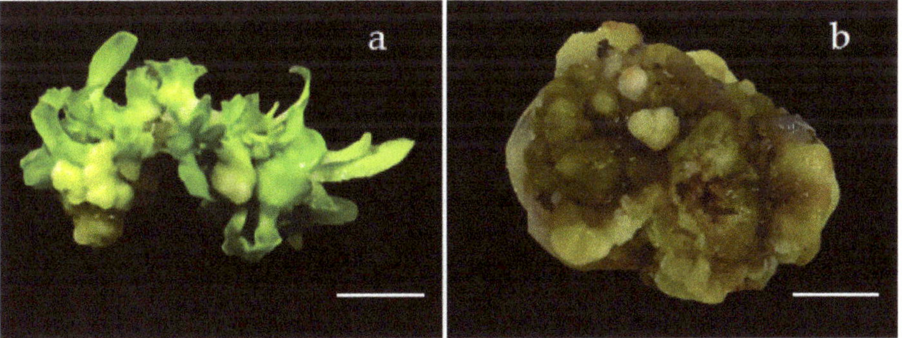

Figure 1. Adventitious shoots induced within 30 days: (**a**) adventitious shoots of callus differentiation; (**b**) browning callus. Bars = 1.0 cm.

3.5. Effects of Different Plant Growth Regulators on Adventitious Shoot Proliferation

After 1 week of culture, adventitious shoots began to proliferate. Low and high concentrations of NAA and ZT produced adventitious shoots with low proliferation coefficients and weak growth shoots with abnormal leaves (Figure 2). The combination of 0.5 mg/L NAA and 2 mg/L ZT resulted in the best adventitious shoot proliferation rate of 95.32% and a proliferation coefficient of 9.42 (Table 5). The shoot leaves were fresh green, the shoot stems were strong, and the growth rate was fast. The deformed adventitious shoots produced by proliferation had a slow growth rate and death during the culture process.

Figure 2. Adventitious shoot proliferation of calluses on different medium combinations: (**a**) adventitious shoots with high proliferation coefficient containing green and normal leaves or (**d**) hyperhydric; (**b**) adventitious shoots with low proliferation coefficient containing green normal leaves or (**c**) yellow-green and abnormal leaves. Scale bars = 1.0 cm.

Table 5. Effects of NAA and ZT on adventitious shoot proliferation.

NAA (mg/L)	ZT (mg/L)	Proliferation Rate (%)	Proliferation Coefficient	Growth Status
0.1	1	66.34 ± 3.06 d	4.67 ± 0.47 f	Leaves were green and healthy, shoots were small
0.2	1	74.03 ± 3.12 c	5.57 ± 0.15 e	Leaves were green and healthy, shoots were small
0.5	1	76.67 ± 3.21 c	7.47 ± 0.36 c	Leaves were green and healthy, shoots were stout
0.1	2	81.47 ± 6.51 b	8.67 ± 0.11 a	Leaves were green and healthy, shoots were stout
0.2	2	88.76 ± 4.51 a	8.97 ± 0.44 a	Leaves were green and healthy, shoots were stout
0.5	2	95.32 ± 2.56 a	9.42 ± 0.27 b	Leaves were green and healthy, shoots were stout
0.1	3	89.32 ± 4.14 a	4.21 ± 0.31 b	Leaves were yellow-green and small, with a few abnormal leaves
0.2	3	84.34 ± 2.65 b	2.64 ± 0.09 c	Leaves were yellow-green and hyperhydric, with many abnormal leaves
0.5	3	74.43 ± 2.52 c	1.95 ± 0.21 e	Leaves were yellow-green and hyperhydric, with many abnormal leaves

Different letters in same column are significantly different at $p < 0.05$ (DMRT).

3.6. Cytological Observation of Shoot Regeneration

Cytological observation at three stages of the adventitious shoot formation from the calluses showed that white and fluffy calluses formed at the bases of the shoots after 20 days of culture (Figure 3a,d). The cells divided rapidly and were closely arranged. After 40 days of culture, the proliferation of the calluses was completed, and the cells continued to divide and expand (Figure 3b,e). After 60 days of culture, adventitious shoots differentiated from the calluses, which are clearly shown on the paraffin section (Figure 3c,f), suggesting that the shoot regeneration from the calluses was indirect.

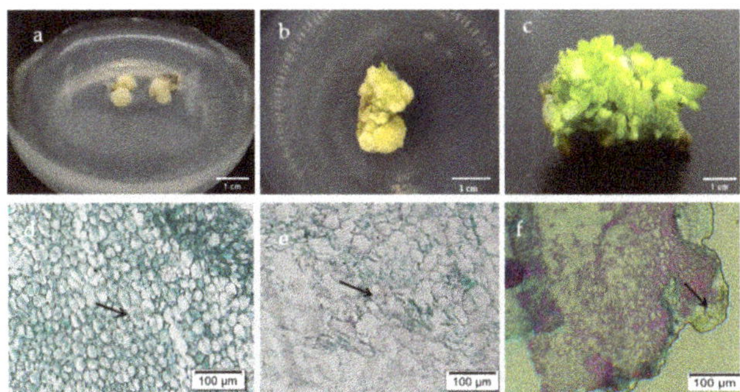

Figure 3. Observation on paraffin section of shoot regeneration of *R. decorum*: (**a**) white fluffy callus; (**b**) proliferation of fluffy calluses; (**c**) differentiation of adventitious shoots from calluses; (**d**) arrow shows cells that divided rapidly after 20 days of culture; (**e**) arrow shows cells that divided and expanded after 40 days of culture; (**f**) arrow shows adventitious shoots that differentiated from callus after 60 days of culture. Bars = 100 μm.

3.7. Rooting and Acclimatization

The adventitious shoots were rooted in a medium containing NAA and IBA. The results are shown in Table 6. Adventitious roots were induced after 30 days of culture (Figure 4). The best medium combination for rooting was WPM + 0.1 mg/L NAA + 1 mg/L IBA, with a rooting rate of 86.9%. A high concentration of IBA and low concentration of NAA induced healthy adventitious roots, the rooting speed was fast, and no calluses appeared at the bases of the adventitious shoots (Figure 4a,b). However, a high concentration of NAA induced slow rooting with fewer roots and a large number of calluses at the base. In addition, after 30 days of transplantation, the survival rates of the rooted plantlets were higher than 90%, and the plantlets were in good growth condition (Figure 4c,d).

Table 6. Effects of NAA and IBA on adventitious shoot rooting.

NAA (mg/L)	IBA (mg/L)	Rooting Rate (%)
0.1	0.5	59.33 ± 4.09 b
0.1	1	86.90 ± 2.97 a
0.5	0.5	21.53 ± 2.49 d
0.5	1	42.67 ± 2.84 c

Different letters in same column are significantly different at $p < 0.05$ (DMRT).

Figure 4. Roots and acclimatization of rooted plantlets: (**a**) roots in WPM medium after 30 days of culture; (**b**) rooted plantlets; (**c,d**) acclimatized plantlets from ex vitro rooting after 30 days. Bars = 1.0 cm.

4. Discussion

There are 8–10 small flower buds in the racemes of *R. decorum* before blossom. Hence, its buds can be used as explants for developing tissue culture systems. Wang et al. [27] reported that the performance of a tissue culture using the flower buds of *Rhododendron hybrids* cn.'Dr. Tjebbe' as explants had a low contamination rate and high induction rate. The use of flower buds as explants neither harms the mother plant nor causes mutation because small flower buds are wrapped in sepals, which are easy to sterilize and do not easily brown [28]. These features may be the reasons that the survival rates of the flower buds subjected to sterilization treatment reached 100% in this study.

In the process of tissue culture, the basal medium is the main source of the nutrients required by the explants, and the appropriate basal medium type is essential for the rapid and healthy growth and development of explants from different plant species [29–31]. Different parts of the same genotype can have different growth responses to a certain basal medium [32–34]. Therefore, the screening of the basal medium is beneficial to the smooth progress of the tissue culture. For example, in the rapid propagation of *Vaccinium ashei* Reade, which is a species belonging to Ericaceae, the basal medium DKW is more suitable for the rapid propagation culture of the stem segments than WPM or MS [35]. By contrast, in the present study, WPM was the most suitable basal medium for culturing the flower buds of *R. decorum*, which is a species that belongs to the same family, followed by DKW and MS. This difference may be caused by the low nitrogen demand of *R. decorum*. WPM, with a low nitrogen content, met this demand [36,37], whereas the DKW and MS media, with relatively high contents of nitrogen, did not [38].

Plant growth regulators are minor natural compounds that are produced in plant metabolism, and they regulate the growth and development processes of plants [39,40]. In tissue culture, plant growth regulators play a key regulatory role in the formation of calluses and the induction and proliferation of adventitious shoots. This effect is affected by the concentration and type of growth regulator, as well as by the interaction between growth regulators. A reasonable ratio of growth regulators is especially crucial for the induction and proliferation of adventitious shoots; thus, the formulas of the growth regulators in tissue culture systems vary among [41,42]. In the tissue culture process of woody plants, the commonly used plant growth regulators are TDZ, NAA, IBA, and kinetin [43,44]. However, different species have different degrees of sensitivity to plant growth regulators. TDZ is currently considered to be one of the most active cytokinins, with good induction effects on calluses and adventitious shoots, and it is widely used in the regeneration processes of *R. calophytum* [45], *R. delavayi* [46], and *Rhododendron* 'Fragrantissimum Improved' [47]. Similarly, in the present study, the plant growth regulator combination 0.2 mg/L NAA + 0.5 mg/L TDZ was the most suitable for the induction of the adventitious shoots of the flower buds of *R. decorum*, and the adventitious shoots grew rapidly and healthily.

Proliferation culture is an indispensable step in the process of tissue culture, and the proliferation coefficient can reflect the speed and efficiency of the propagation in vitro and is an important index for estimating the total production of plantlets. Previous studies have shown that ZT is the most ideal exogenous hormone to induce differentiation and proliferation during the tissue culture process of *Rhododendron* [48]. When the ZT concentration was extremely high, the adventitious shoot proliferation rate and proliferation coefficient were also increased, but the adventitious shoots were prone to elongation, thinness, and deformity, and they were hyperhydric [49].

Rooting induction is another important step in the establishment of a rapid propagation system in vitro. IBA and NAA are commonly used for the rooting induction of *Rhododendrons* species plantlets, which have a high rooting rate and multiple and strong roots, as well as a high survival rate after transplantation. For example, Elmongy et al. [50] found that 2 mg/L of IBA was suitable for the rooting of two azalea cultivars: 'Mingchao' and 'Zihudie'. Almeida found that 1 mg/L of IBA + 2 mg/L of NAA was suitable for the rooting of *R. ponticum* [51]. In this study, NAA and IBA were found to be suitable for inducing the rooting of *R. decorum* adventitious shoots. The results showed that high concentrations of NAA resulted in slow rooting and weak roots, whereas high concentrations of IBA were conducive to the induction of root development and root health. A high concentration of IBA was conducive to the induction of root development, and the root system was developed and strong. The root systems of the regenerated plants were thick, and the survival rate of transplantation was high.

Tissue culture involves the induction, proliferation, rooting, and transplanting of plant materials within a sterile and controlled environment. In this study, the flower buds of *R. decorum* were used for regeneration. The protocol was as follows: (1) the induction of calluses using sterilized flower buds in WPM + 1 mg/L TDZ + 0.2 mg/L NAA in the dark; (2) the induction of adventitious shoots using fluffy calluses in WPM + 0.5 mg/L TDZ + 0.1 mg/L NAA under a 12 h light cycle; (3) the proliferation of adventitious shoots using shoots up to 1 cm in WPM + 2 mg/L ZT + 0.5 mg/L NAA under a 12 h light cycle; (4) the rooting of adventitious shoots using shoots up to 2 cm in WPM + 0.1 mg/L NAA + 1 mg/L IBA under a 12 h light cycle; (5) the transplanting of the rooted plantlets after acclimation under a 16 h light cycle.

5. Conclusions

This study established an in vitro propagation system for *R. decorum* through indirect organogenesis using its flower buds as explants by screening the influencing factors, such as different sterilization method, basal medium, and plant growth regulator combinations. Tissue culture using flower buds as explants has the advantages of thorough sterilization, efficient and rapid regeneration, and it causes little damage to the mother plants. This regeneration system is simple and highly efficient, and it lays a foundation for the plantlet propagation, genetic transformation, and new-variety breeding of *R. decorum*.

Author Contributions: Conceptualization, H.W. and F.L.; methodology, H.W.; validation, Q.A.; formal analysis, H.W., Q.A. and F.L.; investigation, H.W., Q.A., H.L. and F.L.; data curation, H.W. and H.L.; writing, H.W., H.L., Q.A. and H.L.; visualization, Q.A. and H.L.; supervision, H.L.; project administration, H.L.; funding acquisition, H.L. All authors have read and agreed to the published version of the manuscript.

Funding: This research was funded by the National Natural Science Foundation of China (32260415).

Data Availability Statement: Not applicable.

References

1. Flora of China Editorial Committee; Chinese Academy of Sciences. *Flora of China*; Science Press: Beijing, China, 1994; pp. 16–17.
2. Min, T.L. A revision of subgenus Hymenanthes (*Rhododendron* L.) in Yunnan and Xizang. *Plant Divers.* **1984**, *6*, 1.
3. Wang, B.; Zhou, L.Y.; Xia, H.M. Impacts of sucrose, boric acid and Ca+ on pollen germination of *Rhododendron decorum* Franch. *Jiangsu Agric. Sci.* **2021**, *49*, 129–133.
4. Zhu, Y.-X.; Zhang, Z.-X.; Yan, H.-M.; Lu, D.; Zhang, H.-P.; Li, L.; Liu, Y.-B.; Li, Y. Antinociceptive Diterpenoids from the Leaves and Twigs of *Rhododendron decorum*. *J. Nat. Prod.* **2018**, *81*, 1183–1192. [CrossRef] [PubMed]
5. Rateb, M.E.; Hassan, H.; Arafa, E.-S.; Jaspars, M.; Ebel, R. Decorosides A and B, Cytotoxic Flavonoid Glycosides from the Leaves of *Rhododendron decorum*. *Nat. Prod. Commun.* **2014**, *9*, 473–476. [CrossRef] [PubMed]
6. Zhu, Y.-X.; Zhang, Z.-X.; Zhang, H.-P.; Chai, L.-S.; Li, L.; Ma, S.-G.; Li, Y. A new ascorbic acid derivative and two new terpenoids from the leaves and twigs of *Rhododendron decorum*. *J. Asian Nat. Prod. Res.* **2019**, *21*, 579–586. [CrossRef]
7. Shi, Y.; Zhou, M.; Zhang, Y.; Fu, Y.; Li, J.; Yang, X. Poisonous delicacy: Market-oriented surveys of the consumption of *Rhododendron* flowers in Yunnan, China. *J. Ethnopharmacol.* **2021**, *265*, 113320. [CrossRef]
8. Zhang, S.; Dang, Z.; Zhang, L.Y. Research on seeds aseptic germination and seedling growth condition of *Rhododendron decorum* Franch. *North. Hortic.* **2014**, *305*, 77–80. (In Chinese with English Abstract)
9. Lin, L.-C.; Wang, C.-S. Influence of Light Intensity and Photoperiod on the Seed Germination of Four Rhododendron Species in Taiwan. *Pak. J. Biol. Sci.* **2017**, *20*, 253–259. [CrossRef]
10. Giri, C.C.; Shyamkumar, B.; Anjaneyulu, C. Progress in tissue culture, genetic transformation and applications of biotechnology to trees: An overview. *Trees* **2003**, *18*, 115–135. [CrossRef]

11. Yavuz, D.Ö. Optimization of Regeneration Conditions and In Vitro Propagation of Sideritis Stricta Boiss & Heldr. *Int. J. Biol. Macromol.* **2016**, *90*, 59–62. [CrossRef]

12. Nada, S.; Chennareddy, S.; Goldman, S.; Rudrabhatla, S.; Potlakayala, S.D.; Josekutty, P.; Deepkamal, K. Direct Shoot Bud Differentiation and Plantlet Regeneration from Leaf and Petiole Explants of Begonia tuberhybrida. *Hortscience* **2011**, *46*, 759–764. [CrossRef]

13. Zhang, H.P.; Wang, H.B.; Wang, L.Q.; Bao, G.H.; Qin, G.W. A new 1,5-seco grayanotoxane from *Rhododendron decorum*. *J. Asian Nat. Prod. Res.* **2005**, *7*, 87–90. [CrossRef]

14. Long, Y.; Yang, Y.; Pan, G.; Shen, Y. New Insights Into Tissue Culture Plant-Regeneration Mechanisms. *Front. Plant Sci.* **2022**, *13*, 926752. [CrossRef]

15. Xue, Q.W.; Yuan, H.; Chun, L.L. Assessing the genetic consequences of flower-harvesting in *Rhododendron decorum* Franchet (Ericaceae) using microsatellite markers. *Biochem. Syst. Ecol.* **2013**, *50*, 296–303.

16. Wang, X.-Q.; Huang, Y.; Long, C.-L. Isolation and Characterization of Twenty-four Microsatellite Loci for *Rhododendron decorum* Franch. (Ericaceae). *Hortscience* **2009**, *44*, 2028–2030. [CrossRef]

17. Jin, H.Z.; Chen, G.; Li, X.F.; Shen, Y.H.; Yan, S.K.; Zhang, L.; Yang, M.; Zhang, W.D. Flavonoids from *Rhododendron decorum*. *Chem. Nat. Compd.* **2009**, *45*, 85–86. [CrossRef]

18. Zhang, W.; Jin, H.; Chen, G.; Li, X.; Yan, S.; Zhang, L.; Shen, Y.; Yang, M. A new grandame diterpenoid from *Rhododendron decorum*. *Fitoterapia* **2008**, *79*, 602–604. [CrossRef]

19. Zha, H.-G.; Milne, R.I.; Sun, H. Morphological and molecular evidence of natural hybridization between two distantly related *Rhododendron* species from the Sino-Himalaya. *Bot. J. Linn. Soc.* **2008**, *156*, 119–129. [CrossRef]

20. Li, Q.; Li, H.E.; Yang, L.; Guo, Q.; Fu, Y.; Huang, J. Asymmetric hybridization origin of *Rhododendron ageratum* (Ericaceae) in Guizhou, China. *Phytotaxa* **2021**, *510*, 197–212. [CrossRef]

21. Sun, L.; Pei, K.; Wang, F.; Ding, Q.; Bing, Y.; Gao, B.; Zheng, Y.; Liang, Y.; Ma, K. Different distribution patterns between putative ercoid mycorrhizal and other fungal as-semblages in roots of *Rhododendron decorum* in the Southwest of China. *PLoS ONE* **2018**, *7*, e49867.

22. Tian, W.; Zhang, C.; Qiao, P.; Milne, R. Diversity of culturable ericoid mycorrhizal fungi of *Rhododendron decorum* in Yunnan, China. *Mycologia* **2011**, *103*, 703–709. [CrossRef] [PubMed]

23. Murashige, T.; Skoog, F. A Revised Medium for Rapid Growth and Bioassays with Tobacco Tissue Cultures. *Physiol. Plant.* **1962**, *15*, 473–497. [CrossRef]

24. McCown, B.H. Woody Plant Medium (WPM)-a mineral nutrient formulation for microculture for woody plant species. *Hortscience* **1981**, *16*, 453.

25. Driver, J.A.; Kuniyuki, A.H. In Vitro Propagation of Paradox Walnut Rootstock. *Hortscience* **1984**, *19*, 507–509. [CrossRef]

26. Zaytseva, Y.G.; Poluboyarova, T.V.; Novikova, T.I. Effects of thiazine on in vitro morphogenic response of *Rhododendron sichotense* Pojark. and *Rhododendron catawbiense* cv. Grandiflorum leaf explants. *In Vitro Cell. Dev. Biol. Plant* **2016**, *52*, 56–63. [CrossRef]

27. Wang, W.Q.; Xiao, J.Z.; Li, Z.M.; Hu, J.; Bai, X. Study on the bud tissue culture and establishment of optimization system of *Rhododendron*. *J. Hebei Norm. Univ. Sci. Technol.* **2012**, *26*, 17–22. (In Chinese with English Abstract)

28. Tomsone, S.; Gertnere, D. In vitro Shoot Regeneration from Flower and Leaf Explants in *Rhododendron*. *Biol. Plant.* **2003**, *46*, 463–465. [CrossRef]

29. Nobuaki, M.; Shinsaku, T. Effects of medium components and shear conditions on the formation and growth of adventitious bud derived from hairy roots of *Atropa belladonna* L. *Environ. Control Biol.* **2012**, *50*, 393–406.

30. Debnath, S.C. Propagation of Vaccinium in vitro: A review. *Int. J. Fruit Sci.* **2007**, *6*, 47–71. [CrossRef]

31. Mohammed, A.; Chiruvella, K.K.; Namsa, N.D.; Ghanta, R.G. An efficient in vitro shoot regeneration from leaf petiolar explants and ex vitro rooting of *Bixa orellana* L.—A dye yielding plant. *Physiol. Mol. Biol. Plants* **2015**, *21*, 417–424. [CrossRef]

32. Tanmayee, M.; Arvind, G.; Arnab, S. Somatic embryogenesis and genetic fidelity study of micropropagated medicinal species, *Canna indica*. *Horticulture* **2015**, *1*, 3–13.

33. Nowakowska, K.; Pińkowska, A.; Siedlecka, E.; Pacholczak, A. The effect of cytokinins on shoot proliferation, biochemical changes and genetic stability of *Rhododendron* 'Kazimierz Odnowiciel' in the in vitro cultures. *Plant Cell Tissue Organ Cult. PCTOC* **2022**, *149*, 675–684. [CrossRef]

34. Blazich, F.A.; Giles, C.G.; Haemmerle, C.M. Micropropagation of *Rhododendron chapmani*. *J. Environ. Hortic.* **1986**, *4*, 26–29. [CrossRef]

35. Komakech, R.; Kim, Y.-G.; Kim, W.J.; Omujal, F.; Yang, S.; Moon, B.C.; Okello, D.; Rahmat, E.; Kyeyune, G.N.; Matsabisa, M.G.; et al. A Micropropagation Protocol for the Endangered Medicinal Tree *Prunus africana* (Hook f.) Kalkman: Genetic Fidelity and Physiological Parameter Assessment. *Front. Plant Sci.* **2020**, *11*, 548003. [CrossRef]

36. Zhou, J.; Liu, Y.; Wu, L.; Zhao, Y.; Zhang, W.; Yang, G.; Xu, Z. Effects of Plant Growth Regulators on the Rapid Propagation System of *Broussonetia papyrifera* L. Vent Explants. *Forests* **2021**, *12*, 874. [CrossRef]

37. Bell, R.L.; Srinivasan, C.; Lomberk, D. Effect of nutrient media on axillary shoot proliferation and preconditioning for adven-titious shoot regeneration of pears. *In Vitro Cell. Dev. Biol. Plant* **2009**, *45*, 708–721. [CrossRef]

38. Tang, Q.; Guo, X.; Zhang, Y.; Li, Q.; Chen, G.; Sun, H.; Wang, W.; Shen, X. An optimized protocol for indirect organogenesis from root explants of *Agapanthus praecox* subsp. orientalis 'Big Blue'. *Horticulture* **2022**, *8*, 715. [CrossRef]

39. Poothong, S.; Reed, B.M. Modeling the effects of mineral nutrition for improving growth and development of micro propagated red raspberries. *Sci. Hortic.* **2014**, *165*, 132–141. [CrossRef]

40. Carlín, A.P.; Tafoya, F.; Alpuche-Solís, A.; Pérez-Molphe-Balch, E. Effects of different culture media and conditions on biomass production of hairy root cultures in six Mexican cactus species. *In Vitro Cell. Dev. Biol. Plant* **2015**, *51*, 332–339. [CrossRef]

41. Nic-Can, G.I.; Loyola-Vargas, V.M. The role of the auxins during somatic embryogenesis. In *Somatic Embryogenesis: Fundamental Aspects and Applications*; Springer: Cham, Switzerland, 2016; pp. 171–182.

42. Debnath, S.C.; McRae, K.B. An efficient adventitious shoot regeneration system on excised leaves of micro propagated lin-gonberry (*Vaccinium vitisidaea* L.). *J. Hortic. Sci. Biotechnol.* **2002**, *77*, 744–752. [CrossRef]

43. Zheng, M.; Yang, H.; Yang, E.; Zou, X.; Chen, X.; Zhang, J. Efficient in vitro shoot bud proliferation from cotyledonary nodes and apical buds of *Moringa oleifera* Lam. *Ind. Crops Prod.* **2022**, *187*, 115394. [CrossRef]

44. Ahmad, Z.; Yadav, V.; Shahzad, A.; Emamverdian, A.; Ramakrishnan, M.; Ding, Y. Micropropagation, encapsulation, physiological, and genetic homogeneity assessment in *Casuarina equisetifolia*. *Front. Plant Sci.* **2022**, *13*, 905444. [CrossRef] [PubMed]

45. Coste, A.; Vlase, L.; Halmagyi, A.; Deliu, C.; Coldea, G. Effects of plant growth regulators and elicitors on production of secondary metabolites in shoot cultures of *Hypericum hirsutum* and *Hypericum maculatum*. *Plant Cell Tissue Organ Cult. PCTOC* **2011**, *106*, 279–288. [CrossRef]

46. Luo, L.; Bai, J.; Chen, C.; Chen, X.; Chen, K.; Chen, F. Study on plantlet regeneration for blade segments and the physiological trait for heat tolerance in the callus of *Rhododendron calophytum*. *J. Bot. Northwest China* **2014**. *34*, 1377–1382. (In Chinese with English Abstract)

47. Tian, G.; Peng, L.C.; Qu, S.P.; Wang, J.; Zhao, Z.; Li, S.; Jie, W.; Guan, W. Studies on the adventitious bud induction from in vitro leaves of *Rhododendron delavayi* var. delavayi and sociological observation on the bud formation. *J. Hortic.* **2020**, *47*, 2019–2026. (In Chinese with English Abstract)

48. Hebert, C.J.; Touchell, D.H.; Ranney, T.G.; LeBude, A.V. In vitro shoot regeneration and polyploid induction of *Rhododendron* 'Fra-grantissimum Improved'. *Hortic. Sci.* **2010**, *45*, 801–804.

49. Wei, X.; Chen, J.; Zhang, C.; Wang, Z. In vitro shoot culture of *Rhododendron fortunei*: An important plant for bioactive phytochemicals. *Ind. Crops Prod.* **2018**, *126*, 459–465. [CrossRef]

50. Elmongy, M.S.; Cao, Y.; Zhou, H.; Xia, Y. Root development enhanced by using indole-3-butyric acid and naphthalene acetic acid and associated biochemical changes of in vitro Azalea micro shoots. *J. Plant Growth Regul.* **2018**, *37*, 813–825. [CrossRef]

51. Almeida, R.; Gonçalves, S.; Romano, A. In vitro micropropagation of endangered *Rhododendron ponticum* L. subsp. baeticum (Boissier & Reuter) Handel-Mazzetti. *Biodivers. Conserv.* **2005**, *14*, 1059–1069. [CrossRef]

 *horticulturae*

Article

Development of an Improved Micropropagation Protocol for Red-Fleshed Pitaya 'Da Hong' with and without Activated Charcoal and Plant Growth Regulator Combinations

Yu-Chi Lee [1,2] and Jer-Chia Chang [2,*]

[1] Agricultural Management Research Section, Taichung District Agricultural Research and Extension Station, No. 370 Song Hwai Road, Tatsuen Township, Changhua 515, Taiwan; klyc1512@tdais.gov.tw

[2] Department of Horticulture, National Chung-Hsing University, No. 145, Xingda Road, South Dist., Taichung 40227, Taiwan

* Correspondence: jerchiachang@dragon.nchu.edu.tw; Tel.: +886-4-22840340 (ext. 403); Fax: +886-4-22856974

Abstract: Micropropagation protocols for red-fleshed *Hylocereus* species (Cactaceae) have been developed; however, these methods prolong the sprout duration from areoles and produce irregular micro-propagules in 'Da Hong' pitaya. Thus, the present study aimed to establish an improved micropropagation protocol for this cultivar. Shoot regeneration and root induction of self-pollinating seedling segments were evaluated in response to combinations of activated charcoal (AC; 200 mg/L), α-naphthaleneacetic acid (NAA; 0.05, 0.10, and 0.20 mg/L), and 6-benzylaminopurine (BAP; 1.00, 2.00, and 4.00 mg/L). The correlations among plantlet growth characteristics and plantlet survival rate after transplantation under field conditions were calculated. Increasing the NAA concentration increased the number of roots but reduced root length. The addition of AC enhanced shoot length and prevented the regeneration of dried-out, clustered, and abnormal shoots. Plantlets treated with 200 mg/L AC and 0.10 mg/L NAA produced the highest number of shoots, i.e., 4.1 shoots, which however, were shorter and lighter than those cultured with AC alone. Plantlets grown on medium supplemented with BAP showed no advantage in shoot number, shoot weight, plantlet surface area, or plantlet volume. The weight and shoot surface area of plantlets were strongly correlated. All plantlets grew well at 4 weeks post-transplantation. Overall, these results support this improved micropropagation method to regenerate robust ex vitro plantlets.

Keywords: dragon fruit; micropropagation; regeneration; spontaneous rooting; plant growth regulator

Citation: Lee, Y.-C.; Chang, J.-C. Development of an Improved Micropropagation Protocol for Red-Fleshed Pitaya 'Da Hong' with and without Activated Charcoal and Plant Growth Regulator Combinations. *Horticulturae* **2022**, *8*, 104. https://doi.org/10.3390/horticulturae8020104

Academic Editors: Jean Carlos Bettoni, Min-Rui Wang and Qiao-Chun Wang

Received: 4 January 2022
Accepted: 21 January 2022
Published: 25 January 2022

Publisher's Note: MDPI stays neutral with regard to jurisdictional claims in published maps and institutional affiliations.

1. Introduction

Pitaya (*Hylocereus* spp.), which belongs to the family Cactaceae [1], has recently become an important fruit crop in Asia and America due to varietal improvements and increased production during the off-season using night-break/prolonged-daytime techniques [2–6]. *H. polyrhizus* 'Da Hong', also known as 'Big Red', is a red-fleshed pitaya cultivar and is the dominant cultivar grown in China, Vietnam, and Taiwan because of its desirable characteristics, including self-compatibility, large fruit size, high sweetness, and abundant yield [3,6].

However, 'Da Hong' is susceptible to pathogenic infection [*Cactus virus X* [7], *Neoscytalidium dimidiatum* [8], and *Colletotrichum* spp. [9–11]] by penetrating hyphae or mechanical wounds [7,12,13], reducing plant growth vigor, i.e., shoot initiation and flower blooming, and fruit production [9,12]. Agar-based micropropagation has proven to be a valuable method to decrease the risk of infection in tomatoes and cacti, as pathogenic contamination can be visually inspected and prevented [14–16]. Therefore, in vitro propagation may represent a potential strategy to reduce disease risk in 'Da Hong' plantlets and renew infected orchards.

Plant state and environmental conditions during incubation affect pitaya tissue culture and the micro-propagules, including explant type, medium strength (Murashige and Skoog [MS]) [17], i.e., full- or 1/4 strength, and the form and concentration of plant growth regulators (PGRs) [14,18,19]. Culture media are frequently supplemented with the PGRs naphthaleneacetic acid (NAA) and 6-benzylaminopurine (BAP) because of their availability, convenience, and direct shoot (cladode) regeneration properties [20]. *Hylocereus* plantlets cultured on media supplemented with these PGRs show a 2- to 18-fold increase in shoot regeneration rate compared with untreated micro-propagules, depending on the species and cultivars [18,21–25]. For example, plantlets of the white-fleshed pitaya *H. undatus* cultured on MS medium supplemented with appropriate PGRs, i.e., 4 mg/L BAP and 0.1 mg/L NAA, regenerated a maximum of 22 shoots/plantlet; however, the shoot number of the control plantlets was unavailable [26]. Preliminary studies have proposed media for red-fleshed *H. polyrhizus* and its hybrids and reported an 83.2% in vitro germination rate and regeneration of 6–8 shoots on the micro-propagules [25,27,28]. However, the specific cultivars used in these studies are unknown.

Few studies have reported normal regeneration in pitaya micropropagation [18,22]. Shoots regenerated using such PGRs were either semi-transparent or light green [21,28], indicating low shoot vigor, which prolonged the duration of micropropagation [14,29]. In addition, our preliminary results showed that the given media used for red-fleshed pitaya in previous studies were not appropriate for the 'Da Hong' cultivar, which produced abnormal shoots.

Activated charcoal (AC) plays a critical role in tissue cultures, promoting micro-propagule growth and development. Wang and Huang [30] indicated that AC supplementation in the culture medium could darken the medium to simulate the soil conditions and absorb toxic metabolites, thereby increasing shoot and root development in *Phalaenopsis*, *Cymbidium*, and *Dendrobium* species. A reduction of residual PGR side effects in the explant rooting was also shown in Cactaceae species [31]. However, no AC benefit could be observed in shoot initiation in *Vigna radiate* [32] or root induction in *Echinocactus*, *Echiaocereus*, *Mammillaria*, and *Stenocactus* species [33].

Given the advantage of AC in reducing the side effects of PGRs during incubation and the potential to propagate an entire plantlet forming simultaneous shoots and roots from the primordium [32], MS medium supplemented with AC has been used for the micropropagation of white-fleshed pitaya [34]. Despite the importance of 'Da Hong' in the global pitaya industry, designated tissue culture studies of this cultivar are unavailable. The effects of AC on developing red-fleshed pitaya plantlets, particularly 'Da Hong', are thus unknown.

In the present study, in vitro segments of self-pollinating 'Da Hong' progeny were used as explants. To address the knowledge gap, we assessed the effects of AC in combination with different concentrations of PGRs on plantlet characteristics. We then analyzed the relationships among plantlet characteristics and their derivative parameters to construct an essential growth database for the red-fleshed pitaya tissue culture system. Finally, we improved the in vitro culture system based on an existing protocol [26]. The results of this study potentially provide a new in vitro propagation method for producing robust plantlets of *H. polyrhizus* 'Da Hong'.

2. Materials and Methods

2.1. Plant Materials and Culture Establishment

Self-pollinating fruits of *H. polyrhizus* 'Da Hong' were harvested from a commercial orchard in Taichung, central Taiwan (24°14′33.2″ N, 120°48′21.6″ E), on 27 June 2019, 2 November 2019, and 11 July 2020. Seeds were collected from the flesh and sterilized following the method described by De Feria et al. [21], with modifications. In brief, the seeds were immersed in 75% (*v/v*) alcohol for 30 s and then in 1% (*v/v*) sodium hypochlorite containing Tween 20 for 15 min. Next, the seeds were rinsed five times with sterile distilled water. The rinsed seeds were cultured in glass bottles with 1/4 strength MS medium

(M5519, Sigma Chemical Co., St. Louis, MO, USA) supplemented with 3% (*w/v*) sucrose and 0.8% (*w/v*) agar (Kungfei Trading Co., Ltd., Tainan, Taiwan) to shorten by 1.5-fold the germination duration and increase their germination rate by 153% compared to full-strength MS medium [35]. The cultures were incubated in a culture room at 25 °C ± 2 °C with a 16/8 h (L/D) photoperiod under 50 μmol/m^2/s fluorescent light. Before use, the pH of the MS medium was adjusted to 5.8. All experiments were performed in the tissue culture laboratory of the Department of Horticulture at National Chung-Hsing University, Taichung, Taiwan.

2.2. Shoot Regeneration and Root Induction Using AC, NAA, and BAP

We investigated the effects of different concentrations and combinations of PGRs on shoot regeneration and root induction following a previous study [26], with modifications. Briefly, we supplemented the MS medium with either NAA alone (0.05, 0.10, or 0.20 mg/L; Sigma Chemical Co., St. Louis, MO, USA) or 200 mg/L AC in combination with NAA and BAP (1.00, 2.00, and 4.00 mg/L; Sigma Chemical Co., St. Louis, MO, USA) (Table 1).

Table 1. Experimental treatments of Murashige and Skoog (MS) medium supplemented with different combinations and concentrations of activated charcoal (AC) and plant growth regulators (PGRs).

Treatment Label	PGRs [1] (mg/L)		
NAA supplemented in MS medium			
	AC	NAA	BAP
CS	–	–	–
TS01	–	0.05	–
TS02	–	0.10	–
TS03	–	0.20	–
AC, NAA, and BAP supplemented in MS medium			
	AC	NAA	BAP
CM01	–	–	–
CM02	200	–	–
TM01	200	0.05	–
TM02	200	0.05	1.00
TM03	200	0.05	2.00
TM04	200	0.05	4.00
TM05	200	0.10	–
TM06	200	0.10	1.00
TM07	200	0.10	2.00
TM08	200	0.10	4.00
TM09	200	0.20	–
TM10	200	0.20	1.00
TM11	200	0.20	2.00
TM12	200	0.20	4.00

[1] NAA, α-naphthaleneacetic acid; BAP, 6-benzylaminopurine.

Activated charcoal was added to the MS medium supplemented with BAP as the preliminary results showed that shoots regenerated using the MS medium supplemented with BAP without AC dried out, clustered, and became abnormal (Figure 1). We found that adding 200 mg/L of AC into the medium reduced these side effects. The in vitro pitaya seedlings were excised and sub-cultured as mother explants. The shoots regenerated by the mother plants were excised as 15 mm segments. Three replicate bottles, each containing five segments, were prepared for each treatment.

Figure 1. In vitro plantlets of *Hylocereus polyrhizus* 'Da Hong' after 5 months of incubation (**A**) on Murashige and Skoog (MS) medium and (**B**) with 2.00 mg/L 6-benzylaminopurine (BAP). Scale Bar = 1 cm for the plantlets without agar.

Plantlet growth characteristics (i.e., number of shoots, shoot length, root length, and shoot surface area) were measured using ImageJ [36]. Owing to the irregular size and shape of the plantlet shoots, it was difficult to accurately calculate the shoot surface area. In our best effort to measure the full surface area, four sides of the plantlet shoots were photographed, and the photos were transformed into an 8-bit type followed by selection of software functions (i.e., threshold and filter) to match the shape of the shoots. The plantlets were weighed (FW) using an electronic balance (XT220A, Precisa Gravimetrics AG, Dietikon, Switzerland) after 8 weeks of incubation. Archimedes' principle was applied to measure the shoot volume. The shoot surface area per unit length [SAL; mm; Surface area $_{shoots}$ (mm^2)/Length $_{shoots}$ (mm)], shoot weight per unit length [SWL; mg FW/mm; FW $_{shoots}$ (mg)/Length $_{shoots}$ (mm)], root weight per unit length [RWL; mg FW/mm; FW $_{roots}$ (mg)/Length $_{main\ root}$ (mm)], and root/shoot ratio [g FW/g FW; FW $_{roots}$ (g FW)/FW $_{shoots}$ (g FW)] were also calculated, where FW indicated fresh weight. Finally, we analyzed the pairwise Pearson's correlations between plantlet growth characteristics to construct the essential growth database of red-fleshed pitaya in vitro, e.g., the relationship between plantlet weight and surface area and the relationship between (regeneration) shoot length and (regeneration) shoot surface area.

2.3. Acclimatization

After 8 weeks of incubation in the culture room (on 25 April 2020), plantlets were transferred from the bottle, and the agar was gently removed from the roots. The ex vitro plantlets were maintained in a plastic box containing clean water at room temperature (22–25 °C) with a 16/8 h (L/D) photoperiod under 50 $\mu mol/m^2/s$ fluorescent light for 1 week. The plantlets were then planted in plastic pots containing peat soil (pH 5.5–6.0, Stender Peat Substrate, Known-You Seed Co., Ltd., Dashu, Kaohsiung, Taiwan). The plantlets were cultured on the terrace under field conditions with 80 $\mu mol/m^2/s$ solar light for 4 weeks, and the plantlet survival rate was then assessed.

2.4. Data Analysis and Statistics

Data were analyzed using one-way ANOVA in SAS 9.4 (SAS Institute Inc., Cary, NC, USA). The PROC GLM procedure was used to conduct multiple pairwise comparisons using Fisher's protected least significance difference (LSD) test. We considered $p \leq 0.05$ as significant. Figures were constructed using SigmaPlot 10.0 (Systat Software Inc., San Jose, CA, USA) and Powerpoint 2019 (Microsoft Corp., Redmond, WA, USA). A heat map of Pearson's correlation coefficients between pairs of plantlet growth characteristics was created using Python 3.8.0 (Python Software Foundation).

3. Results

3.1. Effects of NAA Alone on Shoot Regeneration and Root Induction

H. polyrhizus 'Da Hong' plantlets grown on MS medium supplemented with various concentrations of NAA are shown in Figure S1. For all concentrations of NAA, there were no significant differences in the shoot regeneration number (3.6 shoots/plantlet), shoot length (57.32 mm), and average shoot length (18.41 mm; Figure 2). Other plantlet growth characteristics, i.e., plantlet weight (Figure S2), plantlet surface area, and plantlet volume (data not shown), were similar for all different concentrations of NAA.

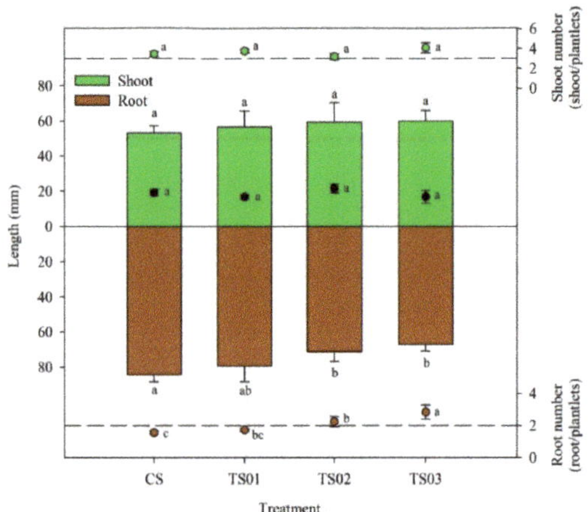

Figure 2. Shoot regeneration number, root induction number, and length of shoots and roots of *Hylocereus polyrhizus* 'Da Hong' grown on Murashige and Skoog (MS) medium supplemented with α-naphthaleneacetic acid (NAA) alone. Treatments: CS, control; TS01, 0.05 mg/L NAA; TS02, 0.10 mg/L NAA; TS03, 0.20 mg/L NAA. Mean ± SE (n = 3, 5 segments per replicate). Means followed by different letters near the column and circle are significantly different at p = 0.05 using Fisher's protected LSD test. The black circle inside the green column indicates the average shoot length.

Plantlets grown on the culture medium supplemented with 0.20 mg/L NAA (treatment TS03) had significantly (p = 0.0024) more roots (2.8 roots/plantlet) than plantlets grown on medium supplemented with 0, 0.05, or 0.10 mg/L NAA (1.6, 1.7, and 2.2 roots/plantlet, respectively; Figure 2). However, plantlets grown on control culture medium (treatment CS) had significantly longer main roots (84.38 mm) than those grown on media supplemented with 0.10 and 0.20 mg/L NAA treatments (71.32 and 67.47 mm, respectively); plantlets grown on control culture medium also had longer main roots than plantlets grown on culture medium supplemented with 0.05 mg/L NAA (79.40 mm), but this difference was not significant. Root weight per unit length was significantly greater in plantlets grown on medium supplemented with 0.20 mg/L NAA (1.09 mg/mm) than in plantlets grown on control medium (0.52 mg/mm; Figure 3).

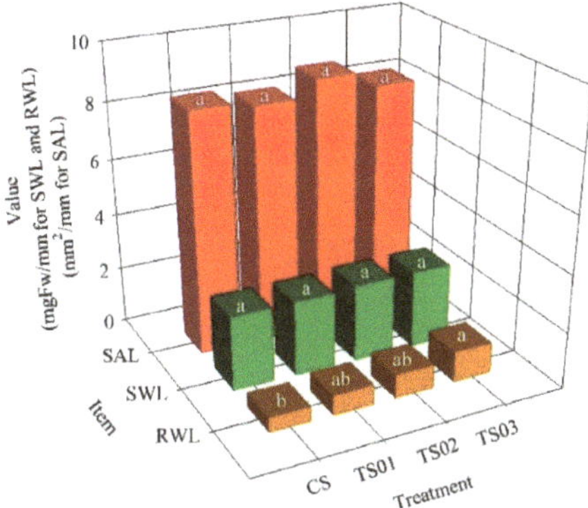

Figure 3. Plantlet growth characteristics of shoot surface area per unit length (SAL; surface area shoots/Length shoots), shoot weight per unit length (SWL; FW shoots/Length shoots), and root weight per unit length (RWL; FW roots/Length main root) of *Hylocereus polyrhizus* 'Da Hong' grown on Murashige and Skoog (MS) medium supplemented with α-naphthaleneacetic acid (NAA) alone. Treatments: CS, control; TS01, 0.05 mg/L NAA; TS02, 0.10 mg/L NAA; TS03, 0.20 mg/L NAA. Mean $\pm$ SE (n = 3, 5 segments per replicate). Means followed by different letters near the column are significantly different at p = 0.05 using Fisher's protected LSD test.

3.2. The Combined Effects of AC, NAA, and BAP on Shoot Regeneration and Root Induction

Shoot appearance was improved by supplementing the MS medium with 200 mg/L AC: the shoots of plantlets grown on AC-supplemented media were dark green, with a succulent structure (Figure 4). The plantlets grown on medium supplemented with AC and 0.10 mg/L NAA (TM05–08) regenerated significantly more shoots (4.1 shoots/plantlet) than plantlets grown on any other media (Figure 5A).

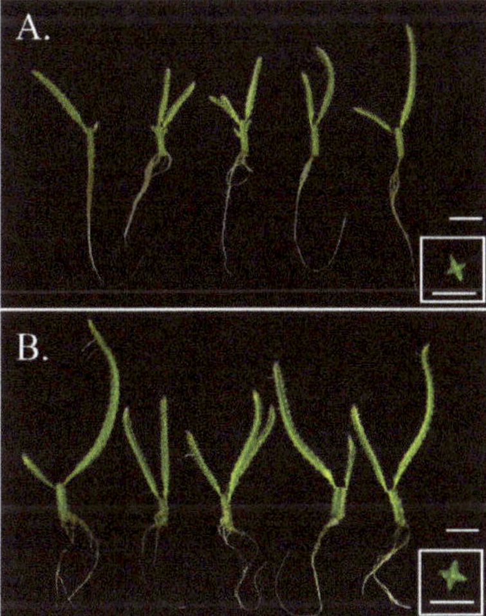

Figure 4. Plantlets of *Hylocereus polyrhizus* 'Da Hong' cultured (**A**) on Murashige and Skoog (MS) medium and (**B**) on activated charcoal (AC)-supplemented MS medium. Profiles in the white square frame are plantlet transections, showing that plantlets grown on the AC-supplemented MS medium were succulent. Bar = 2 cm; bar inside white square = 1 cm.

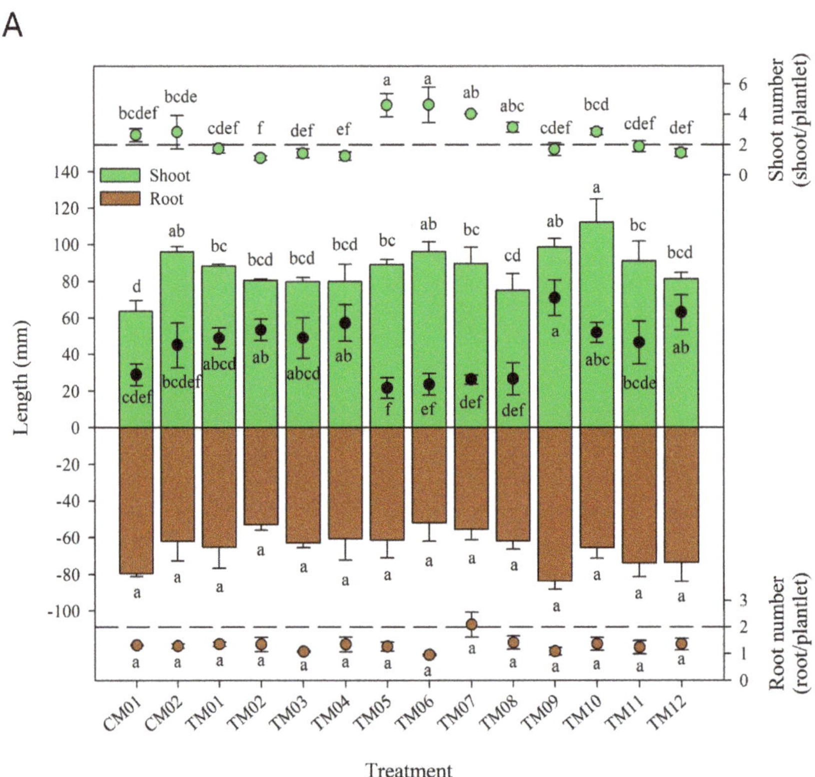

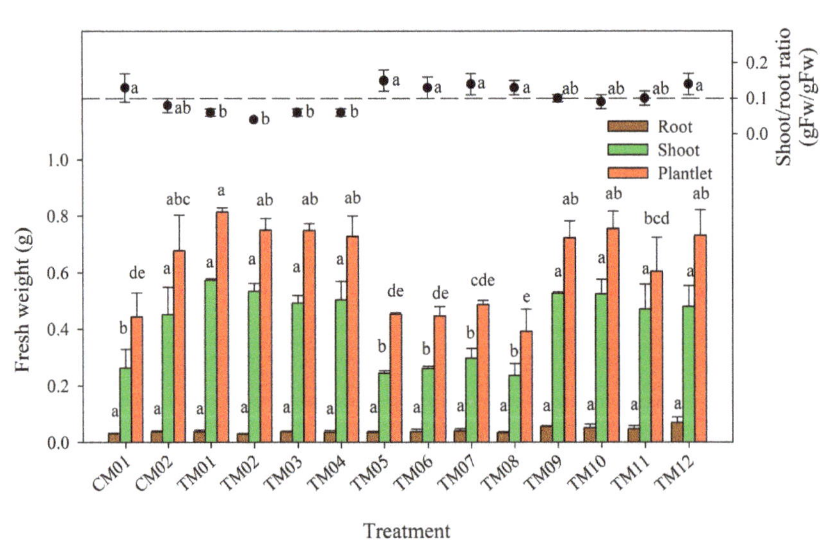

Figure 5. Shoot regeneration number, root induction number, and shoot and root lengths (**A**); fresh weight of plantlets, shoots, and roots, as well as root/shoot ratio (**B**) of *Hylocereus polyrhizus* 'Da Hong'

cultured on Murashige and Skoog (MS) medium supplemented with activated charcoal (AC) combined with α-naphthaleneacetic acid (NAA) and 6-benzylaminopurine (BAP). The black circle inside the green column in plate A indicates the average shoot length. Treatments: CM01, control; CM02, 200 mg/L AC; TM01, 200 mg/L AC and 0.05 mg/L NAA; TM02, 200 mg/L AC, 0.05 mg/L NAA, and 1.00 mg/L BAP; TM03, 200 mg/L AC, 0.05 mg/L NAA, and 2.00 mg/L BAP; TM04, 200 mg/L AC, 0.05 mg/L NAA, and 4.00 mg/L BAP; TM05, 200 mg/L AC and 0.10 mg/L NAA; TM06, 200 mg/L AC, 0.10 mg/L NAA, 1.00 mg/L BAP; TM07, 200 mg/L AC, 0.10 mg/L NAA, and 2.00 mg/L BAP; TM08, 200 mg/L AC, 0.10 mg/L NAA, and 4.00 mg/L BAP; TM09, 200 mg/L AC and 0.20 mg/L NAA; TM10, 200 mg/L AC, 0.20 mg/L NAA, and 1.00 mg/L BAP; TM11, 200 mg/L AC, 0.20 mg/L NAA, and 2.00 mg/L BAP; TM12, 200 mg/L AC, 0.20 mg/L NAA, and 4.00 mg/L BAP. Mean ± SE (n = 3, 5 segments per replicate). Means associated with different letters near the columns are significantly different at p = 0.05 using Fisher's protected LSD test.

The shoot lengths of plantlets grown under the treatment conditions CM02, TM06, TM09, and TM10 were significantly longer (93.97 mm) than those grown under other treatment conditions (63.81–90.96 mm; Figure 5A). The average shoot lengths were significantly greater in treatments TM01–04, TM09, TM10, and TM12 (70.93 mm) compared to those obtained with the controls and other treatments (21.71–46.50 mm; Figure 5A). The addition of 1.00 mg/L BAP tended to increase shoot length, especially in combination with 0.10 and 0.20 mg/L NAA (i.e., TM06 and TM10), although plantlets had similar shoot lengths in medium supplemented with AC alone. However, the average shoot lengths were low for plantlets cultured on 0.10 mg/L NAA, due to the influence of NAA concentration on the shoot number.

Plantlets grown on medium supplemented with 0.05 or 0.20 mg/L NAA had shoots of similar weight to those in treatment CM02 (MS medium supplemented with AC only); shoot weight in these groups was significantly greater than that of plantlets grown on 0.10 mg/L NAA (Figure 5B). In addition, plantlet surface area and plantlet volume were similar across plantlets grown on media supplemented with AC alone (CM02) and AC in combination with 0.05 mg/L NAA or 0.20 mg/L NAA; these values were significantly higher than those obtained with CM01 and the treatments with 0.10 mg/L NAA (Figure 6A). Neither plantlet surface area nor plantlet volume was affected by BAP concentrations (Figure 6A).

Both SWL and SAL were significantly affected by NAA and AC: shoots grown on medium supplemented with AC only (CM02) or medium supplemented with AC in combination with 0.05 or 0.20 mg/L NAA had a heavier SWL and larger SAL than shoots grown on medium supplemented with 0.10 mg/L NAA (Figure 6B).

The root number of the plantlets remained similar (1.3 roots/plantlet) across all combinations of PGR supplements (Figure 5A). Main root length and root weight were also similar across PGR treatments (Figure 5A,B). Irrespective of segment weight, the root/shoot ratios of treatments TM01–04 (0.05 mg/L NAA with 0–4.00 mg/L BAP) were lower than those of CM01 and of treatments TM05–08 and TM12 (0.10–0.20 mg/L NAA with 0–4.00 mg/L BAP; Figure 5B). Notably, the addition of BAP did not affect plantlet fresh weight or root/shoot ratio.

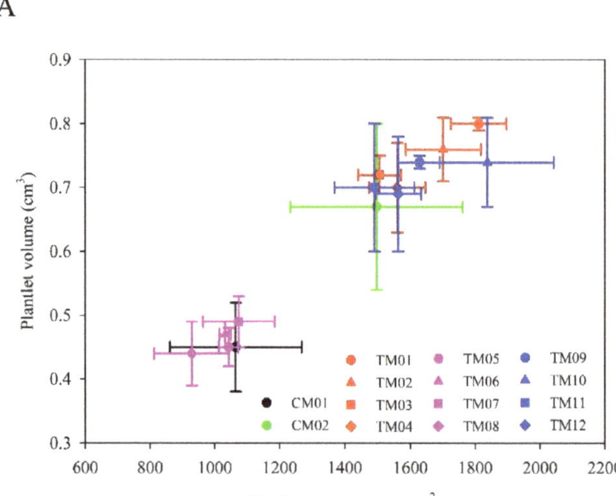

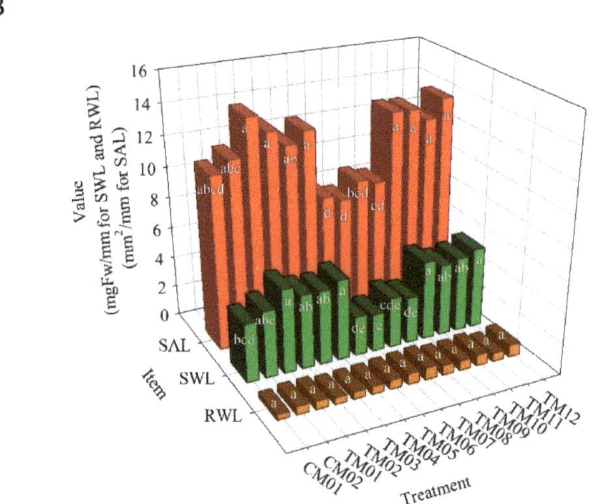

Figure 6. Relationship between plantlet surface area and plantlet volume (**A**) and shoot surface area per unit length (SAL; Surface area $_{shoots}$/Length $_{shoots}$), shoot weight per unit length (SWL; FW $_{shoots}$/Length $_{shoots}$), and root weight per unit length (RWL; FW $_{roots}$/Length $_{main\ root}$) (**B**) of *Hylocereus polyrhizus* 'Da Hong' cultured on Murashige and Skoog (MS) medium supplemented with activated charcoal (AC) combined with α-naphthaleneacetic acid (NAA) and 6-benzylaminopurine (BAP). Treatments: CM01, control; CM02, 200 mg/L AC; TM01, 200 mg/L AC and 0.05 mg/L NAA; TM02, 200 mg/L AC, 0.05 mg/L NAA, and 1.00 mg/L BAP; TM03, 200 mg/L AC, 0.05 mg/L NAA, and 2.00 mg/L BAP; TM04, 200 mg/L AC, 0.05 mg/L NAA, and 4.00 mg/L BAP; TM05, 200 mg/L AC and 0.10 mg/L NAA; TM06, 200 mg/L AC, 0.10 mg/L NAA, 1.00 mg/L BAP; TM07, 200 mg/L AC, 0.10 mg/L NAA, and 2.00 mg/L BAP; TM08, 200 mg/L AC, 0.10 mg/L NAA, and 4.00 mg/L BAP; TM09, 200 mg/L AC and 0.20 mg/L NAA; TM10, 200 mg/L AC, 0.20 mg/L NAA, and 1.00 mg/L BAP; TM11, 200 mg/L AC, 0.20 mg/L NAA, and 2.00 mg/L BAP; TM12, 200 mg/L AC, 0.20 mg/L NAA, and 4.00 mg/L BAP. Mean ± SE (*n* = 3, 5 segments per replicate). Means associated with different letters near the columns are significantly different at *p* = 0.05 using Fisher's protected LSD test.

3.3. Pairwise Correlations between Growth Characteristics of Shoots and Roots

There were no strong correlations between shoot number, root number, and other growth characteristics (Figure 7). However, the shoot lengths of the plantlets were significantly correlated with plantlet surface area (r = 0.647; *p* = 0.0001), and plantlet weight was significantly correlated with plantlet surface area and plantlet volume (r = 0.931 and 0.986, respectively; *p* = 0.0001). A high regression coefficient of 0.9795 was found between regeneration shoot length and surface area (Figure 8A), which coincided with the regression results for plantlet surface area and plantlet volume based on plantlet weight, which were 0.8665 and 0.9767, respectively (Figure 8B,C). Given the intricate measurements of micro-propagules surface area and volume, the regression results provide a well-calculated basis (plantlet weight) for exploring growth characteristics in vitro.

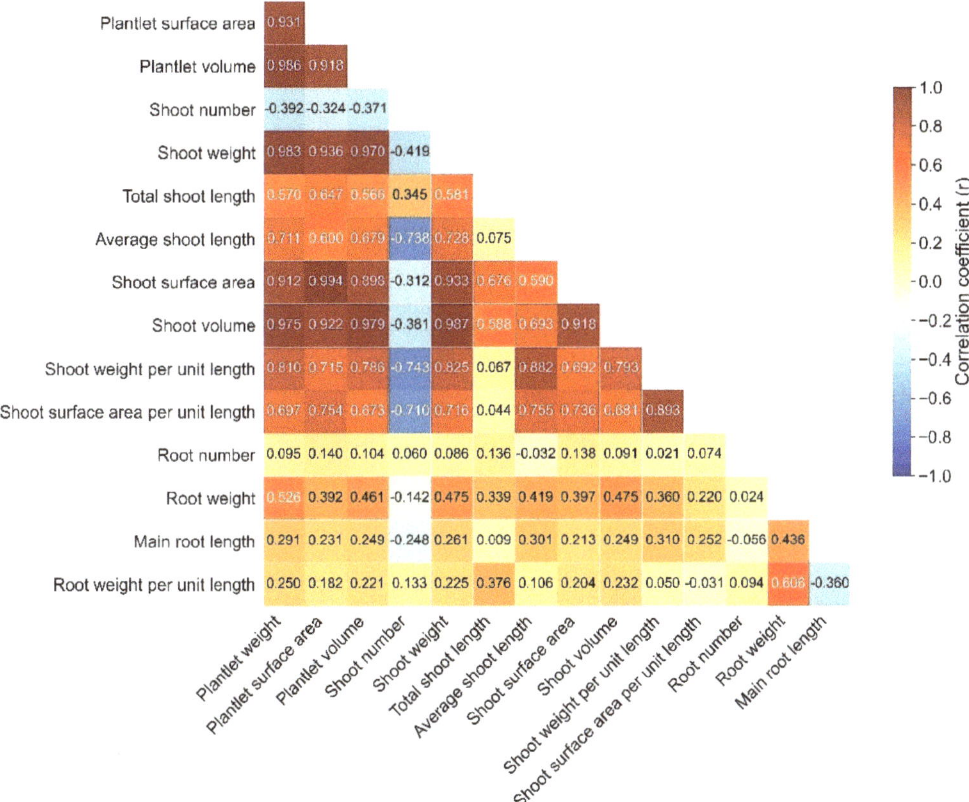

Figure 7. Heat map of Pearson's correlation coefficients for in vitro plantlet growth characteristics of *Hylocereus polyrhizus* 'Da Hong' incubated in a culture room at 25 °C $\pm$ 2 °C with a 16/8 h (L/D) photoperiod under 50 μmol/m^2/s fluorescent light.

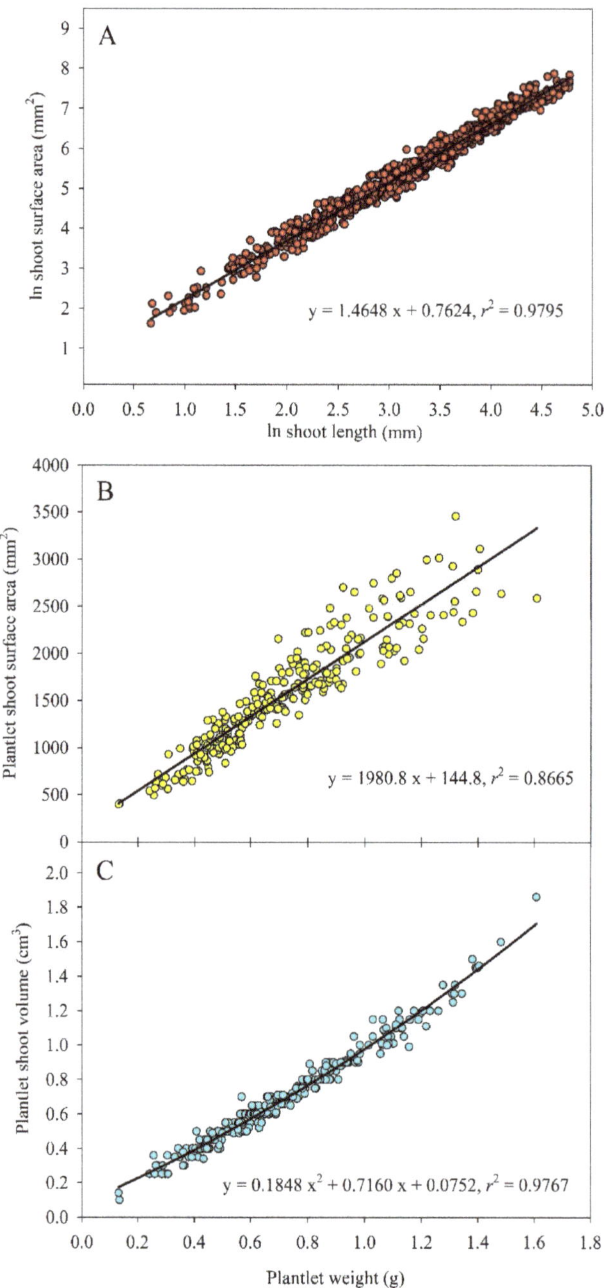

Figure 8. Relationships (**A**) between (regeneration) shoot surface area and (regeneration) shoot length, (**B**) between plantlet weight and plantlet surface area, and (**C**) between plantlet weight and plantlet volume on *Hylocereus polyrhizus* 'Da Hong'. The regression equations were: (**A**) y = 1.4648x + 0.7624 ($n = 808$, $r^2 = 0.9795$, $p < 0.0001$), (**B**) y = 1980.8x + 144.8 ($n = 270$, $r^2 = 0.8665$, $p < 0.0001$), (**C**) y = 0.1848x^2 + 0.7160x + 0.0752 ($n = 270$, $r^2 = 0.9767$, $p < 0.0001$), respectively.

3.4. Cultivated Duration and Acclimatization

Segments excised from the in vitro seedlings regenerated new shoots after 1 week of incubation, irrespective of PGR supplementation. Micro-propagules were transplanted to the field after 4 weeks of in vitro incubation when the shoots were 4 cm long. None of the plantlets died or showed signs of pathogenic infection during the 4 weeks of field cultivation (Figure 9).

Figure 9. Acclimatization of *Hylocereus polyrhizus* 'Da Hong' at room temperature (22–25 °C). (**A**) Plantlets hardened by incubating in water, (**B**) all plantlets grew well at 4 weeks after transplantation.

4. Discussion

4.1. Effects of NAA Alone on Shoot Regeneration and Root Induction

Micro-propagules growth qualities were affected by the concentrations and combinations of PGRs [14,19] and endogenous hormones [37] in the culture medium. The shoot regeneration number of 'Da Hong' pitaya was not affected by single PGR NAA supplementation, in contrast with previous observations of *Mammillaria san-angelensis* (Cactaceae) [38], wherein NAA supplementation increased the number of shoots per explant. The benefits of BAP on in vitro shoot sprouting and development have been reported for certain Cactaceae [37] and *Hylocereus* species [39]. However, BAP supplementation might cause hormonal imbalance in plants. Indeed, supplementation with >1.00 mg/L BAP produced compact, dried-out shoots in 'Da Hong' pitaya in the present study. Our results demonstrate that the *Hylocereus* plantlet response to BAP might be species-specific.

'Da Hong' produced 1.6 roots per plantlet in PGR-free culture medium, consistent with previous results in *H. costaricensis* [39] and other Cactaceae species [38,40,41]. However, root length decreased as the number of roots increased, e.g., for treatment TS03, in contrast to previous observations in *H. undatus* [24]. In addition, root density (RWL) was significantly greater in plantlets cultured on medium supplemented with 0.20 mg/L NAA than in those cultured on PGR-free medium. These results suggest that, in 'Da Hong', the length of the primary adventitious roots decreased, whereas the number/lengths of the lateral roots increased when NAA concentration increased in the culture medium. After exposure to excessively high levels of NAA, this may occur due to root tip enlargement [42] or the synthesis of inappropriate PGRs [43,44]. Both processes would subsequently reduce the vigor and viability of micro-plants [42]. Thus, it is critical to identify appropriate NAA concentrations for optimal healthy root growth in 'Da Hong' pitaya plantlets.

4.2. The Combined Effects of AC, NAA, and BAP on Shoot Regeneration and Root Induction

The addition of AC is known to promote plantlet growth and development [30] and spontaneous rooting during micropropagation [34]. Activated charcoal supplementation facilitated in vitro organogenesis by absorbing the excess substrates and providing a dark environment to simulate soil conditions [30]. AC-related sucrose hydrolysis eased micro-propagule use [45,46]. Moreover, AC promoted both nitrate and ammonium uptake under tissue culture conditions, which facilitated plantlet development [32]. Consistent with this, 'Da Hong' pitaya plantlets grown on medium supplemented with AC exhibited remarkable improvement in several characteristics, including shoot length, shoot weight, shoot surface area, and shoot volume. It is, therefore, possible that AC promotes micro-shoot maturation and accelerates the propagation cycle by increasing shoot length and weight.

The maximum shoot number for regenerated 'Da Hong' (4.1 shoots/plantlet) could be observed in plantlets cultured on medium supplemented with 200 mg/L AC and 0.10 mg/L NAA. This result corresponded to 2.4-fold more shoots (1.7 shoots/plantlet) than those regenerated on medium containing 200 mg/L AC combined with 0.05 or 0.20 mg/L NAA and to 1.5-fold more shoots (2.8 shoots/plantlet) than those regenerated on MS medium containing 200 mg/L AC only. However, the shoot regeneration number (2.6 shoots/plantlet) of 'Da Hong' was 1.4-fold higher on PGR-free MS medium than on MS medium supplemented with multiple PGRs, i.e., 200 mg/L AC, 0.20 mg/L NAA, and 2.00 mg/L BAP. This result was inconsistent with the observation in red/purple-fleshed pitaya *H. purpusii*, in which plantlets produced 5.3-fold more shoots on MS medium supplemented with 0.20 mg/L NAA and 2.00 mg/L BAP than on PGR-free MS medium [21].

Compared to the shoot regeneration results for red-fleshed 'Da Hong' pitaya, plantlets of the white-fleshed pitaya *H. undatus* regenerated 2.2-fold more shoots (2.2 shoots/plantlet) on MS medium supplemented with 1000 mg/L AC combination with 0.10 mg/L NAA and 4.00 mg/L BAP [34] and 18-fold more shoots (18 shoots/plantlet) on MS medium supplemented with 0.01 mg/L NAA and 2.50 mg/L BAP [18] than plantlets grown on PGR-free MS medium (1.0 shoot/plantlet) [24].

Although AC supplementation had several advantages, the negative effects on the non-selective absorption of substances including PGRs should also be considered [32]. The NAA and BAP concentrations used in the present study were similar to those indicated in a previous report [34], but the regeneration number of 'Da Hong' pitaya was higher in combination with 200 mg/L AC than that of white-fleshed pitaya grown in 1000 mg/L AC. Despite the genetic differentiation in species and cultivars, PGR absorption by AC might be a reason behind the reduction in plantlet shoot initiation and regeneration, i.e., the shoot regeneration number of plantlets was higher in AC-free media than in low-AC concentration media, while the media supplemented with AC in high concentration exhibited smaller shoot numbers. PGR concentration after AC supplementation in the culture medium would require investigation, confirming the quality and quantity of medium substances and improving the culture medium.

Consistent with a previous study reporting that similar shoot lengths (18.5 mm) were observed after treatment with varied BAP concentrations [39], our results indicated that BAP supplementation did not affect the average shoot length, considering the shoot number. However, NAA concentration did influence shoot development in 'Da Hong' pitaya, with the average shoot length increasing by 3.21–18.5 mm compared to *H. costaricensis* [39]. NAA supplementation had similar beneficial effects on other shoot characteristics, including weight, surface area, volume, SWL, and SAL. These results are consistent with those of Mauseth and Halperin [37], who suggested that NAA suppresses the effect of BAP on plantlet development and that the types and concentrations of PGRs used affect plantlet organogenesis during tissue culture.

4.3. Pairwise Correlations between Growth Characteristics of Shoots and Roots

Studies on the relationships among plantlet growth characteristics are rare in the genus *Hylocereus*. However, for *Opuntia ficus-indica*, which is another plant of Cactaceae, shoot weight and shoot surface area had a curvilinear relationship under field conditions, with shoot weight increasing much more quickly than shoot surface area due to shoot thickness and succulence increases [47]. That result was partially consistent with our observations in 'Da Hong' pitaya, where plantlet weight and plantlet surface area had a strong linear relationship ($r^2 = 0.8665$). However, this conclusion was incongruent with the hypothesis of Mauseth [48], who suggested that succulent shoot volume should generally not affect surface area due to the shoot rib connection. Notably, the regression equations describing the relationships between (regeneration) shoot length and (regeneration) shoot surface area, and between plantlet weight and plantlet surface area as well as plantlet shoot volume, established in this study, will be helpful for further calculations of gas exchange rate [49] and leaf area index to estimate the best in vitro growth environment for pitaya.

4.4. Cultivated Duration and Acclimatization

Our preliminary results showed that the disinfected seeds of 'Da Hong' emerged after 1 week under aseptic conditions [35], a period that is consistent with previous reports for *H. purpusii* [21] but 3 weeks shorter than that for *H. undatus* [18]. The seedling segments regenerated shoots after 1 week of in vitro cultivation. Therefore, the total duration of 'Da Hong' micropropagation, including seed germination, shoot regeneration, shoot elongation to 4 cm, and root induction, was 8 weeks. That was considerably shorter than the duration of micropropagation in *H. undatus*, in which the shoots took 60 days to develop from 1.5 to 2.0 cm [18,26]. All plantlets grew well without disease symptoms at 4 weeks after transplantation, consistent with observations in other *Hylocereus* species [21,27,39]. These findings suggest that the disease risk of field-transplanted plantlets is rare.

4.5. The Potential to Reduce Disease Risk in Plantlets and Renew Infected Orchards

Pathogens typically have latent infectious characteristics on the host surface or in host organs. The infection pathway depends on the pathogens: both microbial (i.e., bacterial and fungal) and viral pathogens infect succulent shoots after mechanical injury [7,12], whereas microbial hyphae may penetrate plant cell walls and infect hosts when environmental conditions are suitable for mycelium development [11,13]. Host development status (e.g., young shoot, flower, or fruit) may also affect infection susceptibility. For example, *N. dimidiatum* infects young shoots and fruits [8,50], while the symptoms of *Colletotrichum* spp. are manifested in mature shoots and fruits [10,11,51].

Viral and microbial infections can reduce fruit production by up to 40% [9]. Thus, various methodologies that minimize pathogen-associated losses have been proposed [9,13]. Micropropagation performed in aseptic agar media has been a helpful method for minimizing disease risks [14], because microbial infections can typically be identified in seedling explants using the agar plate method [15].

In the present study, robust micro-propagules were produced from disinfected seeds. It is difficult for microbes to invade seeds, and virus transmission through *Hylocereus* seeds and seed-derived seedlings has not been confirmed [12,52]. Therefore, using disinfected seeds for in vitro propagation of *H. polyrhizus* 'Da Hong' reduces disease risk in the resulting micro-propagules [7,53]. Blotter or ELISA methods will be needed for further detection of the predominant microorganisms of *Hylocereus*, including *Cactus virus X*, *N. dimidiatum*, and *Colletotrichum* spp., to confirm disinfection during and after micropropagation [15,54]. Although self-pollinating 'Da Hong' seedlings were used in the present study, there is the possibility of heterogeneity [55]; therefore, identification of genetic differentiation should be investigated in the future.

5. Conclusions

We reported a simple procedure for the in vitro micropropagation of robust 'Da Hong' pitaya. Adding AC to the MS culture medium accelerated plantlet growth and development. Although the shoot regeneration number was not drastically increased, the regenerated shoots were strong, robust, and capable of sustaining multiple subcultures. We also observed spontaneous rooting in 'Da Hong' segments during shoot development, which reduced the duration of the rooting process and shortened the overall culture period. We proved that MS medium supplemented with 200 mg/L AC and 0.10 mg/L NAA was the best for shoot regeneration. In contrast, the best protocol for shoot development involved culture medium supplemented with 200 mg/L AC in combination with 0.20 mg/L NAA and 1.00 mg/L BAP. Successfully micropropagated 'Da Hong' pitaya could be used in orchards, providing new information in this context, valuable to produce robust seedlings of the genus *Hylocereus* tissue culture.

Supplementary Materials: The following supporting information can be downloaded at: https://www.mdpi.com/article/10.3390/horticulturae8020104/s1, Figure S1: In vitro shoot regeneration and root induction of *Hylocereus polyrhizus* 'Da Hong' from Murashige and Skoog (MS) medium supplemented with 0.05–0.20 mg/L α-naphthaleneacetic acid (NAA); Figure S2: Plantlet weight, including shoots and roots, and root/shoot ratio of *Hylocereus polyrhizus* 'Da Hong' grown on Murashige and Skoog (MS) medium supplemented with α-naphthaleneacetic acid (NAA) alone.

Author Contributions: Conceptualization, J.-C.C. and Y.-C.L.; methodology, Y.-C.L.; software, Y.-C.L.; validation, J.-C.C.; formal analysis, Y.-C.L.; investigation, Y.-C.L.; resources, Y.-C.L.; data curation, Y.-C.L.; writing—original draft preparation, Y.-C.L.; writing—review and editing, J.-C.C.; supervision, J.-C.C.; project administration, J.-C.C.; funding acquisition, J.-C.C. All authors have read and agreed to the published version of the manuscript.

Funding: This study was supported by a grant from the Ministry of Science and Technology, Executive Yuan, Taiwan, Republic of China. Project code: MOST-109-2313-B-005-021-MY3 (to J.-C.C.).

Data Availability Statement: Data sets analyzed during the current study are available from the current author on reasonable request.

Acknowledgments: We would like to thank Y.H. Hsu for her valuable suggestions and technical assistance in this work. We are also grateful to M.C. Ho for her support with orchard management.

Conflicts of Interest: The authors declare no conflict of interest.

References

1. Zee, F.; Yen, C.-R.; Nishina, M. Pitaya (dragon fruit, strawberry pear). *Fruits Nuts* **2004**, *9*, 1–3.
2. Chien, Y.-C.; Chang, J.-C. Net houses effects on microclimate, production, and plant protection of white-fleshed pitaya. *HortScience* **2019**, *54*, 692–700. [CrossRef]
3. Chu, Y.-C.; Chang, J.-C. High temperature suppresses fruit/seed set and weight, and cladode regreening in red-fleshed 'Da Hong' pitaya (*Hylocereus polyrhizus*) under controlled conditions. *HortScience* **2020**, *55*, 1259–1264. [CrossRef]
4. Chu, Y.-C.; Chang, J.-C. Regulation of floral bud development and emergence by ambient temperature under a long-day photoperiod in white-fleshed pitaya (*Hylocereus undatus*). *Sci. Hortic.* **2020**, *271*, 109479. [CrossRef]
5. Jiang, Y.-L.; Yang, W.-J. Development of integrated crop management systems for pitaya in Taiwan. In *Improving Pitaya Production and Marketing*; Jiang, Y.-L., Liu, P.-C., Huang, P.H., Eds.; Food Fertilizer Technology Center Press: Taipei, Taiwan, 2015; pp. 73–78.
6. Chiu, Y.C.; Lin, C.P.; Hsu, M.C.; Liu, C.P.; Chen, D.Y.; Liu, P.C. *Cultivation and Management of Pitaya*; Taiwan Agricultural Research Institute: Tainan, Taiwan, 2015; p. 93.
7. Liao, J.Y.; Chang, C.A.; Yan, C.R.; Chen, Y.C.; Deng, T.C. Detection and incidence of *Cactus Virus X* on pitaya in Taiwan. *Plant Pathol. Bull.* **2003**, *12*, 225–234.
8. Chuang, M.F.; Ni, H.F.; Yang, H.R.; Shu, S.L.; Lai, S.Y.; Jiang, Y.L. First report of stem canker disease of pitaya (*Hylocereus undatus* and *H. polyrhizus*) caused by *Neoscytalidium dimidiatum* in Taiwan. *Plant Dis.* **2012**, *96*, 906. [CrossRef]
9. Valencia-Botín, A.J.; Kokubu, H.; Ruiz, D.R. A brief overview on pitahaya (*Hylocereus* spp.) diseases. *J. Prof. Assoc. Cactus* **2013**, *15*, 42–48. [CrossRef]
10. Lin, C.-P.; Ann, P.-J.; Huang, H.-C.; Chang, J.-T.; Tsai, J.-N. Anthracnose of pitaya (*Hylocereus* spp.) caused by *Colletotrichum* spp., a new postharvest disease in Taiwan. *J. Taiwan Agric. Res.* **2017**, *66*, 171–183. [CrossRef]
11. Zakaria, L. Diversity of *colletotrichum* species associated with anthracnose disease in tropical fruit crops—A review. *Agriculture* **2021**, *11*, 297. [CrossRef]
12. Evallo, E.; Taguiam, J.D.; Balendres, M.A. A brief review of plant diseases caused by *Cactus virus X*. *Crop Protect.* **2021**, *143*, 105566. [CrossRef]
13. Zimmermann, H.G.; Granata, G. Insect pests and diseases. In *Cacti: Biology and Uses*; Nobel, P.S., Ed.; University of California Press: Berkeley, CA, USA, 2002; pp. 235–254.
14. Lema-Rumińska, J.; Kulus, D. Micropropagation of cacti—A review. *Haseltonia* **2014**, *19*, 46–63. [CrossRef]
15. Chohan, S.; Perveen, R.; Abid, M.; Naqvi, A.H.; Naz, S. Management of seed borne fungal diseases of tomato: A review. *Pak. J. Phytopathol.* **2017**, *29*, 193–200. [CrossRef]
16. Cassells, A.C. Pathogen and biological contamination management in plant tissue culture: Phytopathogens, vitro pathogens, and vitro pests. In *Plant Cell Culture Protocols*, 3rd ed.; Loyola-Vargas, V.M., Ochoa-Alejo, N., Eds.; Humana Press: Totowa, NJ, USA, 2012; pp. 57–80. [CrossRef]
17. Murashige, T.; Skoog, F. A revised medium for rapid growth and bio assays with tobacco tissue cultures. *Physiol. Plant.* **1962**, *15*, 473–497. [CrossRef]

18. Dahanayake, N.; Ranawake, A.L. Regeneration of dragon fruit (*Hylecereus undatus*) plantlets from leaf and stem explants. *Trop. Agric. Res. Ext.* **2011**, *14*, 85–89. [CrossRef]
19. Sheng, W.K.W.; Sundarasekar, J.; Sathasivam, K.; Subramaniam, S. Effects of plant growth regulators on seed germination and callus induction of *Hylocereus costaricensis*. *Pak. J. Bot.* **2016**, *48*, 977–982.
20. Pérez-Molphe-Balch, E.; del Socorro Santos-Díaz, M.; Ramírez-Malagón, R.; Ochoa-Alejo, N. Tissue culture of ornamental cacti. *Sci. Agric.* **2015**, *72*, 540–561. [CrossRef]
21. De Feria, M.; Rojas, D.; Reyna, M.; Quiala, E.; Solìs, J.; Zurita, F. In vitro propagation of *Hylocereus purpusii* Britton & Rose, a mexican species in danger of extinction. *Biotecnol. Veg.* **2012**, *12*, 77–83.
22. Hua, Q.; Chen, P.; Liu, W.; Ma, Y.; Liang, R.; Wang, L.; Wang, Z.; Hu, G.; Qin, Y. A protocol for rapid in vitro propagation of genetically diverse pitaya. *Plant Cell Tissue Organ. Cult.* **2014**, *120*, 741–745. [CrossRef]
23. Mohamed-Yasseen, Y. Micropropagation of pitaya (*Hylocereus undatus* Britton et Rose). *Vitr. Cell. Dev. Biol. Plant* **2002**, *38*, 427–429. [CrossRef]
24. Suman, K.; Rani, A.R.; Reddy, P.V. Response of dragon fruit (*Hylocereus undatus*) explants on MS media with growth regulators under in vitro for mass multiplication. *Agric. Update* **2017**, *12*, 2371–2375.
25. Trivellini, A.; Lucchesini, M.; Ferrante, A.; Massa, D.; Orlando, M.; Incrocci, L.; Mensuali-Sodi, A. Pitaya, an attractive alternative crop for mediterranean region. *Agronomy* **2020**, *10*, 1065. [CrossRef]
26. Yu, H.-L.; Zhang, W.; Zhu, Y.-Y. Studies on rapid propagation technology of stem segments in vitro of Hongxianmi *Hylocereus undatus*. *J. Anhui Agric. Sci.* **2009**, *37*, 3951–3952. [CrossRef]
27. Kari, R.; Lukman, A.L.; Zainuddin, R.; Ja'afar, H. Basal media for in vitro germination of red-purple dragon fruit *Hylocereus polyrhizus*. *J. Agrobiotechnol.* **2010**, *1*, 87–93.
28. Qin, J.; Wang, Y.; He, G.; Chen, L.; He, H.; Cheng, X.; Xu, K.; Zhang, D. High-efficiency micropropagation of dormant buds in spine base of red pitaya (*Hylocereus polyrhizus*) for industrial breeding. *Int. J. Agric. Biol.* **2017**, *19*, 193–198. [CrossRef]
29. Bidabadi, S.S.; Jain, S.M. Cellular, molecular, and physiological aspects of in vitro plant regeneration. *Plants* **2020**, *9*, 702. [CrossRef]
30. Wang, P.J.; Huang, L.C. Beneficial effects of activated charcoal on plant tissue and organ cultures. *In Vitro* **1976**, *12*, 260–262. [CrossRef]
31. Santos-Díaz, M.S.; Pérez-Molphe, E.; Ramírez-Malagón, R.; Núñez-Palenius, H.G.; Ochoa-Alejo, N. Mexican threatened cacti: Current status and strategies for their conservation. In *Species Diversity and Extinction*; Tepper, G.H., Ed.; Nova Science Publishers, Inc. Press: New York, NY, USA, 2010; pp. 1–60.
32. Thomas, T.D. The role of activated charcoal in plant tissue culture. *Biotechnol. Adv.* **2008**, *26*, 618–631. [CrossRef]
33. Pérez Molphe Balch, E.; Reyes, M.E.P.; Amador, E.V.; Rangel, E.M.; del Rocío Morones Ruiz, L.; Lizalde-Viramontes, H.J. Micropropagation of 21 species of mexican cacti by axillary proliferation. *Vitr. Cell. Dev. Biol. Plant* **1998**, *34*, 131–135. [CrossRef]
34. Peng, L.-C.; Qu, S.-P.; Su, Y.; Zhang, Y.-P.; Wang, L.-H. Study on rapid propagation of *Hylocereus undatus* by two steps. *Southwest China J. Agric. Sci.* **2014**, *27*, 2529–2533.
35. Lee, Y.-C.; Chang, J.-C. Effect of MS strengths and fruit storage duration at 4 °C on 'Da Hong' pitaya seed germination. *HortScience*, 2022; *manuscript in preparation*.
36. Abramoff, M.D.; Magalhães, P.J.; Ram, S.J. Image processing with ImageJ. *Biophotonics Int.* **2004**, *11*, 36–42.
37. Mauseth, J.D.; Halperin, W. Hormonal control of organogenesis in *Opuntia polyacantha* (Cactaceae). *Am. J. Bot.* **1975**, *62*, 869–877. [CrossRef]
38. Rubluo, A.; Marín-Hernández, T.; Duval, K.; Vargas, A.; Márquez-Guzmán, J. Auxin induced morphogenetic responses in long-term in vitro subcultured *Mammillaria san-angelensis* Sánchez-Mejorada (Cactaceae). *Sci. Hortic.* **2002**, *95*, 341–349. [CrossRef]
39. Viñas, M.; Fernández-Brenes, M.; Azofeifa, A.; Jiménez, V.M. In vitro propagation of purple pitahaya (*Hylocereus costaricensis* [F.A.C. Weber] Britton & Rose) cv. Cebra. *Vitr. Cell. Dev. Biol. Plant* **2012**, *48*, 469–477.
40. Clayton, P.W.; Hubstenberger, J.F.; Phillips, G.C.; Butler-Nance, S.A. Micropropagation of members of the Cactaceae subtribe Cactinae. *J. Am. Soc. Hortic. Sci.* **1990**, *115*, 337–343. [CrossRef]
41. Angulo-Bejarano, P.I.; Paredes-López, O. Development of a regeneration protocol through indirect organogenesis in prickly pear cactus (*Opuntia ficus-indica* (L.) Mill). *Sci. Hortic.* **2011**, *128*, 283–288. [CrossRef]
42. Kim, Y.-S.; Hahn, E.-J.; Yeung, E.C.; Paek, K.-Y. Lateral root development and saponin accumulation as affected by IBA or NAA in adventitious root cultures of *Panax ginseng* C.A. Meyer. *Vitr. Cell. Dev. Biol. Plant* **2003**, *39*, 245–249. [CrossRef]
43. Elmongy, M.S.; Cao, Y.; Zhou, H.; Xia, Y. Root development enhanced by using indole-3-butyric acid and naphthalene acetic acid and associated biochemical changes of in vitro azalea microshoots. *J. Plant Growth Regul.* **2018**, *37*, 813–825. [CrossRef]
44. Mendes, A.F.S.; Cidade, L.C.; Otoni, W.C.; Soares-Filho, W.S.; Costa, M.G.C. Role of auxins, polyamines and ethylene in root formation and growth in sweet orange. *Biol. Plant.* **2011**, *55*, 375–378. [CrossRef]
45. Pan, M.J.; van Staden, J. The use of charcoal in in vitro culture—A review. *Plant Growth Regul.* **1998**, *26*, 155–163. [CrossRef]
46. Pan, M.J.; van Staden, J. Effect of activated charcoal, autoclaving and culture media on sucrose hydrolysis. *Plant Growth Regul.* **1999**, *29*, 135–141. [CrossRef]
47. De Cortázar, V.G.; Nobel, P.S. Biomass and fruit production for the prickly pear cactus, *Opuntia ficus-indica*. *J. Am. Soc. Hortic. Sci.* **1992**, *117*, 558–562. [CrossRef]

48. Mauseth, J.D. Theoretical aspects of surface-to-volume ratios and water-storage capacities of succulent shoots. *Am. J. Bot.* **2000**, *87*, 1107–1115. [CrossRef] [PubMed]
49. Malda, G.; Backhaus, R.A.; Martin, C. Alterations in growth and crassulacean acid metabolism (CAM) activity of in vitro cultured cactus. *Plant Cell Tissue Organ. Cult.* **1999**, *58*, 1–9. [CrossRef]
50. Yi, R.H.; Lin, Q.L.; Mo, J.J.; Wu, F.F.; Chen, J. Fruit internal brown rot caused by *Neoscytalidium dimidiatum* on pitahaya in Guangdong province, China. *Australas. Plant Dis. Notes* **2015**, *10*, 13. [CrossRef]
51. Masyahit, M.; Sijam, K.; Awang, Y.; Satar, M.G.M. The first report of the occurrence of anthracnose disease caused by *Colletotrichum gloeosporioides* (Penz.) Penz. & Sacc. on dragon fruit (*Hylocereus* spp.) in peninsular Malaysia. *Am. J. Appl. Sci.* **2009**, *6*, 902–912.
52. Maule, A.J.; Wang, D. Seed transmission of plant viruses: A lesson in biological complexity. *Trends Microbiol.* **1996**, *4*, 153–158. [CrossRef]
53. Ni, H.-F.; Huang, C.-W.; Hsu, S.-L.; Lai, S.-Y.; Yang, H.-R. Pathogen characterization and fungicide screening of stem canker of pitaya. *J. Taiwan Agric. Res.* **2013**, *62*, 225–234.
54. Gebeyaw, M. Review on: Impact of seed-borne pathogens on seed quality. *Am. J. Plant Biol.* **2020**, *5*, 77–81. [CrossRef]
55. Cisneros, A.; Tel-Zur, N. Genomic analysis in three *Hylocereus* species and their progeny: Evidence for introgressive hybridization and gene flow. *Euphytica* **2013**, *194*, 109–124. [CrossRef]

 *horticulturae*

Article

Micropropagation of Plum (*Prunus domestica* L.) in Bioreactors Using Photomixotrophic and Photoautotrophic Conditions

Diego Gago [1,2], Conchi Sánchez [1], Anxela Aldrey [1], Colin Bruce Christie [3], María Ángeles Bernal [2] and Nieves Vidal [1,*]

[1] Misión Biológica de Galicia Sede Santiago de Compostela, Consejo Superior de Investigaciones Científicas, Apdo 122, 15780 Santiago de Compostela, Spain; tresdedecembro@gmail.com (D.G.); conchi@iiag.csic.es (C.S.); anxela.aldrey@iiag.csic.es (A.A.)

[2] Departamento de Biología, Facultad de Ciencias, Universidade da Coruña, Campus da Zapateira s/n, 15071 A Coruña, Spain; angeles.bernal@udc.es

[3] The Greenplant Company, Palmerston North 4410, New Zealand; brucechristie101@gmail.com

* Correspondence: nieves@iiag.csic.es

Abstract: In this study, we propagated two old Galician plum varieties in liquid medium using a temporary immersion system with RITA© bioreactors. Environmental variables including culture system, light intensity, CO_2 enrichment, immersion frequency and sucrose supplementation were evaluated in relation to in vitro proliferation, physiological status and ex vitro performance. Bioreactors were superior to jars for culturing shoots in photomixotrophic conditions, producing up to 2 times more shoot numbers and up to 1.7 times more shoot length (depending on the genotype) using shoot clusters. The number and quality of shoots were positively influenced by the sucrose concentration in the medium, plus by the light and gaseous environment. For individual apical sections the best response occurred with 3% sucrose, 150 μmol m^{-2} s^{-1} photosynthetic photon flux density and 2000 ppm CO_2, averaging 2.5 shoots per explant, 26 mm shoot length and 240 mm^2 leaf area, while with 50 μmol m^{-2} s^{-1} light and ambient CO_2 (400 ppm) values decreased to 1.2 shoots per explant, 14 mm of shoot length and 160 mm^2 of leaf area. Shoots cultured photoautotrophically (without sucrose) were successfully rooted and acclimated despite of showing limited growth, low photosynthetic pigments, carbohydrate, phenolic and antioxidant contents during the multiplication phase.

Keywords: CO_2; liquid medium; local varieties; RITA©; rooting; stress; sucrose; temporary immersion

Citation: Gago, D.; Sánchez, C.; Aldrey, A.; Christie, C.B.; Bernal, M.Á.; Vidal, N. Micropropagation of Plum (*Prunus domestica* L.) in Bioreactors Using Photomixotrophic and Photoautotrophic Conditions. *Horticulturae* **2022**, *8*, 286. https://doi.org/10.3390/horticulturae8040286

Academic Editors: Jean Carlos Bettoni, Min-Rui Wang and Qiao-Chun Wang

Received: 4 March 2022
Accepted: 26 March 2022
Published: 29 March 2022

Publisher's Note: MDPI stays neutral with regard to jurisdictional claims in published maps and institutional affiliations.

1. Introduction

Plums include a large and diverse group of closely related *Prunus* species of the Rosaceae family. European plum (*P. domestica* L.) and Japanese plum (*P. salicina* Lindl) are currently the most globally cultivated plum species. China is the leading producing country, followed by Romania, Serbia, Chile, Iran, the USA, Turkey, Italy, France, Ukraine and Spain [1]. As many other cultivated *Prunus* spp. fruit trees, plums show limited intra-specific genetic variability [2]. The biodiversity loss started with the process of domestication and was exacerbated by clonal propagation, a narrow parentage range for breeding and selection of similar fruit attributes for agronomic, processing and commercial reasons [3]. Reduced genetic variability diminishes the breeding potential and increases the vulnerability to pests, diseases and environmental change [4]. Genotyping projects carried out in several European countries revealed that local cultivars often differ considerably from widespread international cultivars [5], probably because old traditional cultivars and wild relatives have been subjected to less artificial selection pressures [6]. Local varieties may show low yield and not have outstanding agronomic characteristics, but their allelic diversity provides a gene-pool reservoir for breeding for organoleptic characteristics and for adaptation in a changing environment [7–10]. The potential value of most local genotypes for current market needs is largely unknown, since they have not been sufficiently characterized from an genetic, agronomic and consumer point of view [11].

In the present study, we focus on two local plum varieties from Galicia, located in the northwest of Spain. Galicia is one of the regions where the coexistence of forest areas with small but numerous traditional farms favored during the past centuries has contributed to the maintenance of an extensive forest and agricultural and horticultural biodiversity [12–14]. In the last century, large-scale commercialization of food in general and of fruits in particular led to the introduction of new varieties. Plants were selected for larger size and higher yields, which displaced the traditional crops and their associated cultural methodologies. As a result, in many regions local varieties of fruits have a narrow range of distribution and are linked to self-supply practices. Together with the continuous decrease in the population of rural areas, this has promoted a rapid global genetic erosion of agrobiodiversity. There is an urgent need to preserve as much of the current fruit tree diversity as possible to prevent the irretrievable loss of this heritage [13,15]. There are breeding programs in some countries that could use this old genetic material [1,16].

Ex situ conservation of the existing genetic resources is one strategy recommended to alleviate this situation [17]. Within ex situ conservation, micropropagation is the methodology of choice in the case of species or genotypes such as plums, which are propagated vegetatively by budding, grafting or cuttings [18,19]. The first micropropagation protocols in semisolid medium were reviewed by Druart and Gruselle as early as 1986 [20]. In 1991, Druart pointed out the importance of reliable micropropagation protocols for plums to enhance the commercial propagation of virus-free plants [21]. This author also highlighted the need to develop new micropropagation protocols that could be automated. These protocols would make the whole process of in vitro plant production more economically viable and competitive with traditional vegetative propagation. However, the protocols developed afterwards were still based in semisolid medium [22–26]. More recently, liquid medium has been used in bioreactors as an alternative to agar-based protocols, this has the potential to enhance the quality of micropropagated plants and facilitates automation [27]. This technology has been successfully applied to a range of woody species [28], however, to the best of our knowledge it has not been used with European plum (*P. domestica*). In this study, we proliferated plum in a liquid medium by a temporary immersion system using commercial RITA© bioreactors and explored the feasibility of culturing plum shoots without sucrose supplementation. Variables such as culture system, growth conditions (light intensity and CO_2 enrichment), frequency of immersion and sucrose supplementation were evaluated in relation to in vitro proliferation, physiological status and ex vitro performance.

The main goal was to provide information that could be developed into a commercially applicable protocol for rapid propagation of plum trees and may also be useful in rescuing rare or endangered plum germplasm. The application of bioreactors for this work represents a novel approach to plum propagation.

2. Materials and Methods

2.1. Establishment of Shoot Multiplication Cultures

Shoots of *P. domestica*, old Galician landrace varieties "Claudia Blanca País" (CBP) and "Collón de Frade Negro" (CFN) were collected from trees growing in the research orchard of the "Asociacion Galega da Froita Autóctona do Eume", San Sadurniño, A Coruña, Spain (Figure 1). CBP is a productive small to medium-sized plum tree. The fruits ripen during August, having a green-yellowish skin occasionally with a pink tinge and some red spots, the flesh is light yellow, very sweet and tasty. CFN is a vigorous and productive tree with fruit maturing at the end of August. Oval fruits are normally black skinned with greenish flesh that is tasty and a bit sour close to the skin.

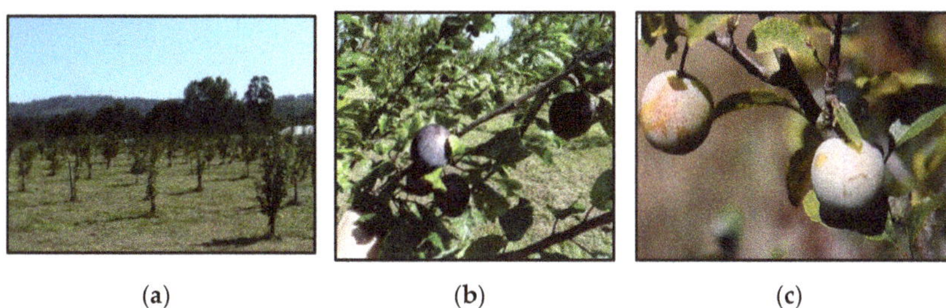

(a)   (b)   (c)

Figure 1. Plant material for establishment of plum cultures. (**a**) Orchard of the "Asociación Galega da Froita Autóctona do Eume", (**b**) Collón de Frade Negro (CFN) genotype and (**c**) Claudia Blanca País (CBP) genotype.

Cultures were initiated from branches with dormant buds collected in early spring from 4-year-old trees following the protocols described in Sánchez and Vieitez [29] with minor nutrient modifications. After fungicide treatment (copper oxychloride), branch segments 25- to 30-cm-long were set upright in moistened perlite and forced axillary buds to flush in a growth cabinet at 25 °C and 80–90% relative humidity with a 16 h photoperiod (90–100 µmol m^{-2} s^{-1} provided by cool-white fluorescent lamps). After 2–3 weeks, recently sprouted shoots (3–5 cm in length) were used as the source of explants. The shoots were stripped of their leaves and surface sterilized by immersion for 30 s in 70% ethanol and for 10 min in a 6 g L^{-1} solution of sodium hypochlorite containing 2–3 drops of Tween 80$^{®}$. The shoots were then rinsed three times in sterile distilled water. Nodal segments (10 mm) were cut from the shoots and inoculated in tubes with semisolid medium (SSM). Culture medium consisted in Murashige and Skoog medium (MS) [30] supplemented with 0.5 mg L^{-1} of N^6-benzyladenine (BA), 0.5 mg L^{-1} of indole-3-butyric acid (IBA), 3% sucrose and 0.65% (*w/v*) agar Vitroagar (Pronadisa, Spain). The medium was adjusted to pH 5.7 before being autoclaved at 120 °C for 20 min. Cultures were incubated under a 16 h photoperiod provided by cool-white fluorescent lamps (photosynthetic photon flux density (PPF) from 50 to 60 µmol m^{-2} s^{-1}) at 25 °C light/20 °C dark (Standard conditions; ST). The explants were transferred every 2 weeks during the first 6 weeks after establishment. Thereafter, the explants were maintained in 50 mL of the medium described above in 300 mL glass jars (8 explants per jar) and were subcultured every 5 weeks. Preliminary work with the two plum clones indicated that they were indistinguishable and could be treated as one and the same.

2.2. Culture System, Hormonal Supplementation and Frequency of Immersion

To study the proposed variables, we performed two experiments. In the first one, clusters (15–20 mm high) with 3–4 shoots (Figure 2a) of CFN genotype were excised from shoots growing in jars in SSM. These initial explants (8 per container) were cultured in jars or RITA© vessels (Vitropic, Saint Mathieu de Tréviers, France) under ST conditions and were subcultured every 5 weeks. In jars, the explants were cultured with 50 mL of SSM, and in RITA© with 150 mL of liquid medium (LM) of the same composition as SSM without agar and immersed for 1 min 3 times per day. The proliferation medium consisted of MS supplemented with 3% sucrose, 0.65% agar and (i) 0.2 mg L^{-1} BA + 0.2 mg L^{-1} IBA, (ii) 0.2 mg L^{-1} BA + 0.5 mg L^{-1} IBA and (iii) 0.5 mg L^{-1} BA + 0.5 mg L^{-1} IBA.

(a)　　　　　　　　(b)　　　　　　　　(c)

Figure 2. (**a**) Cluster of CFN shoots used as initial explant for experiments regarding the culture system and the hormonal concentration of culture media, (**b**) cluster of CBP shoots used as initial explant for experiments regarding the culture system and the immersion frequency and (**c**) apical section of CBP used as initial experiment regarding growth conditions and sucrose supplementation. Bar: 10 mm.

In the second experiment, clusters of similar size as described above were excised from shoots of the CBP genotype growing in jars and were cultured in jars or RITA© vessels with MS supplemented with 0.5 mg L^{-1} BA + 0.5 mg L^{-1} IBA and 3% sucrose (Figure 2b). Agar (0.65%) was added to the jars. In RITA©, shoots were immersed for 1 min 3 or 6 times per day. After 6 weeks of culture under ST conditions, the morphological data were recorded and shoots were used for rooting experiments.

2.3. Growth Conditions and Sucrose Supplementation

Apical sections (20 mm) of CBP genotype (Figure 2c) were cultured in RITA© and were immersed in LM supplemented with 0, 1 or 3% sucrose for 1 min 6 times per day. The explants were cultured under ST conditions and with ambient CO_2 (~400 ppm) or in an experimental unit previously designed for photoautotrophic micropropagation (PAM) of chestnut [31]. In the PAM experimental unit, the cultures grew under high PPF (150 μmol m^{-2} s^{-1}), and CO_2-enriched air (~2000 ppm) was injected to the bioreactors during immersion. The photoperiod and temperature regime were the same as under ST conditions. After 6 weeks of subculture morphological data were recorded and shoots were used for rooting experiments or biochemical analysis (monosaccharides, photosynthetic pigments, soluble phenolic compounds and antioxidant activity).

2.3.1. Biochemical Quantifications
Soluble Monosaccharides

We used the dinitrosalicylic acid (DNS) method [32,33]. Briefly, 100–150 mg of apical leaves was homogenized with 2 mL of distilled water and centrifuged for 5 min at 10,000× *g*. The supernatant was collected, mixed with DNS, heated in a thermoblock at 100 °C for 5 min and placed on ice for 5 min before quantitation in a spectrophotometer at 540 nm. The results were expressed as glucose equivalents on a fresh weight basis.

Total Soluble Sugars

Total soluble carbohydrates were determined by the anthrone method [34]. Fresh leaves (100 mg) were homogenized in 2 mL water and centrifuged at 10,000× *g* for 5 min. Samples (250 μL) were treated with 750 μL of ice-cold anthrone reagent (1 g/L in 96% H_2SO_4). The reaction mixture was heated in a water bath (40 °C) for 40 min and rapidly cooled to 0 °C. Absorbance was measured at 620 nm. The results were expressed as sucrose equivalents on a fresh weight basis.

Photosynthetic Pigments

The two uppermost fully expanded leaves per explant were collected, weighed and extracted with dimethylformamide. Chlorophyll a, b and total carotenoids were quantified using the method described by Wellburn [35].

Total Soluble Phenolic Compounds

The extraction was performed according to [36]. Individual shoots were collected, and leaves were homogenized with methanol 80%. The mixture was centrifuged at 10,000× g for 5 min and the supernatant used for analysis of soluble phenolic compounds and antioxidants.

Total soluble phenolic compounds were assayed using the Folin–Ciocalteu method [37]. The soluble phenols content was calculated from a standard curve based on gallic acid different concentrations and results were expressed as gallic acid equivalents on a fresh weight basis.

Antioxidant Activity

The extraction was performed as described for soluble phenols. Antioxidant activity of plant extracts was determined through spectrophotometry using a 1,1-diphenyl-2-picrylhydrazyl (DPPH) scavenging radical assay [38], and results were expressed as TROLOX mM equivalents per g on a fresh weight basis.

2.4. Rooting and Acclimatization

Shoots of CBP and CFN genotypes (longer than 15 mm) cultured in jars or bioreactors with 3% sucrose were treated with MS with the macronutrients reduced to half strength ($\frac{1}{2}$MS) with 25 mg L^{-1} IBA, 0.7% agar and 3% sucrose. After 24 h, shoots were transferred to jars with SSM of the same composition but without IBA. Alternatively, shoots were inserted in rockwool cubes (2 cm of side) soaked in liquid medium (Figure 3a) and introduced in Plantform™ bioreactors (Plantform, Hjärup, Sweden) without the inner baskets (Figure 3b,c). Plantforms™ were aerated for 1 min at a frequency of 6 times every 24 h for 6 weeks.

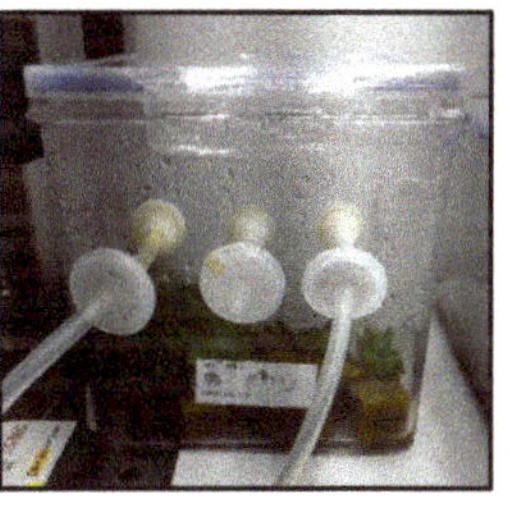

(**a**) (**b**) (**c**)

Figure 3. (**a**,**b**) Shoots of CBP inserted in rockwool cubes for rooting expression and (**c**) shoots inserted in a Plantform™ bioreactor without the inner basket.

Shoots of CBP cultured in bioreactors with different sucrose were treated with 1/2 MS with 25 mg L^{-1} IBA, 0.7% agar and the same sucrose concentration that was present in the multiplication medium. After 24 h, shoots were transferred to IBA-free medium for rooting expression and distributed between the three sucrose treatments. Shoots were inserted in rockwool cubes and placed in Plantform™ bioreactors as described above.

Rooted shoots were transferred to plastic plug trays (size of the plugs 52 × 52 mm by 60 mm height) filled with a peat:perlite (3:1) mixture and placed in a controlled environmental chamber (Fitotron SGC066, Sanyo Gallencamp PLC, Leicestershire, UK) with a photoperiod of 16 h light:8 h dark, a photosynthetic photon flux from 240–250 μmol m^{-2}·s^{-1} and a temperature of 25 °C (day) and 20 °C (night), with a relative humidity of 85%. After 3 weeks in the phytotron, plantlets were transferred to 1.3 L pots and placed into a greenhouse between April and June. The percentage of survival and plantlet growth were recorded when transferred to the greenhouse and 4 weeks later.

2.5. Data Recording and Statistical Analysis

The parameters analyzed were: (a) the number of normal shoots longer than 15 mm produced by each explant (NS); (b) the length of the longest shoot per explant (SL); (c) the length (LL) and width (LW) of the largest leaf per explant; (d) leaf biomass; (e) monosaccharides; (f) total soluble sugars; (g) total soluble phenolic compounds; (h) total antioxidant activity; (i) photosynthetic pigments; (j) rooting and acclimatization percentage. Leaf area was calculated as $1/2$ LL*LW.

For proliferation experiments, three jars or RITA© vessels (24 explants) were used for each treatment and the experiments were repeated twice. For rooting experiments, 18 shoots per treatment were used.

The data were analyzed by Levene's test (to verify the homogeneity of variance) and were then subjected to analysis of variance (ANOVA), followed by comparison of group means (Tukey-b test), or to the Welch ANOVA, followed by Games–Howell post-hoc comparison (when heteroscedasticity was detected). When an interaction between two factors was indicated by the two-way ANOVA, Bonferroni's adjustment was applied to detect the simple main effects in multiple post-hoc comparisons. Statistical analyses were performed using SPSS 26.0 (IBM).

3. Results

3.1. Culture System and Hormonal Composition of the Media

Plum shoots grew successfully in jars and bioreactors (Figure 4), but the culture system and the hormonal composition of the media significantly influenced the performance of shoots. The best results for number and length of shoots and leaf size were obtained in bioreactors (Figure 5). MS medium supplemented with 0.5 mg/L BA and IBA produced more and longer shoots (Figure 5a,b), whereas the size of leaves decreased when more BA and IBA were added to the medium (Figure 5c,d).

(a)   (b)

Figure 4. CFN shoots cultured in jars (**a**) and in bioreactors (**b**) in MS supplemented with 0.5 mg L^{-1} BA, 0.5 mg L^{-1} IBA and 3% sucrose.

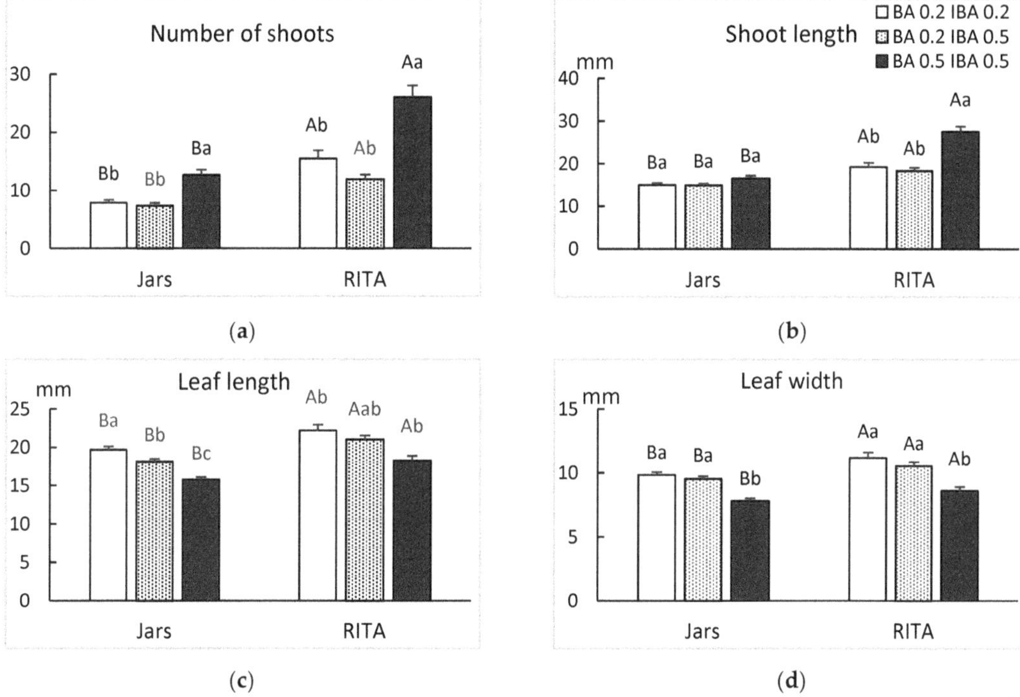

Figure 5. Effect of the hormonal composition of the media and the culture system on the growth of the plum genotype CFN. (**a**) Number of shoots, (**b**) shoot length, (**c**) length and (**d**) width of the largest leaf. Different uppercase letters indicate significant differences between culture system and different lowercase letters indicate significant differences between the hormonal treatments ($p < 0.05$).

3.2. Culture System and Number of Immersions

Table 1 shows the effect of culturing the CBP genotype in jars or in RITA© bioreactors with 3 or 6 immersions per day. The use of bioreactors significantly increased all the growth parameters. The highest values were obtained with six immersions, although significant differences between the two immersion frequencies were only detected in leaf length.

Table 1. Effect of the number of immersions and the culture system on the growth of the plum genotype CBP cultured with 3% sucrose. (a) Shoot length (SL), length (LL) and width (LW) of the largest leaf; (b) Number of shoots (NS). Data are mean values and standard error of 48 explants. Different letters indicate significant differences ($p < 0.05$).

Culture System	NS	SL (mm)	LL (mm)	LW (mm)
Jars	5.3 ± 0.27 b	18.9 ± 0.68 b	15.3 ± 0.63 c	6.9 ± 0.30 b
RITA© (3 immersions/day)	7.7 ± 0.59 a	22.3 ± 0.70 a	21.5 ± 0.51 b	9.3 ± 0.26 a
RITA© (6 immersions/day)	8.7 ± 0.61 a	24.5 ± 0.83 a	23.8 ± 0.69 a	10.1 ± 0.41 a

3.3. Growth Conditions and Sucrose Supplementation

The number and quality of shoots were significantly affected by the sucrose concentration in the medium and by the light and gaseous environment (Figure 6).

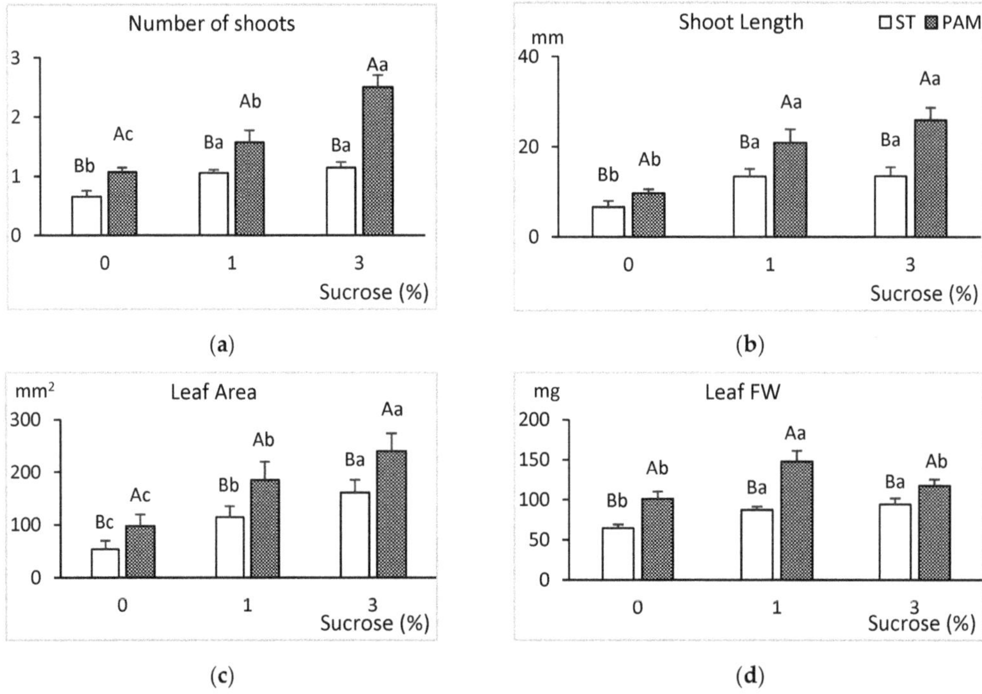

Figure 6. Effect of growth conditions and sucrose supplementation on the growth of the genotype CBP cultured in RITA© with 6 immersions/day. (**a**) Number of shoots, (**b**) shoot length (mm), (**c**) leaf area and (**d**) leaf biomass. Different uppercase letters indicate significant differences in relation to the growth conditions, and different lowercase letters indicate significant differences in relation to the sucrose supplementation ($p < 0.05$). ST: standard conditions, PAM: photoautotrophic conditions and FW: fresh weight.

The best response was obtained in PAM conditions for all the sucrose treatments (Figure 6). Sucrose significantly enhanced the growth of shoots and leaves, and the best response growth responses were obtained with 1–3% sucrose in PAM conditions. Shoots cultured in PAM with 1% sucrose produced more and longer shoots, as well as larger leaves than shoots cultured in ST conditions with 3% sucrose. Explants cultured without sucrose showed limited growth, especially in ST conditions. Frequently, only 20–30% of the explants produced vigorous shoots longer than 15 mm, meaning that shoots suitable for the rooting experiments were barely obtained. For this treatment, the average number of new shoots (0.7) and their short height (7 mm average) were too small for sub-culturing. Leaves of shoots grown without sucrose had the lowest concentration of photosynthetic pigments. Pigments increased which sucrose supplementation but were not significantly influenced by growth conditions (Figure 7).

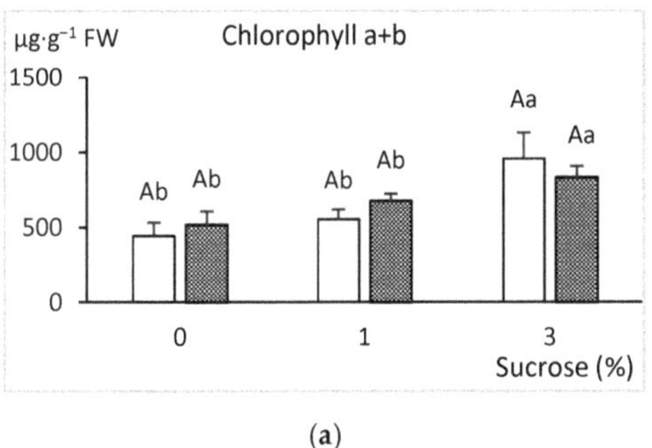
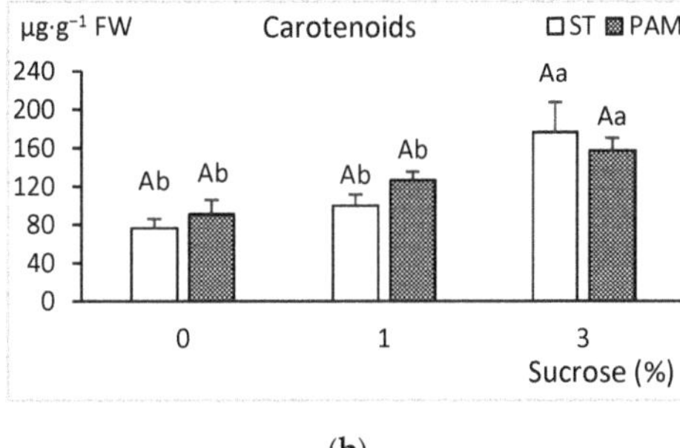

(**a**) (**b**)

Figure 7. Effect of growth conditions and sucrose supplementation on the photosynthetic pigments of leaves of the genotype CBP cultured in RITA© with 6 immersions/day. (**a**) Chlorophyll a+ chlorophyll b and (**b**) total carotenoids. Different uppercase letters indicate significant differences in relation to the growth conditions, and different lowercase letters indicate significant differences in relation to the sucrose supplementation ($p < 0.05$). ST: standard conditions, PAM: photoautotrophic conditions and FW: fresh weight.

Carbohydrate content in leaves was influenced by the sucrose treatment and its interaction with the growth conditions (Figure 8).

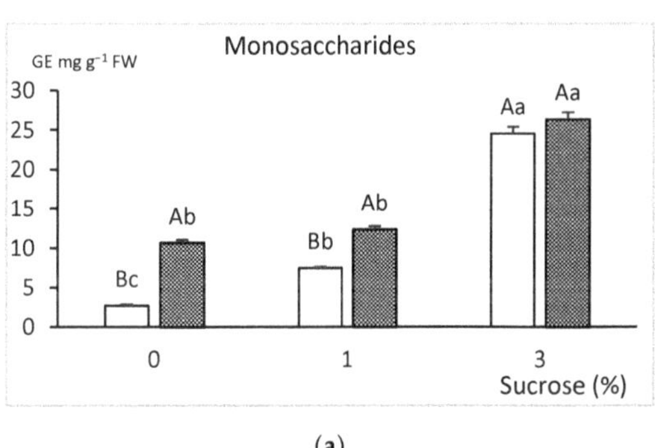
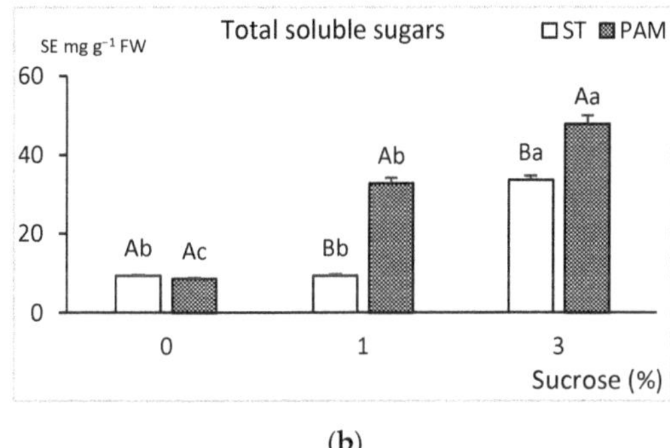

(**a**) (**b**)

Figure 8. Effect of growth conditions and sucrose supplementation on the carbohydrate concentration of leaves of genotype CBP cultured in RITA© with 6 immersions/day. (**a**) Monosaccharides and (**b**) total soluble sugars. Different uppercase letters indicate significant differences in relation to the growth conditions, and different lowercase letters indicate significant differences in relation to the sucrose supplementation ($p < 0.05$). ST: standard conditions, PAM: photoautotrophic conditions and FW: fresh weight. SE: sucrose equivalents and GE: glucose equivalents.

Generally, carbohydrate content (Figure 8a,b) increased with sucrose supplementation and with PAM conditions. There were marked differences between the highest and the lowest concentrations of total soluble sugars and monosaccharides. In monosaccharides, the lowest content was observed in shoots cultured without sucrose in ST conditions. The highest values were for 3% sucrose, with no significant differences between growth conditions (Figure 8a). Total sugars followed a similar trend to the monosaccharides (Figure 8b).

Soluble phenolic compounds and antioxidant activity were quantified to estimate the shoot stress levels in different growth conditions and with different concentrations of sucrose (Figure 9). In ST conditions sugar enhanced phenolic production, but this was not reflected consistently in PAM conditions (Figure 9a). Regarding antioxidants, in the PAM treatment the observed levels were elevated with increased sugar content and were also higher than the corresponding ST treatments (Figure 9b).

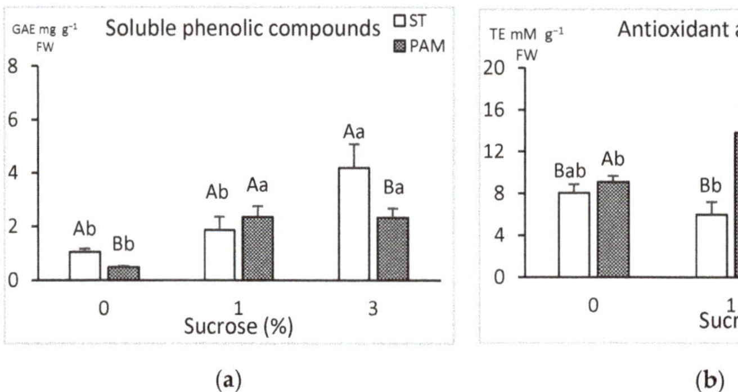

(a)   (b)

Figure 9. Effect of growth conditions and sucrose supplementation on the phenolic and antioxidant concentration of leaves of genotype CBP cultured in RITA© with 6 immersions/day. (a) Total soluble phenolics and (b) total antioxidant activity. Different uppercase letters indicate significant differences in relation to the growth conditions, and different lowercase letters indicate significant differences in relation to the sucrose supplementation ($p < 0.05$). ST: standard conditions, PAM: photoautotrophic conditions and FW: fresh weight. GAE: gallic acid equivalents and TE: Trolox equivalents.

3.4. Effect of the Culture System on Rooting and Acclimation

Shoots of CFN and CBP genotypes cultured in jars or RITA© with 3% sucrose were successfully rooted and acclimated irrespective of the culture system. Six weeks after auxin treatment, the length of the rooted shoots ranged from 30 to 40 mm, and 6 weeks later their height was 10-fold longer, averaging 37 cm (Figures 10 and 11).

(a)   (b)   (c)

Figure 10. Shoots of CBP cultured in jars and RITA© with 3% sucrose and rooted in jars. (a) Shoots proliferated in jars 6 weeks after auxin treatment and (b,c) shoots proliferated in RITA© (6 immersions/day) 12 weeks after auxin treatment.

Rooting and acclimation responses averaged 90%, and we did not observe differences in the percentages of shoots rooted in SSM or in rockwool cubes. Shoots rooted in cubes were easier to handle without compromising the integrity of the roots, whereas some roots were broken when the shoots were extracted from the jars, and cubes were used thereafter.

(a)   (b)   (c)

Figure 11. Shoots of CFN cultured in jars and RITA© with 3% sucrose and rooted in rockwool cubes. (**a**) Shoots proliferated in jars 6 weeks after auxin treatment and (**b**,**c**) plantlets from shoots proliferated in RITA© after being transferred to the phytotron (**b**) and 3 months later in the greenhouse (**c**).

3.5. Effect of Sucrose Supplementation on Rooting and Acclimation

Figures 12 and 13 show the rooting response of CBP shoots cultured in ST or PAM with 0, 1 or 3% sucrose and rooted in rockwool cubes soaked with medium containing 0, 1 or 3% sucrose. Figure 12 shows quantitative data and Figure 13 the appearance of the shoots after rooting and during acclimation. None of the three investigated factors (growth conditions, sucrose during multiplication and rooting) had a significant effect on rooting (p values were 0.238, 0.536 and 0.128 for growth conditions, sucrose during multiplication and sucrose during rooting, respectively), and shoots from all the treatments rooted and acclimated successfully (Figures 12 and 13). However, when shoots were multiplied without sucrose the proportion of rootable shoots was lower than in the rest of treatments, meaning more bioreactors had to be used to obtain enough shoots for the rooting experiments.

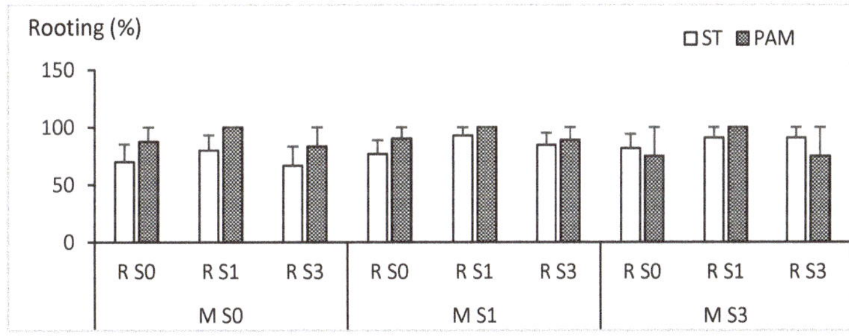

Figure 12. Shoots of CBP cultured in RITA© with 0–3% sucrose in ST or PAM conditions and rooted in rockwool cubes soaked with 0–3% sucrose. ST: standard conditions and PAM: photoautotrophic conditions. M S0, M S1 and M S3: sucrose percentage (0, 1 or 3%) of the multiplication medium in RITA© bioreactors. R S0, R S1 and R S3: sucrose percentage (0, 1 or 3%) of the rooting expression medium.

(a)   (b)   (c)

(d)   (e)

Figure 13. (**a**,**b**) Shoots of CBP cultured in RITA© without sucrose in PAM conditions and rooted in rockwool cubes soaked with 1% (**a**) and 3% sucrose (**b**); (**c**–**e**) shoots cultured in ST with 1% sucrose and rooted with 0, 1 and 3% sucrose 6 weeks after root induction (**c**), after 2 weeks in the phytotron (**d**) and after 2 months in the greenhouse (**e**). ST: standard conditions and PAM: photoautotrophic conditions.

4. Discussion

In the present study, we demonstrated the feasibility of micropropagating two local plum varieties from the northwest of Spain. We obtained high proliferation and rooting rates and the plantlets were successfully acclimated. The multiplication of both clones was more efficient in bioreactors than in jars with semisolid medium. Similar results have been reported for other woody plants such as calabash tree [39], eucalyptus [40], apple [41], teak [42,43], pistachio [44], chestnut [45], hazelnut [46], yerba mate [47], willow [48], olive [49], alder [50] and pear [51].

Most bioreactors include forced ventilation systems, which increase gas exchange [28,52]. It has been reported that this feature promotes the photosynthetic ability of tissues, enabling the decrease or even the elimination of the conventional sugar supplementation [53,54]. For this reason, bioreactors have been used for the photoautotrophic propagation of several plants including eucalyptus [55–58], apple [59], poplar [60], bamboo [61] and willow [62], but to the best of our knowledge there are no reports of photoautotrophic protocols for plum or any other *Prunus* species. In the current study, we explored the feasibility of using commercial RITA© bioreactors for the propagation of plum photoautotrophically (without exogenous sugars) in an experimental unit which provided good results with willow [62]. Previously, we carried out some experiments photomixotrophically (using 3% sucrose) to find the best growing conditions for the plum genotypes under study.

In the photomixotrophic using shoot clusters as initial explants, the shoot number and shoot length were the highest in treatments with BA 0.5 and IBA 0.5, whereas leaf size was negatively correlated. Typically, leaf size decreases with increasing cytokinin supplementation [42,63,64]. Increasing exposure of the plant material to more immersions of up to six per day increased plant growth, whereas using more frequent immersion produced hyperhydric shoots (data not shown). Immersion every 3–6 h has been successfully used for other tree species such as calabash tree [39], *Fraxinus mandshurica* [65], teak, chestnut [45], yerba mate [47], willow [48,62] and apple [66].

For photoautotrophic propagation, we used the best conditions ascertained above, but we selected apical sections as initial explants. This material does not produce as many shoots as clusters, but it is more uniform and maybe less affected by the carry-over as it is more distant from the previous nutritional conditions. In these experiments, we found that sucrose supplementation affected beneficially the number and quality of shoots. Using high light intensity and CO_2 enrichment provided further improvement in growth to explants cultured with 1 and 3% sucrose, but it did not compensate for the absence of sugar supplementation. The explants could be grown photoautotrophically (without sucrose), but the number and size of shoots limited the practical and future value of this option. Increased gaseous exchange enabled shoot growth of *Paulownia fortunei* [67], *Samanea saman* [68], *Eucalyptus spp.* [55–58], *Macadamia tetraphylla* [69], *Bambusa vulgaris* [61] and *Salix viminalis* [62] in media with low sucrose or without any supplementary carbohydrate, whereas the exposure of *Vernonia condensata* [70], *Fraxinus mandshurica* [65], *Juglans regia* [71], *Populus deltoides* [72] and *Pfaffia glomerata* [73] to these environmental conditions resulted in a limited growth in sucrose-free medium and a beneficial effect of sucrose supplementation, as observed in plum.

It has been claimed that culturing plants in low sugar or no sugar media can easily enhance photosynthetic competence [74], but the results of this study do not corroborate this hypothesis. Carbohydrates are the direct products of photosynthesis, and in our study the highest carbohydrate accumulation occurred in leaves of shoots grown with 3% sucrose. Shoots cultured without sucrose showed limited growth and low sugar content, possibly indicating a low photosynthetic competence. In *V. condensata*, [70] *Solanum tuberosum* [74,75] and *Nicotiana tabacum* shoots [74] accumulated more carbohydrates when they were cultured with sucrose. Interestingly, in tobacco and potato, the low content of sugars observed in plants cultured without sucrose was associated with high photosynthetic capacity and successful growth [74], whereas in plum the shoots cultured without sucrose showed the lowest sugar content and limited growth, possibly indicating a low photosynthetic competence. In our experimental conditions, photosynthetic pigments increased with sugar but not with additional light or CO_2 provided by the PAM system and did not correlate exactly with the growth of the explants. Similarly, in *V. condensata*, total chlorophyll content was higher in shoots cultured with sugar but did not follow the same trend of shoot length or biomass accumulation, although in that study photosynthetic pigments did increase with CO_2 exchange [70]. A poor correlation between photosynthetic pigments and growth in addition to this study has also been reported for other plants such as myrtle [76], chestnut [77], apple [66], tobacco, potato, strawberry and rapeseed [74].

Plants produce reactive oxygen species throughout their life cycle. The environmental conditions of micropropagated plants differ from those found in nature and can increase this stress, potentially causing cell damage [78]. Plants counteract reactive oxygen species through enzymatic and non-enzymatic mechanisms [78–80]. Within non-enzymatic antioxidants, phenolics play a major role in defense against reactive oxygen species. Their concentration in explants subjected to different culture conditions may provide useful information about the plant's physiological state [81]. In the current study, we quantified phenolic compounds and antioxidant activity to estimate the influence of sucrose and the growth conditions on the stress levels of plum cultured in bioreactors. Within each growth conditions (ST or PAM), we observed higher phenol content and antioxidant activity with increased sucrose in the culture medium. The addition of sucrose to *V. condensata* leads to an increase in phenolic and flavonoid compounds [70], and in bamboo [82], rose [83], teak [42] and olive [84], the phenol content was associated with treatments producing more and longer shoots.

In plum, growth and biochemical parameters suggest that shoots cultured without sucrose undergo more stress than shoots cultured with this sugar. However, the limited proliferation and the low content in pigments, carbohydrates, phenols and total antioxidants did not hinder rooting and acclimation of shoots propagated without sucrose. We did not observe significant differences between rooting performance of shoots multiplied or rooted with sucrose 0, 1 or 3%, the only inconvenience of the use of sucrose-free medium being the low proportion of vigorous shoots from these treatments. Rooting percentages were high, averaging 81, 89 and 86% for shoots multiplied without sucrose, with 1% and 3% sucrose, respectively. Shoots multiplied with high light intensity and CO_2 enrichment rooted slightly better than those multiplied in ST conditions (89 versus 82%), but these differences were not significant. Likewise, the use of semisolid medium or rockwool cubes soaked in liquid medium did not affect the rooting and acclimation percentages; the best advantage of using rockwool cubes was to facilitate the handling of shoots during transfer to substrate without causing root damage. The use of fibrous or porous support materials for rooting has been recommended as a simple and cost effective means of micropropagation [85,86] and was beneficial for other plants such as sweet potato [87], American and European chestnut [88,89], cannabis [90] and peach [91], and our results demonstrate their applicability to plum as a way to facilitate large-scale propagation.

Reports on *P. domestica* micropropagation include commercial [20,21,24] and local varieties faced with varying degrees of vulnerability [22,23,26,92], and to the best of our knowledge, all these methods have been developed using semisolid medium. Even with a fully prescribed/defined protocol, in these conditions production is limited by the size of containers and physical manipulation restrictions [21] and is not easily scaled up and commercialized. To date, the use of liquid medium for the micropropagation of *Prunus* species has been limited to *P. avium* [93], *P. armeniaca* [94], *P. cerasifera* [95] and a hybrid rootstock of *P. cerasifera* and *P. dulcis* [96]. Published results are consistent with our higher proliferation rates found in temporary immersion bioreactors compared with semisolid medium. In our study, we investigated the effect of high light intensity and CO_2 enrichment, which provided further improvement in growth and enabled us to reduce sucrose concentration. This protocol can be applied to shoot clusters instead of individual apical sections of plum, which increases proliferation, and be extended to other fruit trees and to larger bioreactors used for commercial applications, making the micropropagation process more economically viable. The medium-sized RITA© bioreactors (1L) of this study allowed us to test several treatments with a limited amount of plant material. This approach can be useful for fine-tuning protocols to cope with genotypic differences in micropropagation requirements, which are an ongoing challenge for fruit trees.

5. Conclusions

The use of RITA© bioreactors for plum shoot micropropagation was highly beneficial. The best multiplication rates were obtained with clusters of shoots, which were immersed for 1 min six times per day in MS BA 0.5 mg L^{-1} IBA 0.5 mg L^{-1} with 3% sucrose and cultured with PPF 50 μmol m^{-2} s^{-1} and ambient air. When individual apical sections were used as initial explants, growth was enhanced by using media with 1% or 3% sucrose, PPF 150 μmol m^{-2} s^{-1} and CO_2-enriched air. The absence of sucrose decreased growth and increased the levels of stress indicators, especially when shoots were cultured under 50 μmol m^{-2} s^{-1} PPF and ambient air. However, these shoots reacted positively to rooting and were successfully acclimated.

This is the first report in the photoautotrophic micropropagation of *Prunus domestica*. By using bioreactors, we obtained successful results for rooting and acclimation, whereas the multiplication phase in medium without sugar has to be improved before being more widely applicable.

Author Contributions: Conceptualization, N.V., M.Á.B. and C.S.; formal analysis, D.G. and N.V.; investigation, D.G., N.V. and A.A.; writing—original draft, D.G., C.B.C. and N.V.; writing—review and editing, D.G., C.B.C. and N.V.; funding acquisition, N.V., M.Á.B. and C.S. All authors have read and agreed to the published version of the manuscript.

Funding: This research was funded by Xunta de Galicia (Spain) (project IN607A 2021), by CYTED (P117RT0522), and by CSIC (PIE 202140E015, COOPB20584).

Data Availability Statement: Not applicable.

Acknowledgments: We thank Beatriz Cuenca (Maceda Nursery, TRAGSA) for her advice on photoautotrophic micropropagation, Mª José Cernadas for technical assistance and the Asociación Galega de Froita Autóctona do Eume for providing the plant material. We dedicate this article to the memory of our dear colleague and friend, Brais Bogo Graña, who started the experiments in bioreactors but died prematurely before this research was completed.

Conflicts of Interest: The authors declare no conflict of interest.

References

1. Sottile, F.; Caltagirone, C.; Giacalone, G.; Peano, C. Unlocking Plum Genetic Potential: Where Are We At? *Horticulturae* **2022**, *8*, 128. [CrossRef]
2. Zhebentyayeva, T.; Shankar, V.; Scorza, R.; Callahan, A.; Ravelonandro, M.; Castro, S.; DeJong, T.; Saski, C.A.; Dardick, C. Genetic characterization of worldwide *Prunus domestica* (plum) germplasm using sequence-based genotyping. *Hortic. Res.* **2019**, *6*, 12. [CrossRef]
3. Miller, A.J.; Gross, B.L. From forest to field: Perennial fruit crop domestication. *Am. J. Bot.* **2011**, *98*, 1389–1414. [CrossRef]
4. Goldschmidt, E.E. The Evolution of Fruit Tree Productivity: A Review. *Econ. Bot.* **2013**, *67*, 51–62. [CrossRef] [PubMed]
5. Gaši, F.; Sehic, J.; Grahic, J.; Hjeltnes, S.H.; Ordidge, M.; Benedikova, D.; Blouin-Delmas, M.; Drogoudi, P.; Giovannini, D.; Höfer, M.; et al. Genetic assessment of the pomological classification of plum *Prunus domestica* L. accessions sampled across Europe. *Genet. Resour. Crop Evol.* **2020**, *67*, 1137–1161. [CrossRef]
6. Lansari, A.; Kester, D.E.; Iezzoni, A.F. Inbreeding, coancestry, anf founding clones of almonds of California, Mediterranean shores, and Russia. *J. Am. Soc. Hortic. Sci.* **1994**, *119*, 1279–1285. [CrossRef]
7. Maxted, N.; Kell, S.; Ford-Lloyd, B.; Dulloo, E.; Toledo, Á. Toward the systematic conservation of global crop wild relative diversity. *Crop Sci.* **2012**, *52*, 774–785. [CrossRef]
8. Caballero, A.; García-Dorado, A. Allelic diversity and its implications for the rate of adaptation. *Genetics* **2013**, *195*, 1373–1384. [CrossRef]
9. Marconi, G.; Ferradini, N.; Russi, L.; Concezzi, L.; Veronesi, F.; Albertini, E. Genetic characterization of the apple germplasm collection in central Italy: The value of local varieties. *Front. Plant Sci.* **2018**, *9*, 1460. [CrossRef] [PubMed]
10. Aranzana, M.J.; Decroocq, V.; Dirlewanger, E.; Eduardo, I.; Gao, Z.S.; Gasic, K.; Iezzoni, A.; Jung, S.; Peace, C.; Prieto, H.; et al. Prunus genetics and applications after de novo genome sequencing: Achievements and prospects. *Hortic. Res.* **2019**, *6*, 58. [CrossRef]
11. Urrestarazu, J.; Errea, P.; Miranda, C.; Santesteban, L.G.; Pina, A. Genetic diversity of Spanish *Prunus domestica* L. germplasm reveals a complex genetic structure underlying. *PLoS ONE* **2018**, *13*, e0195591. [CrossRef] [PubMed]
12. Ferrás-Sexto, C.; O'Flanagan, P. Small-holdings and sustainable family farming in Galicia and Ireland. A comparative case study. *Norois* **2012**, *224*, 61–76. [CrossRef]
13. Pereira-Lorenzo, S.; Urrestarazu, J.; Ramos-Cabrer, A.M.; Miranda, C.; Pina, A.; Dapena, E.; Moreno, M.A.; Errea, P.; Llamero, N.; Díaz-Hernández, M.B.; et al. Analysis of the genetic diversity and structure of the Spanish apple genetic resources suggests the existence of an Iberian genepool. *Ann. Appl. Biol.* **2017**, *171*, 424–440. [CrossRef]
14. Goded, S.; Ekroos, J.; Domínguez, J.; Guitián, J.A.; Smith, H.G. Effects of organic farming on bird diversity in North-West Spain. *Agric. Ecosyst. Environ.* **2018**, *257*, 60–67. [CrossRef]
15. FAO. *Contributing to Food Security and Sustainability in a Changing World*; FAO: Rome, Italy, 2011; ISBN 9789251067482.
16. Miloševic, T.; Miloševic, N. Plum (*Prunus* spp.) Breeding. In *Advances in Plant Breeding Strategies: Fruits*; Al-Khayri, J.M., Jain, S.M., Johnson, D.V., Eds.; Springer: Berlin/Heidelberg, Germany, 2018; pp. 165–215.
17. Dulloo, M.E.; Hunter, D.; Borelli, T. Ex situ and in situ conservation of agricultural biodiversity: Major advances and research needs. *Not. Bot. Horti Agrobot. Cluj-Napoca* **2010**, *38*, 123–135. [CrossRef]
18. Postman, J.; Hummer, K.; Stover, E.; Krueger, R.; Forsline, P.; Grauke, L.J.; Zee, F.; Ayala-Silva, T.; Irish, B. Fruit and Nut Genebanks in the U.S. National Plant Germplasm System. *HortScience* **2006**, *41*, 1188–1194. [CrossRef]
19. Pence, V.C. In Vitro Methods and the Challenge of Exceptional Species for Target 8 of the Global Strategy for Plant Conservation 1. *Ann. Mo. Bot. Gard.* **2013**, *99*, 214–220. [CrossRef]
20. Druart, P.; Gruselle, R. Plum (*Prunus domestica*). In *Biotechnology in Agriculture and Forestry, Vol 1: Trees I*; Bajaj, Y., Ed.; Springer: Berlin/Heidelberg, Germany, 1986; pp. 130–154. ISBN 3-540-15581-3.
21. Druart, P. In Vitro Culture and Micropropagation of Plum (*Prunus* spp.). In *High-Tech and Micropropagation II Biotechnology in Agriculture and Forestry, Vol 18*; Bajaj, Y.P.S., Ed.; Springer: Berlin/Heidelberg, Germany, 1992; pp. 279–303.
22. Andreu, P.; Marín, J.A. In vitro culture establishment and multiplication of the Prunus rootstock 'Adesoto 101' (*P. insititia* L.) as affected by the type of propagation of the donor plant and by the culture medium composition. *Sci. Hortic.* **2005**, *106*, 258–267. [CrossRef]

23. Ruzic, D.; Vujovic, T.; Cerovic, R. In vitro preservation of autochthonous plum genotypes. *Bulg. J. Agric. Sci.* **2012**, *18*, 55–62.
24. Wolella, E.K. Surface sterilization and in vitro propagation of *Prunus domestica* L. cv. Stanley using axillary buds as explants. *J. Biotech Res.* **2017**, *8*, 18–26.
25. Alburquerque, N.; Faize, L.; Burgos, L. Silencing of Agrobacterium tumefaciens oncogenes ipt and iaaM induces resistance to crown gall disease in plum but not in apricot. *Pest Manag. Sci.* **2017**, *73*, 2163–2173. [CrossRef] [PubMed]
26. Vujović, T.; Jevremović, D.; Marjanović, T.; Glišić, I. In vitro propagation and medium-term conservation of autochthonous plum cultivar "Crvena Ranka". *Acta Agric. Serbica* **2020**, *25*, 141–147. [CrossRef]
27. Etienne, H.; Berthouly, M. Temporary immersion systems in plant micropropagation. *Plant Cell Tissue Organ Cult.* **2002**, *69*, 215–231. [CrossRef]
28. Vidal, N.; Sánchez, C. Use of bioreactor systems in the propagation of forest trees. *Eng. Life Sci.* **2019**, *19*, 896–915. [CrossRef]
29. Sanchez, M.C.; Vieitez, A.M. In vitro morphogenetic competence of basal sprouts and crown branches of mature chestnut. *Tree Physiol.* **1991**, *8*, 59–70. [CrossRef]
30. Murashige, T.; Skoog, F. A Revised Medium for Rapid Growth and Bio Assays with Tobacco Tissue Cultures. *Physiol. Plant.* **1962**, *15*, 473–497. [CrossRef]
31. Cuenca Valera, B.; Aldrey Villar, A.; Blanco Beiro, B.; Vidal González, N. Use of a Continuous Immersion System (CIS) for micropropagation of chestnut in photoautotrophic and photomixotrophic conditions. In *Woody Plant Production Integrating Genetic and Vegetative Propagation Technologies, Proceedings of the 3rd International Conference of the IUFRO Unit 2.09.02, Vitoria-Gasteiz, Spain, 8–12 September 2014*; Park, Y.S., Bonga, J.M., Eds.; IUFRO: Vienna, Austria, 2015; pp. 112–116.
32. Miller, G.L. Use of Dinitrosalicylic Acid Reagent for Determination of Reducing Sugar. *Anal. Chem.* **1959**, *31*, 426–428. [CrossRef]
33. Lindsay, H. A colorimetric estimation of reducing sugars in potatoes with 3,5-dinitrosalicylic acid. *Potato Res.* **1973**, *16*, 176–179. [CrossRef]
34. Yemm, E.W.; Willis, A.J. The Estimation of Carbohydrates in Plant Extracts by Anthrone. *Biochem. J.* **1954**, *57*, 508–514. [CrossRef]
35. Wellburn, A.R. The Spectral Determination of Chlorophylls a and b, as well as Total Carotenoids, Using Various Solvents with Spectrophotometers of Different Resolution. *J. Plant Physiol.* **1994**, *144*, 307–313. [CrossRef]
36. Díaz, J.; Bernal, A.; Pomar, F.; Merino, F. Induction of shikimate dehydrogenase and peroxidase in pepper (*Capsicum annuum* L.) seedlings in response to copper stress and its relation to lignification. *Plant Sci.* **2001**, *161*, 179–188. [CrossRef]
37. Singleton, V.; Rossi, J. Colorimetry of Total Phenolics with Phosphomolybdic-Phosphotungstic Acid Reagents. *Am. J. Enol. Vitic.* **1965**, *16*, 144–168.
38. Blois, M.S. Antioxidant Determinations by the Use of a Stable Free Radical. *Nature* **1958**, *181*, 1199–1200. [CrossRef]
39. Murch, S.J.; Liu, C.; Romero, R.M.; Saxena, P.K. In vitro Culture and Temporary Immersion Bioreactor Production of Crescentia cujete. *Plant Cell Tissue Organ Cult.* **2004**, *78*, 63–68. [CrossRef]
40. McAlister, B.; Finnie, J.; Watt, M.; Blakeway, F. Use of the temporary immersion bioreactor system (RITA®) for production of commercial Eucalyptus clones in Mondi Forests (SA). *Plant Cell Tissue Organ Cult.* **2005**, *81*, 347–358. [CrossRef]
41. Zhu, L.-H.; Li, X.-Y.; Welander, M. Optimisation of growing conditions for the apple rootstock M26 grown in RITA containers using temporary immersion principle. *Plant Cell Tissue Organ Cult.* **2005**, *81*, 313–318. [CrossRef]
42. Quiala, E.; Cañal, M.J.; Meijón, M.; Rodríguez, R.; Chávez, M.; Valledor, L.; de Feria, M.; Barbón, R. Morphological and physiological responses of proliferating shoots of teak to temporary immersion and BA treatments. *Plant Cell Tissue Organ Cult.* **2012**, *109*, 223–234. [CrossRef]
43. Aguilar, M.E.; Garita, K.; Kim, Y.W.; Kim, J.-A.; Moon, H.K. Simple Protocol for the Micropropagation of Teak (Tectona grandis Linn.) in Semi-Solid and Liquid Media in RITA Bioreactors and ex Vitro Rooting. *Am. J. Plant Sci.* **2019**, *10*, 1121–1141. [CrossRef]
44. Akdemir, H.; Süzerer, V.; Onay, A.; Tilkat, E.; Ersali, Y.; Çiftçi, Y.O. Micropropagation of the pistachio and its rootstocks by temporary immersion system. *Plant Cell Tissue Organ Cult.* **2014**, *117*, 65–76. [CrossRef]
45. Vidal, N.; Blanco, B.; Cuenca, B. A temporary immersion system for micropropagation of axillary shoots of hybrid chestnut. *Plant Cell Tissue Organ Cult.* **2015**, *123*, 229–243. [CrossRef]
46. Latawa, J.; Shukla, M.R.; Saxena, P.K. An efficient temporary immersion system for micropropagation of hybrid hazelnut. *Botany* **2016**, *94*, 1–8. [CrossRef]
47. Luna, C.V.; Gonzalez, A.M.; Mroginski, L.A.; Sansberro, P.A. Anatomical and histological features of Ilex paraguariensis leaves under different in vitro shoot culture systems. *Plant Cell Tissue Organ Cult.* **2017**, *129*, 457–467. [CrossRef]
48. Regueira, M.; Rial, E.; Blanco, B.; Bogo, B.; Aldrey, A.; Correa, B.; Varas, E.; Sánchez, C.; Vidal, N. Micropropagation of axillary shoots of Salix viminalis using a temporary immersion system. *Trees* **2018**, *32*, 61–71. [CrossRef]
49. Benelli, C.; De Carlo, A. In vitro multiplication and growth improvement of *Olea europaea* L. cv Canino with temporary immersion system (Plantform™). *3 Biotech* **2018**, *8*, 317. [CrossRef]
50. San José, M.C.; Blázquez, N.; Cernadas, M.J.; Janeiro, L.V.; Cuenca, B.; Sánchez, C.; Vidal, N. Temporary immersion systems to improve alder micropropagation. *Plant Cell Tissue Organ Cult.* **2020**, *143*, 265–275. [CrossRef]
51. Lotfi, M.; Bayoudh, C.; Werbrouck, S.; Mars, M. Effects of meta–topolin derivatives and temporary immersion on hyperhydricity and in vitro shoot proliferation in Pyrus communis. *Plant Cell Tissue Organ Cult.* **2020**, *143*, 499–505. [CrossRef]
52. Georgiev, V.; Schumann, A.; Pavlov, A.; Bley, T. Temporary immersion systems in plant biotechnology. *Eng. Life Sci.* **2014**, *14*, 607–621. [CrossRef]

53. Xiao, Y.; Niu, G.; Kozai, T. Development and application of photoautotrophic micropropagation plant system. *Plant Cell Tissue Organ Cult.* **2011**, *105*, 149–158. [CrossRef]
54. Watt, M.P. The status of temporary immersion system (TIS) technology for plant micropropagation. *Afr. J. Biotechnol.* **2012**, *11*, 14036–14043. [CrossRef]
55. Kirdmanee, C.; Kitaya, Y.; Kozai, T. Effects of CO_2 enrichment and supporting materialin vitro on photoautotrophic growth ofEucalyptus plantletsin vitro andex vitro. *Vitr. Cell. Dev. Biol.-Plant* **1995**, *31*, 144–149. [CrossRef]
56. Zobayed, S.M.A.; Afreen, F.; Kozai, T. Physiology of Eucalyptus plantlets grown photoautotrophically in a scaled-up vessel. *Vitr. Cell. Dev. Biol.-Plant* **2001**, *37*, 807–813. [CrossRef]
57. Zobayed, S. Mass Propagation of Eucalyptus camaldulensis in a Scaled-up Vessel Under In Vitro Photoautotrophic Condition. *Ann. Bot.* **2000**, *85*, 587–592. [CrossRef]
58. Tanaka, M.; Giang, D.T.T.; Murakami, A. Application of a novel disposable film culture system to photoautotrophic micropropagation of Eucalyptus uro-grandis (Urophylia × grandis). *Vitr. Cell. Dev. Biol.-Plant* **2005**, *41*, 173–180. [CrossRef]
59. Fuljahn, S.; Tantau, H.-J. Process engineering as a means of regulating the microclimate in a photoautotrophic in vitro culture. *Acta Hortic.* **2009**, *817*, 143–150. [CrossRef]
60. Arencibia, A.D.; Gómez, A.; Poblete, M.; Vergara, C. High-performance micropropagation of dendroenergetic poplar hybrids in photomixotrophic Temporary Immersion Bioreactors (TIBs). *Ind. Crops Prod.* **2017**, *96*, 102–109. [CrossRef]
61. García-Ramírez, Y.; Barrera, G.P.; Freire-Seijo, M.; Barbón, R.; Concepción-Hernández, M.; Mendoza-Rodríguez, M.F.; Torres-García, S. Effect of sucrose on physiological and biochemical changes of proliferated shoots of Bambusa vulgaris Schrad. Ex Wendl in temporary immersion. *Plant Cell Tissue Organ Cult.* **2019**, *137*, 239–247. [CrossRef]
62. Gago, D.; Vilavert, S.; Bernal, M.Á.; Sánchez, C.; Aldrey, A.; Vidal, N. The Effect of Sucrose Supplementation on the Micropropagation of *Salix viminalis* L. Shoots in Semisolid Medium and Temporary Immersion Bioreactors. *Forests* **2021**, *12*, 1408. [CrossRef]
63. Greenboim-Wainberg, Y.; Maymon, I.; Borochov, R.; Alvarez, J.; Olszewski, N.; Ori, N.; Eshed, Y.; Weiss, D. Cross Talk between Gibberellin and Cytokinin: The Arabidopsis GA Response Inhibitor SPINDLY Plays a Positive Role in Cytokinin Signaling. *Plant Cell* **2005**, *17*, 92–102. [CrossRef]
64. Silva, L.A.S.; Costa, A.D.O.; Batista, D.S.; Silva, M.L.D.; Costa Netto, A.P.D.; Rocha, D.I. Exogenous gibberellin and cytokinin in a novel system for in vitro germination and development of African iris (*Dietes bicolor*). *Rev. Ceres* **2020**, *67*, 402–409. [CrossRef]
65. Deng, Z.; Chu, J.; Wang, Q.; Wang, L. Effect of different carbon sources on the accumulation of carbohydrate, nutrient absorption and the survival rate of Chinese Ash (*Fraxinus mandshurica*) explants in vitro. *Afr. J. Agric. Res.* **2012**, *7*, 3111–3119. [CrossRef]
66. Kim, N.-Y.; Hwang, H.-D.; Kim, J.-H.; Kwon, B.-M.; Kim, D.; Park, S.-Y. Efficient production of virus-free apple plantlets using the temporary immersion bioreactor system. *Hortic. Environ. Biotechnol.* **2020**, *61*, 779–785. [CrossRef]
67. Sha Valli Khan, P.S.; Kozai, T.; Nguyen, Q.T.; Kubota, C.; Dhawan, V. Growth and Water Relations of Paulownia fortunei Under Photomixotrophic and Photoautotrophic Conditions. *Biol. Plant.* **2003**, *46*, 161–166. [CrossRef]
68. Mosaleeyanon, K.; Cha-um, S.; Kirdmanee, C. Enhanced growth and photosynthesis of rain tree (Samanea saman Merr.) plantlets in vitro under a CO_2-enriched condition with decreased sucrose concentrations in the medium. *Sci. Hortic.* **2004**, *103*, 51–63. [CrossRef]
69. Cha-um, S.; Chanseetis, C.; Chintakovid, W.; Pichakum, A.; Supaibulwatana, K. Promoting root induction and growth of in vitro macadamia (*Macadamia tetraphylla* L. 'Keaau') plantlets using CO_2-enriched photoautotrophic conditions. *Plant Cell Tissue Organ Cult.* **2011**, *106*, 435–444. [CrossRef]
70. Fortini, E.A.; Batista, D.S.; Mamedes-Rodrigues, T.C.; Felipe, S.H.S.; Correia, L.N.F.; Chagas, K.; Silva, P.O.; Rocha, D.I.; Otoni, W.C. Gas exchange rates and sucrose concentrations affect plant growth and production of flavonoids in Vernonia condensata grown in vitro. *Plant Cell Tissue Organ Cult.* **2021**, *144*, 593–605. [CrossRef]
71. Hassankhah, A.; Vahdati, K.; Lotfi, M.; Mirmasoumi, M.; Preece, J.; Assareh, M.-H. Effects of Ventilation and Sucrose Concentrations on the Growth and Plantlet Anatomy of Micropropagated Persian Walnut Plants. *Int. J. Hort. Sci. Technol.* **2014**, *1*, 111–120.
72. Mingozzi, M.; Montello, P.; Merkle, S. Adventitious shoot regeneration from leaf explants of eastern cottonwood (*Populus deltoides*) cultured under photoautotrophic conditions. *Tree Physiol.* **2009**, *29*, 333–343. [CrossRef]
73. Saldanha, C.W.; Otoni, C.G.; Notini, M.M.; Kuki, K.N.; da Cruz, A.C.F.; Neto, A.R.; Dias, L.L.C.; Otoni, W.C. A CO_2-enriched atmosphere improves in vitro growth of Brazilian ginseng [*Pfaffia glomerata* (Spreng.) Pedersen]. *Vitr. Cell. Dev. Biol.-Plant* **2013**, *49*, 433–444. [CrossRef]
74. Ševčíková, H.; Lhotáková, Z.; Hamet, J.; Lipavská, H. Mixotrophic in vitro cultivations: The way to go astray in plant physiology. *Physiol. Plant.* **2019**, *167*, 365–377. [CrossRef]
75. Badr, A.; Angers, P.; Desjardins, Y. Metabolic profiling of photoautotrophic and photomixotrophic potato plantlets (Solanum tuberosum) provides new insights into acclimatization. *Plant Cell Tissue Organ Cult.* **2011**, *107*, 13–24. [CrossRef]
76. Lucchesini, M.; Monteforti, G.; Mensuali-Sodi, A.; Serra, G. Leaf ultrastructure, photosynthetic rate and growth of myrtle plantlets under different in vitro culture conditions. *Biol. Plant.* **2006**, *50*, 161–168. [CrossRef]
77. Sáez, P.L.; Bravo, L.A.; Latsague, M.I.; Sánchez, M.E.; Ríos, D.G. Increased light intensity during in vitro culture improves water loss control and photosynthetic performance of *Castanea sativa* grown in ventilated vessels. *Sci. Hortic.* **2012**, *138*, 7–16. [CrossRef]

78. Gaspar, T.; Franck, T.; Bisbis, B.; Kevers, C.; Jouve, L.; Hausman, J.F.; Dommes, J. Concepts in plant stress physiology. Application to plant tissue cultures. *Plant Growth Regul.* **2002**, *37*, 263–285. [CrossRef]

79. Ahmad, P.; Jaleel, C.A.; Salem, M.A.; Nabi, G.; Sharma, S. Roles of enzymatic and nonenzymatic antioxidants in plants during abiotic stress. *Crit. Rev. Biotechnol.* **2010**, *30*, 161–175. [CrossRef]

80. Sharma, P.; Jha, A.B.; Dubey, R.S.; Pessarakli, M. Reactive Oxygen Species, Oxidative Damage, and Antioxidative Defense Mechanism in Plants under Stressful Conditions. *J. Bot.* **2012**, *2012*, 217037. [CrossRef]

81. Gharibi, S.; Tabatabaei, B.E.S.; Saeidi, G.; Goli, S.A.H. Effect of Drought Stress on Total Phenolic, Lipid Peroxidation, and Antioxidant Activity of Achillea Species. *Appl. Biochem. Biotechnol.* **2016**, *178*, 796–809. [CrossRef]

82. García-Ramírez, Y.; Gonzáles, M.G.; Mendoza, E.Q.; Seijo, M.F.; Cárdenas, M.L.O.; Moreno-Bermúdez, L.J.; Ribalta, O.H. Effect of BA Treatments on Morphology and Physiology of Proliferated Shoots of Bambusa vulgaris Schrad. Ex Wendl in Temporary Immersion. *Am. J. Plant Sci.* **2014**, *05*, 205–211. [CrossRef]

83. Malik, M.; Warchoł, M.; Pawłowska, B. Liquid culture systems affect morphological and biochemical parameters during Rosa canina plantlets in vitro production. *Not. Bot. Horti Agrobot. Cluj-Napoca* **2018**, *46*, 58–64. [CrossRef]

84. Regni, L.; Del Buono, D.; Micheli, M.; Facchin, S.L.; Tolisano, C.; Proietti, P. Effects of Biogenic ZnO Nanoparticles on Growth, Physiological, Biochemical Traits and Antioxidants on Olive Tree In Vitro. *Horticulturae* **2022**, *8*, 161. [CrossRef]

85. Newel, C.; Growns, D.; McComb, J. The influence of medium aeration on in vitro rooting of Australian plant microcuttings. *Plant Cell Tissue Organ Cult.* **2003**, *75*, 131–142. [CrossRef]

86. Dutta Gupta, S.; Prasad, V.S.S. Matrix-Supported Liquid Culture Systems for Efficient Micropropagation of Floricultural Plants. In *Floriculture, Ornamental and Plant Biotechnology*; Teixeira da Silva, J., Ed.; Global Science Books: Kagawa, Japan, 2006; pp. 487–495.

87. Afreen-Zobayed, F.; Zobayed, S.M.A.; Kubota, C.; Kozai, T.; Hasegawa, O. Supporting material affects the growth and development of in vitro sweet potato plantlets cultured photoautotrophically. *Vitr. Cell. Dev. Biol.-Plant* **1999**, *35*, 470–474. [CrossRef]

88. Maner, L.; Merkle, S. Polymerized peat plugs improve American chestnut somatic embryo germination in vitro. *J. Am. Chestnut Found.* **2010**, *24*, 16.

89. Cuenca, B.; Sánchez, C.; Aldrey, A.; Bogo, B.; Blanco, B.; Correa, B.; Vidal, N. Micropropagation of axillary shoots of hybrid chestnut (*Castanea sativa* × *C. crenata*) in liquid medium in a continuous immersion system. *Plant Cell Tissue Organ Cult.* **2017**, *131*, 307–320. [CrossRef]

90. Kodym, A.; Leeb, C.J. Back to the roots: Protocol for the photoautotrophic micropropagation of medicinal Cannabis. *Plant Cell Tissue Organ Cult.* **2019**, *138*, 399–402. [CrossRef] [PubMed]

91. Adelberg, J.; Naylor-Adelberg, J.; Miller, S.; Gasic, K.; Schnabel, G.; Bryson, P.; Saski, C.; Parris, S.; Reighard, G. In vitro co-culture system for *Prunus* spp. and *Armillaria mellea* in phenolic foam rooting matric. *Vitr. Cell. Dev. Biol.-Plant* **2021**, *57*, 387–397. [CrossRef]

92. Gianní, S.; Sottile, F. In vitro storage of plum germplasm by slow growth. *Hortic. Sci.* **2016**, *42*, 61–69. [CrossRef]

93. Godoy, S.; Tapia, E.; Seit, P.; Andrade, D.; Sánchez, E.; Andrade, P.; Almeida, A.M.; Prieto, H. Temporary immersion systems for the mass propagation of sweet cherry cultivars and cherry rootstocks: Development of a micropropagation procedure and effect of culture conditions on plant quality. *Vitr. Cell. Dev. Biol.-Plant* **2017**, *53*, 494–504. [CrossRef]

94. Zare Khafri, A.; Solouki, M.; Zarghami, R.; Fakheri, B.; Mahdinezhad, N.; Naderpour, M. In vitro propagation of three Iranian apricot cultivars. *Vitr. Cell. Dev. Biol.-Plant* **2021**, *57*, 102–117. [CrossRef]

95. Damiano, C.; Monticelli, S.; La Starza, S.R.; Gentile, A.; Frattarelli, A. Temperate fruit plant propagation through temporary immersion. *Acta Hortic.* **2003**, *625*, 193–200. [CrossRef]

96. Cantabella, D.; Mendoza, C.R.; Teixidó, N.; Vilaró, F.; Torres, R.; Dolcet-Sanjuan, R. GreenTray® TIS bioreactor as an effective in vitro culture system for the micropropagation of *Prunus* spp. rootstocks and analysis of the plant-PGPMs interactions. *Sci. Hortic.* **2022**, *291*, 110622. [CrossRef]

 *horticulturae*

MDPI

Article

How to Get More Silver? Culture Media Adjustment Targeting Surge of Silver Nanoparticle Penetration in Apricot Tissue during in Vitro Micropropagation

Cristian Pérez-Caselles [1], Nuria Alburquerque [1], Lydia Faize [1], Nina Bogdanchikova [2], Juan Carlos García-Ramos [3], Ana G. Rodríguez-Hernández [4], Alexey Pestryakov [5] and Lorenzo Burgos [1,*]

1 Group of Fruit Tree Biotechnology, Department of Plant Breeding, CEBAS-CSIC, Campus de Espinardo, Edif. 25, 30100 Murcia, Spain
2 Centro de Nanociencias y Nanotecnología, Universidad Nacional Autónoma de México, Ensenada 22860, Mexico
3 Escuela de Ciencias de la Salud, Campus Ensenada, Universidad Autónoma de Baja California (UABC), Ensenada 22890, Mexico
4 Center for Nanosciences and Nanotechnology, Universidad Nacional Autónoma de México, CATEDRA CONACyT Researcher at CNyN–UNAM, Ensenada, B.C, Mexico City 04510, Mexico
5 Research School of Chemistry and Applied Biomedical Sciences, Tomsk Polytechnic University, 634050 Tomsk, Russia
* Correspondence: burgos@cebas.csic.es; Tel.: +34-968396200 (ext. 445317)

Citation: Pérez-Caselles, C.; Alburquerque, N.; Faize, L.; Bogdanchikova, N.; García-Ramos, J.C.; Rodríguez-Hernández, A.G.; Pestryakov, A.; Burgos, L. How to Get More Silver? Culture Media Adjustment Targeting Surge of Silver Nanoparticle Penetration in Apricot Tissue during in Vitro Micropropagation. *Horticulturae* 2022, 8, 855. https://doi.org/10.3390/horticulturae8100855

Academic Editors: Jean Carlos Bettoni, Min-Rui Wang and Qiao-Chun Wang

Received: 26 July 2022
Accepted: 15 September 2022
Published: 20 September 2022

Publisher's Note: MDPI stays neutral with regard to jurisdictional claims in published maps and institutional affiliations.

Abstract: The use of silver nanoparticles (AgNPs) is increasing nowadays due to their applications against phytopathogens. Temporary Immersion Systems (TIS) allow the micropropagation of plants in liquid media. This work aims to develop an effective protocol for apricot micropropagation in TIS and to study the necessary conditions to introduce AgNPs in apricot plants, as well as the effect of its application on proliferation-related parameters. AgNPs were introduced in different media at a concentration of 100 mg L^{-1} to test the incorporation of silver to plant tissues. Silver content analysis was made by Inductively Coupled Plasma-Optical Emission Spectroscopy (ICP-OES). The effect of initial shoot density and the addition of AgNPs on micropropagation were evaluated after four weeks in culture on TIS. Productivity, proliferation, shoot-length and leave surface were measured. The better micropropagation rate was obtained with 40 initial shoots, 2 min of immersion every 6 h and 3 min of aeration every 3 h. To introduce AgNPs in apricot plants it is necessary to culture them in liquid media without chloride in its composition. These results will contribute to the development of an in vitro protocol for virus inhibition by AgNPs application. This depends on the introduction of Ag nanoparticles within the plant tissues, and this is not possible if AgNPs after interaction with Cl$^-$ ions precipitate as silver chloride salts.

Keywords: TIS; AgNPs; liquid media; virucide; micropropagation

1. Introduction

Emerging and re-emerging viruses are considered a continuous threat to human, animal and plant health as they are able to adapt to their current host, migrate to new hosts, as well as develop strategies to escape antiviral measures [1–3].

Diseases caused by plant viruses can cause severe damage in crops and the magnitude of losses in crop yield or product quality can be devastating [4]. Among control strategies for the protection of plants against viral diseases, using pesticides against virus vectors is the most common [5]. However, due to environmental damages caused by an excessive use of pesticides, it is necessary to fight plant diseases without harming natural ecosystems [5,6]. Nanotechnology may be a useful tool in protecting crops against viral diseases [7].

Silver nanoparticles (AgNPs) have been used because of their antimicrobial activity. They are able to control different types of pathogens such as bacteria, virus and fungus, without affecting the development and functionality of plants [8,9]. Additionally, it has been described that they may improve, in vitro, the growing of plants when added to the media [10,11].

Apricot (*Prunus armeniaca* L.) is one of the most economically important species of stone fruits. World production in 2020 was 3.7 million tons in 562.475 hectares [12]. Apricots have been traditionally cultured in Spain because in this country, there are many areas with appropriate environmental conditions for cultivation of different apricot cultivars.

In vitro propagation allows for the production of abundant clonal plant material and has been used for germplasm conservation in controlled and aseptic conditions. The fruit biotechnology group at CEBAS-CSIC, Murcia (Spain) has developed efficient protocols for the micropropagation of apricot cultivars in semisolid media [13]. Additionally, there is the possibility of propagating plant material in liquid media by using Temporal Immersion Systems (TIS). These are bioreactors where filtered air can be introduced to displace the liquid medium to the part of the reactor where the plant material is placed and therefore immersing it for a certain time. Later air can be introduced to return the liquid medium to its original location or just to dry out the plants to avoid excessive humidity [14]. By optimizing immersion and aeration times plants can be proliferated without producing detrimental physiological changes. TIS allows for the introduction, transport and accumulation of specific compounds, such as AgNPs, in plants [15].

To the best of our knowledge, AgNPs have been used to control plant viruses by spraying leaves [16–18] but have never been added to culture media during in vitro propagation to study their effect on virus or other pathogens within the plants. Here, we have used Argovit®, a commercial product of AgNPs with an antimicrobial activity previously demonstrated [19]. It was also revealed the virucidal properties of Argovit® AgNPs in different biological systems: infectious bronchitis coronavirus in chickens [20], *canine distemper virus* in dogs [21], white spot syndrome virus in shrimp [22], Rift Valley fever virus in vitro and in vivo [23] and SARS-CoV-2 coronavirus in vitro and in health workers [24].

The objectives of this work are: (1) to develop the micropropagation protocol of apricot in TIS by searching the conditions providing penetration of Argovit® AgNPs into the tissues of in vitro propagated apricot plants by adding AgNPs to the media and (2) to determine if the addition of AgNPs has any effect on the apricot proliferation. This work will serve as the first stage for a long-term objective which is to produce virus and viroid-free apricot plants during micropropagation. For this long term objective, the penetration of AgNPs with antiviral properties into plan tissue is essential, which is the short term objective of the present publication.

2. Materials and Methods

2.1. Plan Material, Maintenance, and Proliferation in Semisolid or Liquid Medium

The apricot cultivars 'Canino' and 'Mirlo Rojo' are part of the apricot germplasm maintained in vitro at the Fruit Biotechnology Group laboratory in CEBAS-CSIC and were used for the experiments in this work. In vitro plants are maintained in a culture chamber at temperature of 23 ± 1 °C, 16/8 h light/dark photoperiod (56 µmol m^{-2}s^{-1} of light intensity) and are transferred to fresh medium every four weeks.

Micro-shoots grow and multiply in 500 mL glass jars, 12 shoots per jar with initial approximated length of 1.5 cm (Figure 1). The semisolid culture medium (SSM) used in this work has been described by Wang et al. [25]. Basal salt composition can be found in Table 1 but briefly SSM has QL [26] macronutrients, DKW [27] micronutrients and vitamins plus the addition of 3 mM calcium chloride, 0.8 mM phloroglucinol, 3% sucrose, 0.7% agar, 1.12 mM 6-benzylaminopurine riboside (BAPr), 0.05 mM IBA, 2.1 mM Meta-Topolin (6-(3-hydroxybenzylamino) purine) and 29.6 mM adenine. Medium pH is adjusted to 5.7 by adding NaOH 0.1 N before autoclaving at 121 °C for 20 min.

Figure 1. Temporal Immersion System (SIT) in Plantform™ bioreactors. Initial stage of 'Canino' shoots (**A,B**), and proliferation after four weeks (**C,D**).

Table 1. Culture media used for apricot micropropagation and modifications to study AgNPs incorporation in plant tissues.

Media Codes	Basal Salts	CaCl$_2$ (3 mM)	Other Components [z]	Agar (7 gL^{-1})
SSM	QL + DKW	Yes	a	Yes
LM1	QL + DKW	Yes	a	No
LM2	QL + DKW	No	b	No
LM3	QL + DKW	No	a	No
LM4	QL modified [y] + DKW	No	a	No

[y] QL macronutrients was modified by adding 1,908 mg L^{-1} Ca(NO$_3$)$_2$ · 4 H$_2$O and 160 mg L^{-1} NH$_4$NO$_3$ instead of 1,200 mg L^{-1} and 400 mg L^{-1}, respectively. [z] a: 0.8 mM phloroglucinol, 3% sucrose, 1.12 mM BAPr, 0.05 mM IBA, 2.1 mM Meta-Topolin and 29.6 mM adenine; b: 3% sucrose, 1.96 mM BAP and 0.2 mM IBA.

For apricot micropropagation in liquid media 4 L Plantform™ bioreactors were used. To maintain sterility inside the bioreactor after autoclaving, all manipulation was performed inside the hood and pressure air was introduced through 0.22 μm air filters. Liquid medium with the same composition as SSM was used, but without agar (LM1 in Table 1).

In each bioreactor, 500 mL of LM1 was poured and every four weeks, the medium was refreshed. Based on the results from previous experiments, the following optimal protocol was applied: 2 min of immersion every 6 h and 3 min of aeration every 3 h. To improve micropropagation in bioreactors with liquid media the effect of initial density of shoots was evaluated starting the culture with 10, 20, 30 or 40 'Canino' shoots.

2.2. Physicochemical Characteristics of AgNPs

Argovit®-7 AgNPs used in this work were kindly donated by the Scientific-Production Centre Vector-Vita Ltd., Novosibirsk, Russia. According to the manufacturer's data Argovit-7 is an aqueous solution with 1.2% (w/w) of metallic silver and 18.8% (w/w) of polyvinylpirrolidone (PVP), with integral AgNPs concentration (metallic silver and PVP) of 200 g L^{-1}.

High-Resolution Transmission Electron Microscope (HR-TEM) a JEM-2010 (JEOL©) was used to determine AgNPs morphology. NPs average diameter (φAg) and size distribution analysis was determined from TEM micrographs (n = 150) using ImageJ Software [28]. Thermogravimetric analysis (TGA) and differential scanning calorimetry (DSC) measurements were performed with a DSC/TGA thermal analysis system (SDT Q600, Newtown (PA), USA) under argon atmosphere from 30 to 750 °C at a rate of 10 °C min^{-1}. Surface plasmon analysis was performed using AgNPs dilutions on distilled water acquired in a Thermo-Scientific Genesys 30 UV/Vis spectrophotometer with a 1 cm optical path quartz cell. Zeta potential and hydrodynamic diameter were determined by dynamic light scattering (DLS) using a Zetasizer Nano-ZS with capillary cuvettes (Malvern Instruments®, Malvern, UK).

2.3. Addition of AgNPs to Different Media Compositions

To test the incorporation of silver to plant tissues, AgNPs were introduced in different media at a concentration of 100 mg L^{-1} (metallic silver plus PVP). AgNPs were added after autoclaving and cooling down the media in the semisolid (SSM) and liquid medium (LM1) with a similar composition. Additionally, other basal salt combinations were tested as well as modifications of LM1. The media tested are detailed in Table 1 but briefly medium LM2 is based in QL [26] macronutrients and DKV [27] micronutrients and vitamins, 3% sucrose, 1.96 mM BAP and 0.2 mM IBA. LM3 is the medium LM1 without calcium chloride. Finally, medium LM4 is LM1 without calcium chloride, but to compensate for a possible calcium deficiency calcium nitrate was added and ammonium nitrate was reduced to avoid excessive nitrogen in the medium (Table 1).

2.4. Micropropagation Evaluation

The effect of initial shoot density and the addition of AgNPs on micropropagation were evaluated after four weeks in culture. Recorded dependent variables were proliferation (number of new growths longer than 1 cm from each initial shoot), mean length of new shoots longer than 1 cm, productivity (the product of proliferation and mean new shoot length) and leaf area (which was measured for the fourth leaf on the main shoot). Twenty shoots from each treatment were measured.

2.5. Determination of Silver Content in the Plant Tissues

To determine silver content 200 mg of dry material, obtained from whole shoots without basal calli, were used. Fresh samples were dried out at 60 °C for 72 h. Dried powder was subjected to acid digestion in a HNO_3/H_2O_2 solution (4:1) using an UltraCLAVE microwave. Silver concentration was determined by Inductively Coupled Plasma-Optical Emission Spectroscopy (ICP-OES) at CEBAS Ionomic Service using iCAP 7000 (Thermo Fisher Scientific, Waltham, MA, USA). Each silver content is the average from three values obtained for independent samples.

2.6. Statistical Analyses

Results were analyzed using SAS version 9.4 (SAS Institute, Cary, NC, USA). A variance analysis determined if there were significant differences between treatments (initial shoot density or media). The effect of the initial shoot density was evaluated by a trend analysis and orthogonal contrasts.

Data of silver content in plant material were transformed with the natural logarithm before the analyses to homogenize variances. Results obtained from each culture medium were compared with the control, SSM without AgNPs addition, by means of one-tail Dunnett test.

Proliferation variables evaluated for plants grown in media without calcium chloride were compared after the ANOVA with a Duncan post-hoc test.

3. Results

3.1. Physicochemical Characterization

AgNPs formulation used in this work was characterized by TEM, UV-is, DCS/TGA and DLS analysis. TEM images exhibit spherical AgNPs, size distribution range between 22.1 and 44.8 nm, with an average size of 31.9 nm, and with tendency of these nanoparticles to agglomerates into groups with the size more than 100 nm, as shown in Figure 2A. UV-vis spectra show three absorption maxima at 407, 439 and 526 nm, associated with the size distribution of the sample (Figure 2B). DSC/TGA analysis allows for the determination of the composition of the AgNPs formulation, with 75.82% of water (loss at 85 °C), 22.21% of K-17 polyvinylpyrrolidone (PVP), and 1.96% of metallic silver. The sharp endothermic process found at 447 °C let a pure coating agent be identified (Figure 2C). Hydrodynamic diameter and zeta potential measured wit DLS are 56.1 nm and −3.85 mV, respectively.

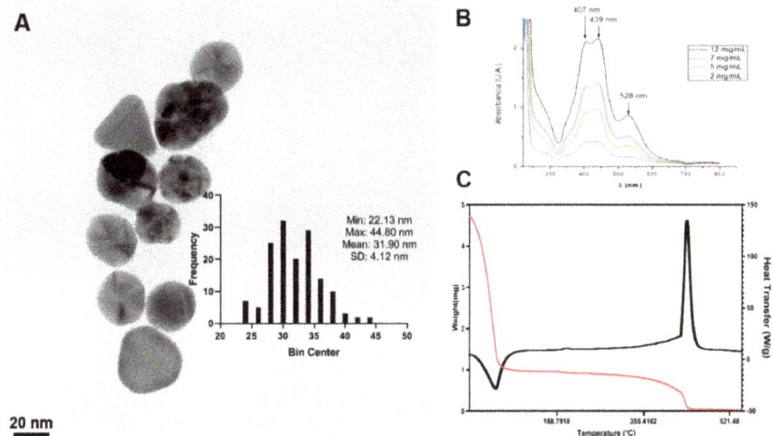

Figure 2. Physicochemical properties of AgNPs. (**A**) TEM images showing AgNPs morphology and size distribution (inset histogram), (**B**) surface plasmon registered at different concentrations in distilled water, and (**C**) DSC/TGA analysis showing the mass loss as a function of temperature increase (red line) and the endothermic/exothermic processes (black line).

3.2. Effect of Initial Shoot Density

The number of shoots in the bioreactor at the beginning of the culture cycle affects different parameters related to micropropagation in the TIS. 'Canino' micropropagation was studied with different initial shoot density (10, 20, 30 and 40 shoots). Figure 3 shows a trend analysis for the micropropagation variables with the increase in initial shoot density in TIS. Moreover, a prediction equation was calculated for those variables where significant differences between treatments were found.

Statistical analyses showed significant differences ($p < 0.01$) in productivity which rises with the increase in the initial shoot number (Figure 3). No significant differences were found for the number of new shoots longer than 1 cm when the initial shoot density increased. However, mean shoot length was significantly affected ($p < 0.001$) with a maximum mean length of 2.26 cm for the highest initial shoot density (40 shoots). Leaf area was also significantly affected ($p < 0.001$) by initial density and drastically increased, reaching a maximum value of 2.62 cm^2 at the highest initial shoot number (Figure 3). Considering all parameters studied, the best results were obtained with initial shoot number of 40 shoots, and this was the shoot density in following experiments. With the same immersion and aeration conditions and an initial density of 40 shoots we tested the effect of TIS on 'Mirlo Rojo' (Table 2, medium LM1 without AgNPs). Therefore, 'Mirlo Rojo' had lower proliferation rates than 'Canino' in these culture conditions.

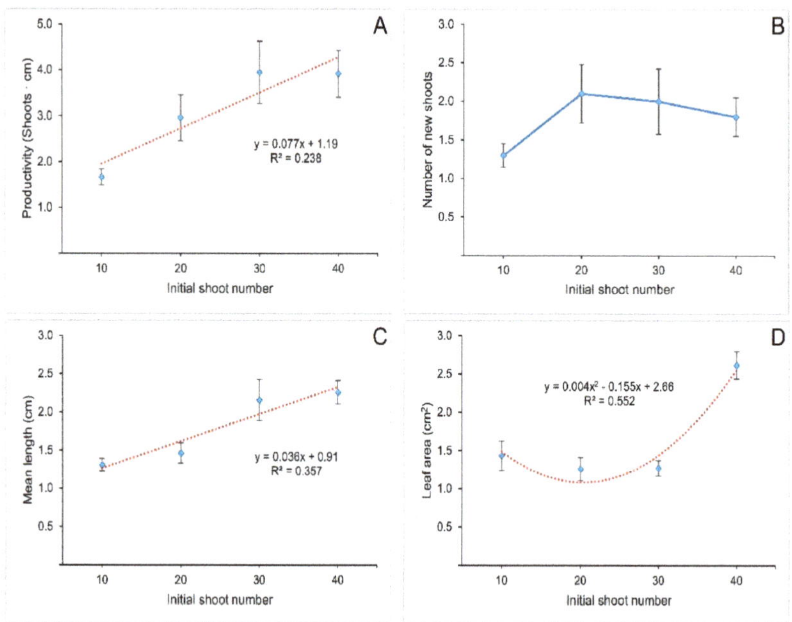

Figure 3. Effect of initial shoot density in TIS on productivity (**A**), number of new shoots (**B**), mean length (**C**) and leaf surface (**D**) of 'Canino' shoots. Discontinuous line represents the results of trend analysis. Bars are the standard error of mean values.

3.3. Silver Content within the Plants

The long-term objective of this study is to study virucidal activity of AgNPs in plant tissue. It is paramount that AgNPs penetrate the plant tissues. Therefore, the effect of media composition containing 100 mg L^{-1} AgNPs (Table 1) on the silver content in the plant tissue after four weeks of cultivation was studied. Figure 4 shows silver content found in 'Mirlo Rojo' shoots cultured in each of the media described in Table 1. Additionally, silver content in 'Canino' shoots cultured in control, SMM and LM3 media are shown.

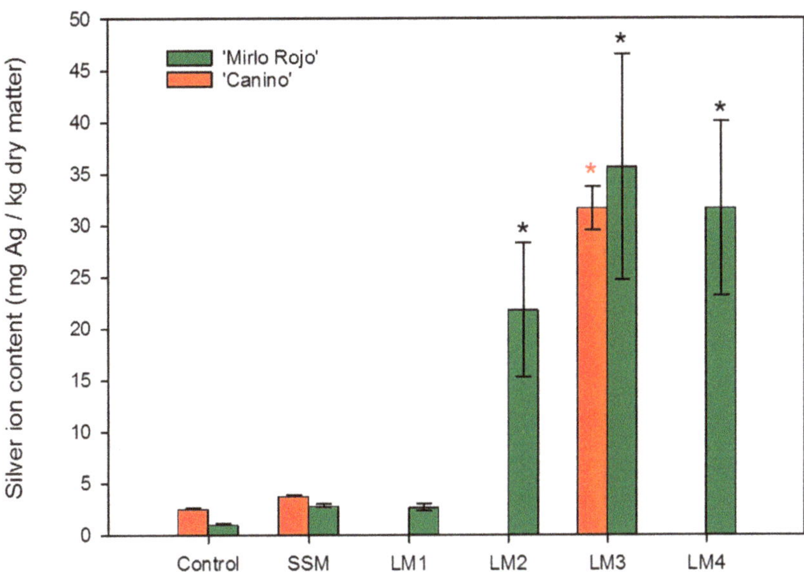

Figure 4. Silver concentration in 'Mirlo Rojo' and 'Canino' shoots cultured in different media containing 100 mg L^{-1} AgNPs. Control medium is SSM without AgNPs. Other media contents are described in Table 1. Bars represent standard errors. Asterisks indicate treatments significantly different from the controls for each cultivar ($p < 0.05$) according to a one-tail Dunnett test.

Control treatment corresponds to the shoots cultured in semisolid medium without AgNPs addition (Figure 4). SSM (containing agar and $CaCl_2$) and LM1 (without agar and with $CaCl_2$) with AgNPs showed a slight, but not significant, increase in silver concentration compared with control. However, for LM2, LM3 and LM4 (all of them without agar and $CaCl_2$) a significant increase ($p < 0.05$) in the silver content compared with the control was recorded. LM3 with 35.6 mg of silver per kg of dry plant was the medium where the highest content of silver was found (Figure 4). When the silver concentration was studied in 'Canino' shoots cultured in SMM and in LM3, 3.71 and 31.6 mg/kg of silver were found, respectively. These are values similar to those found for 'Mirlo Rojo' in these same media. Only silver content in LM3 significantly differed from the control of 'Canino' shoots cultured on SMM without AgNPs.

3.4. Effects of AgNPs on Apricot Micropropagation

For the short- and long-term objectives of this work, it is essential to determine if the addition of AgNPs has any detrimental effect on the micropropagation of apricot cultivars. Therefore, different variables related to micropropagation were measured after exposure of 'Mirlo Rojo' apricot shoots to media with AgNPs. Since the effect of initial number of shoots, proliferation, mean shoot length, productivity and leaf area (Figure 3) were studied without AgNPs, the same micropropagation variables were measured with AgNPs at the concentration of 100 mg L^{-1} after culturing 'Mirlo Rojo' shoots in LM1 (control medium with $CaCl_2$) and LM2, LM3 and LM4 (without $CaCl_2$) media (Table 2).

A specific contrast was performed in order to determine the effect of AgNPs addition on micropropagation of 'Mirlo Rojo' by comparing LM1 with the average of values recorded in the rest of liquid media with AgNPs (Table 2, treatment LM is the average of all media containing AgNPs). Treatments with AgNPs increased productivity when compared with the control LM1 (Table 2) due mainly to an increase in the number of new shoots (proliferation). On the other hand, there were not significant differences in mean shoot length. Leaf area was significantly larger in the media with AgNPs than in the control medium (Table 2).

Since the addition of AgNPs had a significant effect improving productivity we decided to study which composition of liquid media can provide the best micropropagation-related variables for 'Mirlo Rojo' shoots (Table 2). ANOVA found significant differences among the three AgNPs-containing media (LM2, LM3 and LM4) ($p < 0.05$) for all variables studied. Productivity and the number of new shoots were significantly better in media LM3, but mean shoot length and leaf area were larger in LM2 (Table 2).

Table 2. Effect of AgNPs addition on micropropagation of 'Mirlo Rojo' apricot in different liquid media.

Treatment [z]	Productivity Shoots x cm	Proliferation Shoot Number	Shoot Length cm	Leaf Area cm^2
LM1 (no AgNPs)	1.92 ± 0.41	1.40 ± 0.22	1.35 ± 0.11	2.53 ± 0.17
LM (+AgNPs)	4.95 ± 0.34 *	3.57 ± 0.27 *	1.50 ± 0.10	3.79 ± 0.50 *
LM2 (+ AgNPs)	5.33 ± 0.65 [ab]	3.08 ± 0.48 [b]	2.06 ± 0.27 [a]	7.21 ± 1.16 [a]
LM3 (+ AgNPs)	6.10 ± 0.61 [a]	4.58 ± 0.48 [a]	1.35 ± 0.05 [b]	2.50 ± 0.31 [b]
LM4 (+ AgNPs)	3.92 ± 0.46 [b]	3.22 ± 0.38 [b]	1.23 ± 0.04 [b]	2.38 ± 0.33 [b]

[z] Asterisks represent significant differences ($p < 0.05$) between the media without AgNPs addition (LM1) and the average of AgNPs-containing media (LM+AgNPs) according to a specific contrast. Different letters within the AgNPs-containing media represent significant differences ($p < 0.05$) between treatments according to a Duncan test.

The best medium found for 'Mirlo Rojo' (LM3) was tested for 'Canino' that produced a ratio of 6.1 ± 1.36 shoots with an average length of 1.51 ± 0.34 and leaf area of 2.57 ± 0.57, which gives a productivity of 9.12 ± 2.04 even higher than in 'Mirlo Rojo'.

4. Discussion

The use of AgNPs is increasing due to their ability to eliminate phytopathogens, bacteria, virus and fungus. AgNPs have been successfully used to disinfect explants when establishing plants in vitro [29,30]. The antivirus efficiency of Argovit[@] AgNPs has been demonstrated in different biological systems. Bogdanchikova et al. (2016) described that, after application of Argovit AgNPs, symptoms of *Canine distemper virus* (CDV) disappeared in dogs without neurological symptoms of the disease. Argovit® AgNPs were effective in the treatment or prophylaxis of infectious bronchitis coronavirus in chickens [20], white spot syndrome virus in shrimp [22,31], as well as in in vitro and in vivo models of Rift Valley fever virus [23] and SARS-CoV-2 coronavirus [24].

When applied on guar leaf surface along with *Sunhemp rosette virus* (SHRV), AgNPs have a preventive effect on the presence of symptoms on leaves (Jain & Kothari, 2014). Similarly, a post-infection treatment by spraying with AgNPs prevented the destructive symptoms caused by the *Bean yellow mosaic virus* (BYMV) in faba bean plants [17]. Recently, in the same species, leaf application of AgNPs have been demonstrated effective in preventing infection with BYMV and reducing virus replication [16]. Some authors found that AgNPs reduce infection ability of viruses when applied as a preventive treatment [32]. However, to the best of our knowledge, there is not a methodology for producing virus-free plants by adding AgNPs to the in vitro culture media.

The Fruit Biotechnology Group at CEBAS-CSIC has a collection of apricot cultivars maintained in vitro culture in semisolid medium. A protocol for the micropropagation of apricot cultivars was already available. Accordingly, 'Mirlo Rojo' shoots were cultured in SSM medium (Table 1) with the addition of AgNPs at 100 mg L^{-1} (silver and PVP) to study the capacity of AgNPs to penetrate into plant tissues. Bello-Bello et al. [11] described that using Argovit® between 50 and 100 mg L^{-1} allowed for the micropropagation of sugar cane (*Saccharum spp*) but resulted toxic at concentrations over 200 mg L^{-1}. Therefore, we decided to choose 100 mg L^{-1} Argovit-7 AgNPs concentration since the objective was to use the highest possible concentration without detrimental effect on micropropagation. After four weeks of culture in SSM, silver was not detected in 'Mirlo Rojo' or 'Canino' shoot tissues (Figure 4). This may be due to slow migration of AgNPs inside the agar matrix. Agar at concentration used in this work (7 g L^{-1}) have a pore size of around 100 nm [33]. This could be close to 56.1 nm of hydrodynamic diameter of AgNPs nanoparticles studied here which would impede the mobility of nanoparticles and therefore the possibility to be absorbed by the plant. Moreover, the results of TEM study showed that AgNPs agglomerates into groups with a size more of than 100 nm that does not allow their diffusion through 100 nm agar pores. Apparently low negative charge of the particles (zeta potential is -3.85 mV) facilitate agglomeration of AgNPs into the groups.

A methodological approach to address this problem could be application of liquid medium to culture apricot shoots. Castro-González et al. [15] found that silver from Argovit$^@$ formulation was transported from culture medium and was accumulated within plant tissues when Stevia (*Stevia rebaudiana* B.) was cultured in liquid medium in TIS. Therefore, conditions including the application of liquid medium to culture apricot in TIS were also applied in the present work.

There are several studies on the micropropagation of woody plants in TIS [14,34]. However, it is necessary to modify protocols for each species, and even for each cultivar.

The main parameters affecting efficiency of micropropagation in TIS are immersion time, aeration time, bioreactor volume and culture medium volume [35]. Immersion and aeration times determine nutrient absorption and control of hyperhydration [35]. Hyperhydration is one of the main limiting factors of micropropagation in TIS due to the apparition of physiological and morphological disorders, such as irregular shoots growth or production of anomalous leaves [36]. Our previous experience with apricot revealed that this species is very susceptible to hyperhydration [37] and therefore short immersion times and frequent aeration periods are necessary. These results showed that a combination of 2 min of immersion every 6 h and 3 min of aeration every 3 h leaded to effective proliferation of apricot in TIS.

Vegetative propagation in bioreactors is directly related to the bioreactor volume. Propagation of grapes in bioreactors with different volumes demonstrated that the larger volume the higher number of longer new shoots [38]. In the present work we found that increased number of initial shoots had a positive effect in apricot micropropagation. Figure 3 shows that productivity of 'Canino' raised with the increase in the initial shoot number. This effect could be explained by the beneficial conditions effect of our experiments (limited space, Figure 1C,D), with shoots located close to each other, which can stimulate vertical growth of the main stems. When there is a large space between shoots, they tend to grow horizontally which limits their length.

Medium volume depends on the type of bioreactor used. To produce efficient plant propagation in TIS, medium should completely cover the explants during the immersion phase. For the 4 L Plantform^{TM} bioreactors used here, 500 mL of medium gave good results immersing completely the explants.

Culture conditions were established for 'Canino' in TIS because its behavior in semisolid medium is well known. The best determined conditions in TIS were tested for 'Mirlo Rojo'. We found expected differences between genotypes but both cultivars were successfully propagated in TIS. We paid special attention to 'Mirlo Rojo' in the experiments with addition of AgNPs because we have this cultivar infected with *Hop stunt Viroid*.

'Mirlo Rojo' shoots were cultured in liquid medium LM1 supplemented with 100 mg L^{-1} AgNPs. Results indicate that silver did not penetrate the plant tissues (Figure 4). There was a slight non-significant increase, as compared with the control medium without AgNPs, but it could be due to silver precipitated on the shoots surface. At the end of the culture cycle, we observed solid precipitate at the bottom of the bioreactors, probably a precipitated salt. Major and minor nutrients for the plant in culture media are in the form of salts that are dissociated at the pH of the media and then plants absorb the dissociated ions. AgNPs may react with different salts components of the media. For instance, chloride, sulfate, or nitrate anions can react with silver cations (which slowly are released from AgNPs) forming salts that have different degrees of solubility. We decided to check if chloride anions, dissociated from $CaCl_2$, (a major component of SSM and therefore of LM1) can react with Ag^+ forming the highly insoluble silver chloride salt [39]. Although we found precipitate at the bottom of the bioreactors after using media without calcium chloride, it was in much lower amounts than in medium with $CaCl_2$, which supports the hypothesis that silver may precipitate in the same way after reacting with other salts but in less amounts than with $CaCl_2$. Modifying culture media is always a risk since they usually are very well balanced, and the modification of any component may strongly affect micropropagation rates and plant behavior in vitro. Nevertheless, we decided to start testing modifications. The choice fell on $CaCl_2$ since it is in larger amounts than sulfates or nitrates and solubility of AgCl is much lower than the other silver salts. On one hand LM2 medium formulation is frequently used, after adding agar, to micropropagate some apricot cultivars, as a semisolid medium. Moreover, on the other hand, LM2 does not have $CaCl_2$ in its composition. Therefore, we included it, just as a liquid medium without modifications. Additionally, we completely removed $CaCl_2$ in LM3. In LM4, $CaCl_2$ was also removed substituting Ca cation by $Ca(NO3)_2$ and reducing NH_4NO_3 to compensate for a possible calcium deficiency and avoid excess of nitrogen, respectively.

After culturing 'Mirlo Rojo' shoots in three different liquid media without chloride ions (LM2, LM3 and LM4) and with the addition of 100 mg L^{-1} AgNPs, silver was found in plant tissues from the shoots cultured in these three media in concentrations significantly higher than in shoots cultured in the control medium (Figure 4). LM3 promoted the penetration of the highest silver concentration to shoot tissues from culture medium compared with LM2 and LM4. Silver introduction in 'Canino' shoots cultured in LM3 was also significantly larger than in control shoots. Therefore, culture in liquid medium with a simple change consisting in removing $CaCl_2$ avoided immobilization of AgNPs in agar matrix and precipitation of silver chloride salts. These modifications facilitated the penetration of silver into plant tissue. This is the first, inevitable and therefore paramount, step ensuring the interaction of AgNPs with pathogenic microorganisms inside shoot tissues.

As indicated above, the modification of culture media may be risky and, therefore, the influence of AgNPs addition to media needs to be evaluated to discard a negative effect on apricot micropropagation. Table 2 shows a comparison between micropropagation parameters of 'Mirlo Rojo' shoots in LM1, liquid medium whithout AgNPs addition, with those in LM2, LM3 and LM4, where AgNPs were added, and silver is entering plant tissues. A positive effect of AgNPs addition on 'Mirlo Rojo' and 'Canino' micropropagation was found, increasing productivity, mainly due to a higher number of new shoots. This positive effect was also described in sugarcane micropropagation [11], vanilla (*Vanilla planifolia* Jacks. ex Andrews) [10] and stevia [15] when using Argovit® in TIS. Biostimulants are compounds that induce a physiological response in the plant improving growth and development [40]. AgNPs have a biostimulation effect when they are used in TIS.

In order to maximize the efficiency of AgNPs addition on apricot micropropagation it is necessary to use the optimal medium composition. From the media tested here, LM2 and LM3 lead to higher productivity. In LM3 the improvement was due to the production of a higher number of new shoots and a higher proliferation, than in the other two media without calcium chloride, whereas LM2 induced the production of significantly longer shoots with a larger leaf surface (Table 2). It should be noted that high leaf surface leads to dominant use of energy to produce biomass instead of shoot growth. When the effect of LM3 was tested on 'Canino' it also improved micropropagation rates. Therefore, LM3 medium, traditionally used for apricot cultivar micropropagation, after elimination of calcium chloride proved to be the most effective medium for AgNPs penetration into the plant tissue.

These results can undoubtedly be used in further studies targeting AgNPs application for pathogen microorganism growth inhibition in plant tissue during micropropagation in TIS.

5. Conclusions

The long-term objective of this work is the application of AgNPs treatments to prevent or cure microbe's contamination in plant tissues during micropropagation. However, problems related with immobilization of nanoparticles in the agar matrix and/or precipitation of silver salts prevented the migration of silver into plant tissues. By using liquid media in Temporal Immersion Systems and a simple medium modification consisting in removal of calcium chloride we have demonstrated that silver penetration into plant tissues increases, as average for both cultivars, 23-fold and without any detrimental effect on apricot productivity, that on the contrary, was improved by 2.8-fold as average for both cultivars, due to the addition of AgNPs in the culture media.

Author Contributions: Conceptualization, L.B.; methodology, L.F., C.P.-C., A.G.R.-H. and J.C.G.-R.; investigation, C.P.-C., N.A. and L.B.; resources, N.B. and A.P.; writing-original draft preparation, L.B. and C.P.-C.; writing-review and editing, N.B. and N.A.; supervision, L.B. All authors have read and agreed to the published version of the manuscript.

Funding: This research was funded by Project CARM-Agroalimentario Ref: CA20565 "OSIRIS".

Institutional Review Board Statement: Not applicable.

Informed Consent Statement: Not applicable.

Data Availability Statement: Not applicable.

Acknowledgments: Authors wish to thank Spanish Ministry of Education for a FPU fellowship granted to C.P.-C. and Russian Science Foundation and Tomsk region for grant 22-13-20032. The authors thank Amelia Olivas from CNyN-UNAM for TGA/DGS spectra acquisition.

Conflicts of Interest: The authors declare no conflict of interest.

References

1. Galdiero, S.; Falanga, A.; Vitiello, M.; Cantisani, M.; Marra, V.; Galdiero, M. Silver Nanoparticles as Potential Antiviral Agents. *Molecules* **2011**, *16*, 8894–8918. [CrossRef] [PubMed]
2. Rai, M.; Kon, K.; Ingle, A.; Duran, N.; Galdiero, S.; Galdiero, M. Broad-Spectrum Bioactivities of Silver Nanoparticles: The Emerging Trends and Future Prospects. *Appl. Microbiol. Biotechnol.* **2014**, *98*, 1951–1961. [CrossRef] [PubMed]
3. Saravanan, M.; Mostafavi, E.; Vincent, S.; Negash, H.; Andavar, R.; Perumal, V.; Chandra, N.; Narayanasamy, S.; Kalimuthu, K.; Barabadi, H. Nanotechnology-Based Approaches for Emerging and Re-Emerging Viruses: Special Emphasis on COVID-19. *Microb. Pathog.* **2021**, *156*, 104908. [CrossRef] [PubMed]
4. Jones, R.A.C. Global Plant Virus Disease Pandemics and Epidemics. *Plants* **2021**, *10*, 233. [CrossRef] [PubMed]
5. Gupta, N.; Upadhyaya, C.P.; Singh, A.; Abd-Elsalam, K.A.; Prasad, R. Applications of Silver Nanoparticles in Plant Protection. In *Nanobiotechnology Applications in Plant Protection*; Springer: Cham, Switzerland, 2018; pp. 247–265. [CrossRef]
6. Jones, R.A.C.; Naidu, R.A. Global Dimensions of Plant Virus Diseases: Current Status and Future Perspectives. *Annu. Rev. Virol.* **2019**, *6*, 387–409. [CrossRef] [PubMed]
7. Farooq, T.; Adeel, M.; He, Z.; Umar, M.; Shakoor, N.; Da Silva, W.; Elmer, W.; White, J.C.; Rui, Y. Nanotechnology and Plant Viruses: An Emerging Disease Management Approach for Resistant Pathogens. *ACS Nano* **2021**, *15*, 6030–6037. [CrossRef] [PubMed]
8. Avila-Quezada, G.D.; Golinska, P.; Rai, M. Engineered Nanomaterials in Plant Diseases: Can We Combat Phytopathogens? *Appl. Microbiol. Biotechnol.* **2022**, *106*, 117–129. [CrossRef] [PubMed]
9. Kim, D.H.; Gopal, J.; Sivanesan, I. Nanomaterials in Plant Tissue Culture: The Disclosed and Undisclosed. *RSC Adv.* **2017**, *7*, 36492–36505. [CrossRef]
10. Spinoso-Castillo, J.L.; Chavez-Santoscoy, R.A.; Bogdanchikova, N.; Pérez-Sato, J.A.; Morales-Ramos, V.; Bello-Bello, J.J. Antimicrobial and Hormetic Effects of Silver Nanoparticles on in Vitro Regeneration of Vanilla (*Vanilla Planifolia* Jacks. Ex Andrews) Using a Temporary Immersion System. *Plant Cell Tissue Organ Cult.* **2017**, *129*, 195–207. [CrossRef]
11. Bello-Bello, J.J.; Chavez-Santoscoy, R.A.; Lecona-Guzmán, C.A.; Bogdanchikova, N.; Salinas-Ruíz, J.; Carlos Gómez-Merino, F.; Pestryakov, A. Hormetic Response by Silver Nanoparticles on In Vitro Multiplication of Sugarcane (*Saccharum* Spp. Cv. Mex 69-290) Using a Temporary Immersion System. *Dose-Response* **2017**, *15*. [CrossRef] [PubMed]
12. FAOSTAT. Food and Agriculture Organization of the United Nations 2022. Available online: https://www.fao.org/faostat/es/#data/QCL (accessed on 15 May 2022).
13. Pérez-Tornero, O.; Burgos, L. Apricot Micropropagation. In *Protocols for Micropropagation of Woody Trees and Fruits*; Jain, S.M., Häggman, H., Eds.; Springer: Dordrecht, The Netherlands, 2007; pp. 267–278.
14. Vidal, N.; Sánchez, C. Use of Bioreactor Systems in the Propagation of Forest Trees. *Eng. Life Sci.* **2019**, *19*, 896–915. [CrossRef] [PubMed]
15. Castro-González, C.G.; Sánchez-Segura, L.; Gómez-Merino, F.C.; Bello-Bello, J.J. Exposure of Stevia (*Stevia Rebaudiana* B.) to Silver Nanoparticles in Vitro: Transport and Accumulation. *Sci. Rep.* **2019**, *9*, 1–10. [CrossRef]
16. El-Gamal, A.Y.; Tohamy, M.R.; Abou-Zaid, M.I.; Atia, M.M.; El Sayed, T.; Farroh, K.Y. Silver Nanoparticles as a Viricidal Agent to Inhibit Plant-Infecting Viruses and Disrupt Their Acquisition and Transmission by Their Aphid Vector. *Arch. Virol.* **2022**, *167*, 85–97. [CrossRef]
17. Elbeshehy, E.K.F.; Elazzazy, A.M.; Aggelis, G. Silver Nanoparticles Synthesis Mediated by New Isolates of *Bacillus* Spp., Nanoparticle Characterization and Their Activity against Bean Yellow Mosaic Virus and Human Pathogens. *Front. Microbiol.* **2015**, *6*, 453. [CrossRef] [PubMed]
18. Jain, D.; Kothari, S.L. Green Synthesis of Silver Nanoparticles and Their Application in Plant Virus Inhibition. *J. Mycol. Plant Pathol.* **2014**, *44*, 21–24.

19. Podkopaev, D.O.; Shaburova, L.N.; Balandin, G.V.; Kraineva, O.V.; Labutina, N.V.; Suvorov, O.A.; Sidorenko, Y.I. Comparative Evaluation of Antimicrobial Activity of Silver Nanoparticles. *Nanotechnologies Russ.* **2014**, *9*, 93–97. [CrossRef]
20. Nefedova, E.; Koptev, V.; Bobikova, A.S.; Cherepushkina, V.; Mironova, T.; Afonyushkin, V.; Shkil, N.; Donchenko, N.; Kozlova, Y.; Sigareva, N.; et al. The Infectious Bronchitis Coronavirus Pneumonia Model Presenting a Novel Insight for the Sars-Cov-2 Dissemination Route. *Vet. Sci.* **2021**, *8*, 239. [CrossRef] [PubMed]
21. Bogdanchikova, N.; Vázquez-Muñoz, R.; Huerta-Saquero, A.; Pena-Jasso, A.; Aguilar-Uzcanga, G.; Picos-díaz, P.L.; Pestryakov, A. Silver Nanoparticles Composition for Treatment of Distemper in Dogs. *Int. J. Nanotechnol.* **2016**, *13*, 227–237. [CrossRef]
22. Romo-Quiñonez, C.R.; Álvarez-Sánchez, A.R.; Álvarez-Ruiz, P.; Chávez-Sánchez, M.C.; Bogdanchikova, N.; Pestryakov, A.; Mejia-Ruiz, C.H. Evaluation of a New Argovit as an Antiviral Agent Included in Feed to Protect the Shrimp *Litopenaeus Vannamei* against White Spot Syndrome Virus Infection. *PeerJ* **2020**, *8*, e8446. [CrossRef] [PubMed]
23. Borrego, B.; Lorenzo, G.; Mota-Morales, J.D.; Almanza-Reyes, H.; Mateos, F.; López-Gil, E.; de la Losa, N.; Burmistrov, V.A.; Pestryakov, A.N.; Brun, A.; et al. Potential Application of Silver Nanoparticles to Control the Infectivity of Rift Valley Fever Virus in Vitro and in Vivo. *Nanomed. Nanotechnol. Biol. Med.* **2016**, *12*, 1185–1192. [CrossRef] [PubMed]
24. Almanza-Reyes, H.; Moreno, S.; Plascencia-López, I.; Alvarado-Vera, M.; Patrón-Romero, L.; Borrego, B.; Reyes-Escamilla, A.; Valencia-Manzo, D.; Brun, A.; Pestryakov, A.; et al. Evaluation of Silver Nanoparticles for the Prevention of SARS-CoV-2 Infection in Health Workers: In Vitro and in Vivo. *PLoS ONE* **2021**, *16*, e0256401. [CrossRef] [PubMed]
25. Wang, H.; Petri, C.; Burgos, L.; Alburquerque, N. Phosphomannose-Isomerase as a Selectable Marker for Transgenic Plum (*Prunus Domestica* L.). *Plant Cell. Tissue Organ Cult.* **2013**, *113*, 189–197. [CrossRef]
26. Quoirin, M.; Lepoivre, P. Improved Media for in Vitro Culture of *Prunus* Sp. *Acta Hortic.* **1977**, *78*, 437–442. [CrossRef]
27. Driver, J.A.; Kuniyuki, A.H. In Vitro Propagation of Paradox Walnut Rootstock. *HortScience* **1984**, *19*, 507–509.
28. Schneider, C.A.; Rasband, W.S.; Eliceiri, K.W. NIH Image to ImageJ: 25 Years of Image Analysis. *Nat. Methods* **2012**, *9*, 671–675. [CrossRef]
29. Arab, M.M.; Yadollahi, A. Effects of Antimicrobial Activity of Silver Nanoparticles on in Vitro Establishment of G x N15 (Hybrid of Almond x Peach) Rootstock. *J. Genet. Eng. Biotechnol.* **2014**, *12*, 103–110. [CrossRef]
30. Moradpour, M.; Aziz, M.A.; Abdullah, S.N.A. Establishment of in Vitro Culture of Rubber (*Hevea Brasiliensis*) from Field-Derived Explants: Effective Role of Silver Nanoparticles in Reducing Contamination and Browning. *J. Nanomed. Nanotechnol.* **2016**, *7*, 2. [CrossRef]
31. Ochoa-Meza, A.R.; Álvarez-Sánchez, A.R.; Romo-Quiñonez, C.R.; Barraza, A.; Magallón-Barajas, F.J.; Chávez-Sánchez, A.; García-Ramos, J.C.; Toledano-Magaña, Y.; Bogdanchikova, N.; Pestryakov, A.; et al. Silver Nanoparticles Enhance Survival of White Spot Syndrome Virus Infected *Penaeus Vannamei* Shrimps by Activation of Its Immunological System. *Fish Shellfish Immunol.* **2019**, *84*, 1083–1089. [CrossRef]
32. El-Dougdoug, N.K.; Bondok, A.M.; El-Dougdoug, K.A. Evaluation of Silver Nanoparticles as Antiviral Agent against ToMV and PVY in Tomato Plants. *Middle East J. Appl. Sci.* **2018**, *8*, 100–111.
33. Narayanan, J.; Xiong, J.Y.; Liu, X.Y. Determination of Agarose Gel Pore Size: Absorbance Measurements Vis a Vis Other Techniques. *J. Phys. Conf. Ser.* **2006**, *28*, 83–86. [CrossRef]
34. Carvalho, L.S.O.; Ozudogru, E.A.; Lambardi, M.; Paiva, L.V. Temporary Immersion System for Micropropagation of Tree Species: A Bibliographic and Systematic Review. *Not. Bot. Horti Agrobot. Cluj-Napoca* **2019**, *47*, 269–277. [CrossRef]
35. Etienne, H.; Berthouly, M. Temporary Immersion Systems in Plant Micropropagation. *Plant Cell. Tissue Organ Cult.* **2002**, *69*, 215–231. [CrossRef]
36. Rojas-Martínez, L.; Visser, R.G.F.; Klerk, G. De The Hyperhydricity Syndrome: Waterlogging of Plant Tissues as a Major Cause. *Propag. Ornam. Plants* **2010**, *10*, 169–175.
37. Pérez-Tornero, O.; Egea, J.; Olmos, E.; Burgos, L. Control of Hyperhydricity in Micropropagated Apricot Cultivars. *In Vitro Cell. Dev. Biol. Plant.* **2001**, *37*, 250–254. [CrossRef]
38. Monette, P.L. Influence of Size of Culture Vessel on in Vitro Proliferation of Grape in a Liquid Medium. *Plant Cell Tissue Organ Cult.* **1983**, *2*, 327–332. [CrossRef]
39. Syafiuddin, A.; Fulazzaky, M.A.; Salmiati, S.; Kueh, A.B.H.; Fulazzaky, M.; Salim, M.R. Silver Nanoparticles Adsorption by the Synthetic and Natural Adsorbent Materials: An Exclusive Review. *Nanotechnol. Environ. Eng.* **2020**, *5*, 1. [CrossRef]
40. Du Jardin, P. Plant Biostimulants: Definition, Concept, Main Categories and Regulation. *Sci. Hortic.* **2015**, *196*, 3–14. [CrossRef]

 *horticulturae*

Article

In Vitro Techniques for Shipping of Micropropagated Plant Materials

Jingwei Li [1,2,†], Min He [1,2,†], Xiuhong Xu [1,2], Tingmin Huang [1,2], Huan Tian [3] and Wanping Zhang [1,2,*]

1   Vegetable Research Institute, Guizhou University, Guiyang 550025, China; jwli3@gzu.edu.cn (J.L.);
    hemin97@yeah.net (M.H.); xuxiuhong_2000@163.com (X.X.); huangtm182@gmail.com (T.H.)
2   College of Agriculture, Guizhou University, Guiyang 550025, China
3   Guizhou Modern Seed Industry Group Co., Ltd., Guiyang International Financial Center,
    Guiyang 550081, China; huantian0510@163.com
*   Correspondence: drwpzhang@163.com
†   These authors contributed equally to this work.

Abstract: Shipping of in vitro micro-cuttings in tubes or jars is a frequently used method as the plants are more likely to quickly reproduce and comply with quarantine regulations in plant germplasm distribution. However, these containers are fragile during transportation. To diminish the risk associated with the long-distance shipping of in vitro plants, a safe and widely applicable packing and conservation technique based on microplate and slow growth was developed in this study. Potato cultivar ZHB and ginger cultivar G-2 were used to optimize the system with microplates (96 wells), vacuum-sealed packaging, and slow-growth techniques. Under regular culture conditions, packing in vacuum-sealed microplates reduced the survival of ZHB and G-2 micro-cuttings to 85.8% and 20.0%, respectively, and regeneration to 61.8% and 0%, respectively. Reducing the temperature to 10 °C maintained the survival of ZHB and G-2 micro-cuttings in the range of 83.3–100% after 60 days. Exposure to darkness decreased the survival of G-2 and inhibited regrowth. Thus, conservation in darkness at 10 °C is suggested. The effects of iron concentration and plant growth retardants were further assessed. The addition of 1/4 MS medium combined with 100 mg/L chlormequat chloride (CCC) resulted in full survival and growth inhibition of plantlets, without malformation identified. Finally, incubation with 1/4 MS medium supplemented with 100 mg/L CCC in vacuum-sealed microplates at 10 °C in the dark resulted in high survival and suppressed germination. Sweet potato HXS was incubated as well to test the broad-spectrum applications of the technique; 100% survival and 6.7% germination was gained. Morphological indices of released cuttings recovered to control levels after two cycles of subculture in MS medium. A 0.1–0.2% genetic variation was detected by SSR and ISSR, suggesting genetic stability of the conserved samples. Finally, micro-cuttings were safely transported to cities located thousands of kilometers away without package and sample damage. Our results enable easy distribution of in vitro plant germplasms.

Keywords: plant conservation; long distance shipping; slow growth; microplate; vacuum package

Citation: Li, J.; He, M.; Xu, X.; Huang, T.; Tian, H.; Zhang, W. In Vitro Techniques for Shipping of Micropropagated Plant Materials. *Horticulturae* 2022, 8, 609. https://doi.org/10.3390/horticulturae8070609

Academic Editor: Sergey V. Dolgov

Received: 9 June 2022
Accepted: 1 July 2022
Published: 6 July 2022

1. Introduction

Availability of and easy access to diverse plant germplasm, including cultivation crops, their wild relatives, and wild species, are of great importance to human survival and contribute to people's livelihood via staple and cash crop breeding [1], pharmaceuticals [2], rehabilitation [3], environmental beautification [4,5], and ecological governance and stabilization [3]. Vegetatively propagated plants shipped by wrapping the plant or vegetative mass in express containers for long-distance transportation increases the risk of quarantine and environmental contamination [6]. Furthermore, seasonal availability of scion wood or rooted cuttings may limit their usefulness in germplasm distribution. Thus, a safe and efficient method for plant germplasm transportation, especially for vegetatively propagating plants, is needed. Micropropagation, which exploits the totipotent nature of plant cells to generate new individuals from protoplasts, cells, undifferentiated masses of cells (callus), small pieces of tissue, and/or excised organs, is a time-tested and practical ex situ technique for the short- and medium-term conservation of plant germplasm [2,7,8].

Plant germplasm resources are currently exchanged using in vitro micro-cuttings, which are more likely to massively reproduce and comply with quarantine regulations [9–12]. In vitro samples are subcultured after arriving at the destination and carry no superficial pathogens or insects. Shipping of micro-cuttings can be a challenge as well. Once the containers or culture bags are damaged during transportation or travel, the germplasms cannot be recovered. Biosafety problems still exist if the plant resources carry obligate pathogens, such as bacteria, viruses, and viroids, and the released micro-cuttings can infect other healthy plants [9,10,13]. Medium liquefaction or combination with explants due to shaking or changes in cabin pressure and failure to maintain sterility within the container, in addition to neglect on shipping docks for extended periods, are additional challenges [14]. According to a report from the National Clonal Germplasm Repository in Corvallis, OR, USA, in vitro micro-cuttings are transported and exchanged in sealed and semi-permeable plastic bags containing firm medium (7–8 g/L agar) and they are folded and packed in crushproof containers, which minimizes the shifting of plants and medium in transit. Weather conditions and arrival date also contribute to loss of shipments. However, appropriate packaging and shipping can increase the viability of the transported cultures for a month or longer under normal conditions [14].

Decreasing the cellular metabolism of micro-cuttings and prolonging the intervals between subcultures by slow-growth techniques [15] may minimize the adverse effects on plant viability associated with delays in customs or quarantine. Slow growth of plant germplasm via medium-term conservation based on in vitro micropropagation reduces costs [16]. This method is usually conducted by reducing the culture temperature, supplying osmotic agents in culture medium, and adding growth inhibitors to or removing growth promoters from the medium [17,18]. To date, several plant germplasms have been successfully conserved using this method. For instance, seven genotypes of wild and elite plants were preserved with seven slow-growth media for 12 months; Tavazza et al. (2015) confirmed that treatments resulted in 65% to 85% of survival and 100% of regrowth in surviving plants [19]. SSR (simple sequence repeats) and ISSR (inter-simple sequence repeats) were used to analyze the genetic stability of slow-growing conserved samples. Tahtamouni et al. (2016) conserved *Thymbra spicata* supplemented with 0.2 M sucrose storage medium for 3 months, resulting in 100% survival [20]. The growth, oil yield, and carvacrol content of recovered plantlets remained unchanged. *Eustoma grandiflorum* was successfully conserved in vitro for 90 days by Ramírez-Pérez et al. (2020) [16], without significant changes in vitality. The technique of slow growth prolonged the interval between subcultures significantly and did not alter the genetic stability of conserved samples, which facilitates long-distance shipping and exchange of plant micro-cuttings.

The microplate (96 sample wells) offers limited space for culture and biochemical analysis of organisms. It allows less than 250 μL of medium loading and shorter than 0.5 cm of micro-cutting culture. Space-limited culture ensures less thrashing during shipping and smaller explants, resulting in a slower growth rate. Vacuum sealing can effectively reduce mechanical damage, bumping, and microbe germination of items during transportation [21,22]. It has been widely used in the long-distance shipping of fresh plant fruit [23] or semi-finished bioproducts [22]. In addition, reducing air pressure around plants is another important aspect of slow-growth conservation. When the air pressure is reduced in the culture environment, the growth of in vitro plantlets is reduced as well. The initiation of regrowth under normal conditions revealed no phenotypic modification of the plantlets developing from the inoculums [24]. Space-limited incubation and vacuum-sealed packaging stabilize the substances in the container while reducing the microbe germination and plantlet growth, and thus has a high potency for application in long-distance shipping of plant germplasm.

The present study developed an in vitro incubation system for the long-distance transportation of plant germplasm. Potato and ginger, which are two globally important vegetatively propagated agricultural and horticultural crops, were used to optimize the incubation system based on slow-growth techniques, microplates (96 sample wells), and vacuum-sealed packaging. Recovery and regeneration of the packed micro-cuttings were analyzed under normal in vitro culture conditions. The genetic stability was analyzed using molecular markers SSR and ISSR. The transportation resistance of the present system was tested using automobile transportation and traveling by train. The application of this method was tested in sweet potato in vitro micro-cuttings as well.

2. Materials and Methods

2.1. Plant Materials

In vitro materials of 3 cultivated varieties, potato cultivar 'Zihuabai' (ZHB), ginger 'Guizhouxiaohuangjiang-2' (G-2) (Figure 1A), and sweet potato 'Hongxinshu' (HXS), were employed in this study to establish a stable method for plant germplasm medium-term conservation and long-distance shipping. Stock plantlets were cultured in Murashige and Skoog (1962) medium (MS medium) supplemented with 30 g/L sucrose and 7 g/L agar (pH = 5.8) [25]. The cultures were grown at a consistent temperature of 24 $\pm$ 2 °C for a 16 h photoperiod with a light intensity of 50 μM s^{-1} m^{-2}. Subculturing occurred every 3 weeks (Figure 1B).

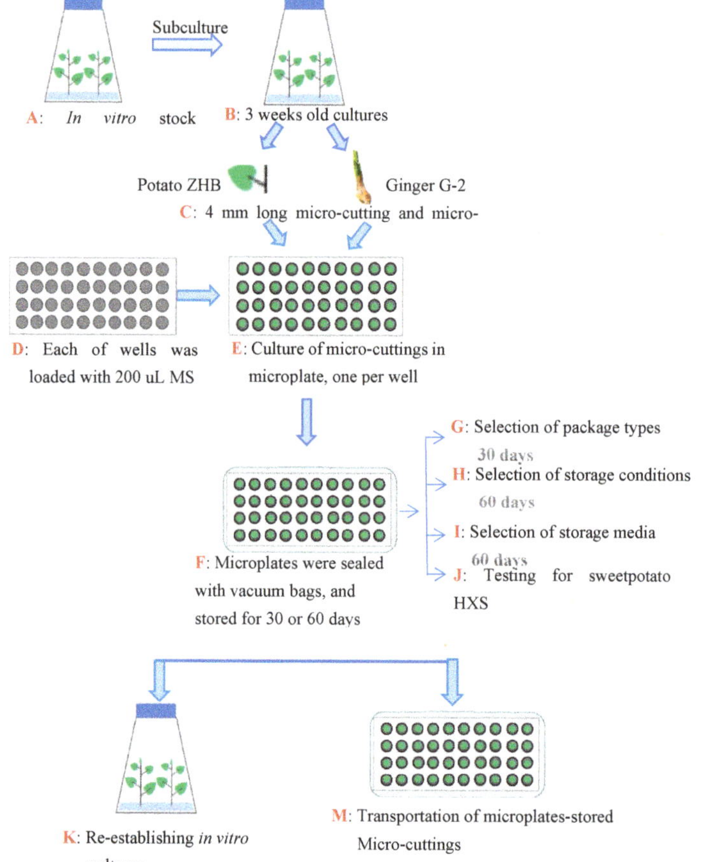

Figure 1. A flow chart of use of microplates for storage and transportation of in vitro micropropagated plant materials.

2.2. Assessment of Packaging Type

Micro-cuttings, which consisted of one leaf on a 0.4 mm shoot of ZHB and a 0.4 mm long micro-tiller containing one terminal bud of G-2, were harvested (Figure 1C). Each well of sterilized microplates (96 wells, Virya, Shanghai, China) was loaded with 200 μL of MS medium and covered with a lid (Figure 1D). Each microplate was used to ship both samples, one sample per well (Figure 1E). Plates with plant samples were packed in a transparent vacuum-sealing packing bag (sterilized or non-sterilized) (Figure 1F). Vacuum treatment was conducted to test whether evacuating air outside the plate and fastening the plate limited the growth of plant samples. Samples were incubated in a normal culture environment for 30 d as described above. The survival rate (%) and new tissue (newly developed tissues or organs form the original explants; example see results. A newly elongated shoot of ZHB) germination rate (%) were thereafter measured. The survival rate was calculated as the number of samples with living tissue, or total number of samples * 100%. The regrowth of new tissue was measured by dividing the number of samples with newly generated tissue or organs by the total number of samples and multiplied by 100%. Micro-cuttings were then transplanted into MS medium under normal culture conditions for 10 d to test for possible microbial contamination of the unsterilized package (Figure 1G).

2.3. Assessment of Incubation Condition

The plant samples shipped in vacuum-sealed microplates were incubated in a growth chamber under conditions of controlled temperature with photoperiod and light intensity. The cultures were maintained under the following conditions: (1) 25 °C for a 16 h photoperiod with a light intensity of 50 μM s^{-1} m^{-2} (25 °C + 16 h light); (2) 25 °C in the dark (25 °C + dark); (3) 10 °C for a 16 h photoperiod with a light intensity of 50 μM s^{-1} m^{-2} (10 °C + 16 h light), (4) 10 °C in the dark (10 °C + dark); (5) 4 °C for a 16 h photoperiod with a light intensity of 50 μM/m^2/s (4 °C + 16 h light); and (6) 4 °C in the dark (4 °C + dark). Periods were 60 culture days and the survival rate (%) and new tissue germination rate (%) were assessed as described above (Figure 1H).

2.4. Assessment of Culture Medium

The growth of in vitro samples was slowed down to extend the duration of germplasm conservation and long-distance shipping. The effect of ion levels in the culture medium was assessed by decreasing the concentration of major and minor ions to 25%. Then, 1/4 MS medium without or with either daminozide (B9; 60, 80, 100, 120 or 140 mg/L), chlormequat chloride (CCC; 25, 50, 75 or 100 mg/L), paclobutrazol (PP333; 1, 2, 3 and 4 mg/L), or abscisic acid (ABA; 1, 2, 3 and 4 mg/L) was tested. The application of 38 treatments resulted in a combination of 2 plants with 19 media. ZHB and G-2 explants measuring approximately 0.4 cm in length from micropropagated plants were maintained in 300 mL glass jars with 40 mL of medium and stored for 60 d under standard culture conditions as noted above. The survival rate (%) and new tissue regrowth (%) were measured after incubation. Then, the morphology indices, such as plant height (cm) and the number of newly formed leaves, roots, and tillers (for ginger) were also calculated to evaluate the effects of ion density and plant growth regulation (Figure 1I). The most efficient media were then loaded in microplates and used in the following experiments.

2.5. Regrowth Capacity Assessment of Maintained Plants

The micro-cuttings of ZHB and G-2 derived from routinely propagated in vitro plants were maintained in 1/4 MS medium supplemented with 100 mg/L CCC (1/4MS 100 CCC medium) in sterilized microplates (Figure 2a–c). The plates were then vacuum-sealed with non-sterile packages (Figure 2d,e). The cultures were maintained at 10 °C in the dark (Figure 2f). Micro-cuttings were transplanted into MS medium after 60 days of incubation, and incubated under normal culture conditions. The regrowth capacity in terms of percentage of shoots resuming normal growth was evaluated 30 days after sample release (the 1st cycle of culture). The subculture and incubation of recovered ZHB and G-2 plants were repeated (the 2nd cycle of culture). The regrowth capacity and morphology indices were measured as well. In order to test the feasibility of the present method in other plant species, sweet potato HXS in vitro cuttings (0.4 mm long shoot with one 0.2 mm petiole) were harvested and transplanted in sterilized microplates supplemented with 1/4 MS 100 CCC medium. The plates were vacuumized and maintained at 10 °C in the dark for 60 d, followed by analysis of the survival and regeneration (Figure 1J,K).

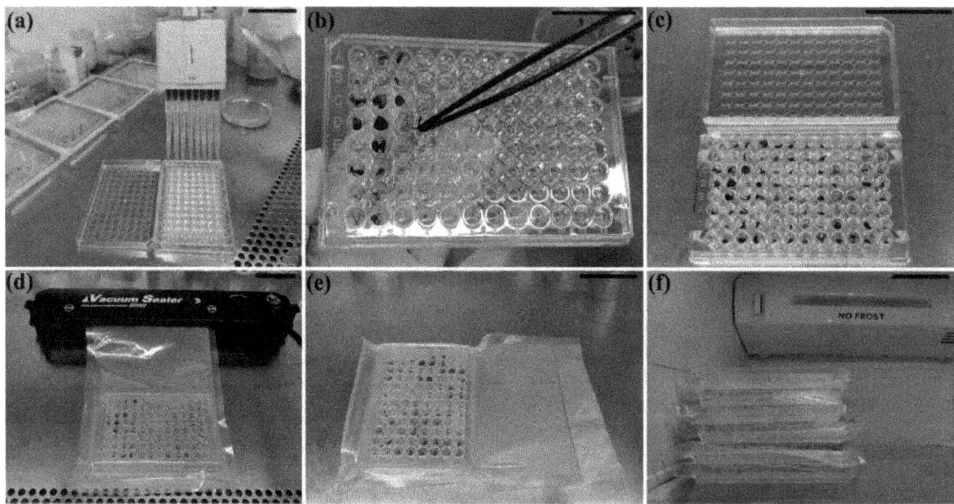

Figure 2. Manufacturing procedures of microplate shipping of vacuum-packed plant micro-cuttings. (**a**) Loading each of microplate well with 200 uL 1/4MS 100 CCC medium; (**b**) micro-cutting culture; (**c**) cutting cultured plate; (**d**) vacuumization; (**e**) vacuum-packed plate; (**f**) packaged plates stored at 10 °C in the dark (light on state in picture). Bar indicates 5 cm.

2.6. Assessment of Genetic Stability

The morphology indices, including plant height, number of leaves (tillers), length of roots, and number of roots, were measured to assess the generic stability on a morphological level. For the molecular marker test, total genomic DNA was extracted from 100 mg of leaves obtained from ZHB and G-2 plants, which was released from a microplate and subcultured twice using the GeneJET Plant Genomic DNA Purification Mini Kit (Thermo Scientific, Waltham, MA, USA). DNA quality was checked using electrophoresis of the samples on 1% agarose gel and stained with StarStain Red Plus (GenStar, Beijing, China). DNA concentration was determined via spectrophotometry. The molecular analysis was performed using 5 SSR and 5 ISSR primers (Table S1). The amplifications were carried out in 20 µL volumes, containing 2 ng genomic DNA, 1X PCR buffer (Biotools, Madrid, Spain), 200 µM of each dNTP (Roche, South San Francisco, CA, USA), 0.25 U of Taq DNA polymerase (Roche, Basilea, Suiza), and 0.2 µM of forward and reverse primers. The PCRs were carried out using a gene amplification instrument (FastAmp-T96, BIO-DL, Shanghai, China) with initial denaturation at 94 °C for 5 min, followed by 40 cycles of 94 °C for 30 s (94 °C for 50 s for ISSR), annealing temperature at 56 °C for 30 s (58 °C for 50 s for ISSR), 72 °C for 2 min (72 °C for 1 min for ISSR), and a final extension at 72 °C for 10 min. Automatic acquisition and reading was performed using ChampChemi610 (BeijingSaizhi, Beijing, China). All fragments in the size range of 100–2000 bp generated from ZHB were assumed to represent a single dominant locus. Fragments in the range of 100–3000 bp were considered and registered as a single codominant locus. All the reactions were performed in triplicate with 10 plantlets selected randomly (Figure 1L).

2.7. Transportation Tolerance Test

The vacuum-packaged plates carrying ZHB, G-2, and HXS samples were delivered by express mail through automobile transportation without any cushioning materials to test their transportation tolerance. The plates were delivered to four destinations from the starting point at Guizhou University, Guizhou City, Guiyang Province, China. The first location was an ornamental plant germplasm resources nursery on Huangzhuang South Road in Baiyun District, Guangzhou City, Guangdong Province, which is 925 km away from the starting point. The highest temperature at the arrival date was 20 °C and the temperature in the car ranged from around 10–22 °C during a 2-day trip (courier No. YT6303669102525). The next destination was 99-1 Yingchengzi Street, Hunnan District, Shenyang City, Liaoning Province, which is 2293.6 km away from the starting point. The highest temperature at the arrival date was 9 °C and the temperature in the car ranged from around 9–10 °C during a 3.5-day trip (courier No. 75853845991528). The third location was Beijing Agricultural College, Changping District, Beijing, which is 2214 km away from the starting point. The highest temperature at the arrival date was 7 °C and the temperature in the car ranged from around 7–10 °C during a 7-day trip (courier No. 9886429762337). The final destination by car was to Jingfengjiayuan residential quarters, Taoshan District, Qitaihe City, Heilongjiang Province, which is 3681 km away from the starting point. The highest temperature at the arrival date was 7 °C and the temperature in the car ranged from around 7–10 °C during an 8-day trip (courier No. 9886429709918). Furthermore, plates were carried in a coat pocket and suitcase and transported 1426 km by train to Hefei City, Anhui Province to test their ability to withstand shipping stress. The temperature in carriage and car were around 12–16 °C and it was taken 7.5 h by train. The integrity of packaged plates and micro-cuttings was immediately checked after arrival (Figure 1M).

2.8. Experiment Design and Statistical Analysis

For the detailed experiment design, see Figure 1. A complete randomized design was used. At least 10 samples were included in each of the experiments. All experiments were performed in triplicate and conducted at least twice. Data were subjected to one-way ANOVA and the least significant difference (LSD) was calculated at $p < 0.05$.

3. Results

3.1. The Plant Growth and Contamination Response to Package Types

Vacuum-sealed packaging significantly influenced the survival and new tissue germination of the micro-cuttings of both ZHB and G-2. ZHB plantlets in vacuumized microplates showed a significantly lower survival rate (85.8–87.7%) and regrowth rate (61.8–62.5%) compared to the non-vacuumized counterparts. Similarly, in ginger G-2, vacuumization resulted in low survival (20–30%) and no new tissue germination. When vacuumization was not conducted, all of the ZHB micro-cuttings and 72.4–83.3% G-2 cuttings survived after 30 days of culture under normal subculture conditions, and 94.0–100% of ZHB and 67.0–83.3% of G-2 germinated new tissue (Table 1). Increased germination of in vitro cuttings during long-distance storage and transportation depletes the growth resources of plants, leading to aging. We found that the germinated and elongated potato seedlings pushed the cover and separated the lid and plate (data not shown), which increased the potential risk of contamination, suggesting that vacuum packing was needed. Unsterilized packages did not result in the contamination of the samples or medium if the mother plants were in a sanitary condition (Table 1). Furthermore, high temperature and pressure damaged the plastic membrane. Thus, in the following study, vacuum packages with unsterilized bags were used.

3.2. Plant Growth Response to Incubation Conditions

The variation of the culture period and conservation conditions significantly affect the survival and regeneration of micro-cuttings in vacuum-packed plates. Compared to 30-day incubated samples (Table 1), normal culture conditions for 60 d resulted in a significant decrease in the vitality of ZHB (35.7% survival and 20.8% regeneration) and G-2 (0% survival and regrowth), which was even worse in the dark (Table 2). Regardless of illumination conditions, nearly 80% of micro-cuttings developed new tissue before they lost vitality at 25 °C. Such samples were still counted as non-survival and regrowth as those without living cells or new tissue formed on cuttings. In potato ZHB, the conservation resistance of cuttings under encapsulated conditions was significantly improved by low temperature treatments ranging from 4 to 10 °C, and the survival rate was elevated to 81.4% to 100%. Furthermore, the darkness enhanced the stability of the cuttings and significantly less new tissue regrowth was detected. A slight difference was observed with ginger G-2, as the survival declined when samples were incubated in the dark compared to light, but this difference was not significant, and similar tissue regrowth was observed under both treatments. However, no increased survival rate of G-2 occurred at 4 °C (Table 2). The morphology of surviving, regenerated, and dead micro-cuttings is shown in Figure 3. It is worth noting that, during conservation, very limited (less than 0.5%, data not shown) ZHB cuttings developed a callus (Figure 3(b1)), while the majority of regrowing samples developed an elongated stem (Figure 3b). In contrast to potato, the main shoot terminal tended to germinate tillers instead of elongated stems (Figure 3e). Considering that both ZHB and G-2 only survived at 10 °C, the 16 h photoperiod induced an increase in tissue germination in ZHB, which prevented germplasm storage and transportation. Darkness was similar to the real transportation environment. Therefore, the conservation conditions of 10 °C and darkness were selected for the following experiment.

Table 1. The effect of package types on plant survival, new tissue regrowth, and medium contamination.

Sample	Vacuum Package	Package Sterilization	Survival (%)	New Tissue Regrowth (%)	Contamination
Potato ZHB	Vacuumized	+	87.7 ± 8.4 b	61.8 ± 6.3 z	−
		−	85.8 ± 7.1 b	62.5 ± 8.8 z	−
	Non-vacuumized	+	100.0 ± 0.0 a	100.0 ± 0.0 x	−
		−	100.0 ± 0.0 a	94.0 ± 3.5 y	−
Ginger G-2	Vacuumized	+	30.0 ± 0.0 b	0.0 ± 0.0 z	−
		−	20.0 ± 0.0 c	0.0 ± 0.0 z	−
	Non-vacuumized	+	83.3 ± 3.3 a	83.3 ± 3.3 x	−
		−	72.4 ± 10.5 a	67. 0 ± 15.4 y	−

Microplates were placed at a consistent temperature of 24 ± 2 °C under 16 h photoperiod conditions. Survival and regrowth were recorded 30 days after culture, while contamination was measured 10 days after culture. "+" in "Package sterilization" indicates package was sterilized; "−" indicates package was not sterilized. "+" in "Contamination" indicates samples were contaminated; "−" indicates not contaminated. Data are presented by means ± SE, with different letters indicating significant differences analyzed by one-way ANOVA at $p < 0.05$.

Table 2. Survival and new tissue regrowth of cuttings encapsulated in microplates wrapped in vacuum package under various incubation conditions.

Conditions	Potato ZHB		Ginger G-2	
	Survival Rate (%)	New Tissue Regrowth (%)	Survival Rate (%)	New Tissue Regrowth (%)
25 °C + 16 h light	35.7 ± 17.1 c	20.8 ± 8.8 y	0.0 ± 0.0 d	0.0 ± 0.0 y
25 °C + dark	18.0 ± 10.5 c	0.0 ± 0.0 z	0.0 ± 0.0 d	0.0 ± 0.0 y
10 °C + 16 h light	100.0 ± 0.0 a	83.1 ± 8.4 x	100.0 ± 0.0 a	2.6 ± 0.2 x
10 °C + dark	100.0 ± 0.0 a	27.7 ± 11.6 y	83.3 ± 16.7 a	2.0 ± 0.2 x
4 °C + 16 h light	81.4 ± 8.2 b	14.3 ± 5.0 y	14.6 ± 3.3 c	0.0 ± 0.0 y
4 °C + dark	100.0 ± 0.0 a	0.0 ± 0.0 z	35.0 ± 8.3 b	0.0 ± 0.0 y

Survival and regrowth were measured 60 days post-culture. Data are presented as means ± SE. Letters indicate significant differences, a–d indicate significant differences among survival rate, while x, y, z indicate significant differences among new tissue regrowth rate. Significant differences were analyzed within ZHB or G-2 via one-way ANOVA at $p < 0.05$.

Figure 3. Survival and new tissue regeneration by micro-cuttings or loss of vitality. (**a**) Potato ZHB survived cutting without new tissue germination; (**b**) new tissue germinated ZHB cuttings, (**b1**) ZHB micro-cutting with callus and bud germination; (**c**) dead ZHB cuttings; (**d**–**f**) ginger G-2 survived cutting, new tissue germinated, and dead cutting, respectively. Images were acquired post 60 d of conservation. Bar indicates 2 mm.

3.3. Plant Growth Response to Slow-Growth Culture Media

Both MS and 1/4 MS media showed 100% survival and total regrowth after 60 days in storage on both potato ZHB and ginger G-2. For ZHB, B9, PP333, and ABA caused partial death of samples in general. CCC-supplemented plants showed the highest survival rate in potato ZHB. Except for MS medium treated with 75 mg/L CCC medium, 25, 50, and 100 mg/L CCC did not decrease the survival (Table 3). In addition, all the plantlets incubated in 25 mg/L and 100 mg/L CCC-supplemented MS medium reformed new tissues. Malformation including leaf bleaching, leathery leaves, tissue necrosis, or swelling of samples were seen on a majority of media except those treated with 75 and 100 mg/L CCC (data not shown). Generally, B9, PP333, and ABA caused malformation on G-2. For G-2, samples incubated in 1/4 MS medium supplemented with 25 and 100 mg/L CCC showed nearly 100% survival and no malformation of micro-cuttings. The addition of ABA led to a significant decrease in survival or regrowth of samples compared to the control (Table 3).

Morphology indices of in vitro plants under each treatment were measured. For ZHB, compared to MS medium, 1/4 MS medium significantly increased the roots but decreased the shoot growth; no obvious changes on other morphological indexes were observed. Following 1/4 MS-medium-based treatment, 60–100 mg of B9 did not significantly affect the regeneration of leaf and root, while 120–140 mg of B9 inhibited it; however, the elongation of the shoot was reduced regardless of dose. Treatment with 25 and 100 mg of CCC suppressed potato growth whereas 50 and 75 mg did not, while 100 mg of CCC generated the shortest plantlets among all CCC treatments. ABA addition resulted in the lowest average number of leaf, root, and stem height on potato ZHB in general, followed by PP333 supplementation (Figure 4). For ginger G-2, decreasing the iron content to 1/4 inhibited rooting and tillering. Regeneration and growth were enhanced by B9 and PP333 depending on the dose. Exposure to 50 and 70 mg of CCC improved the development of roots, leaves, and stems, while 25 and 100 mg doses significantly inhibited all indices, except for the number of tillers. The low regrowth rate under ABA treatment resulted in the shortest and the most undeveloped G-2 plantlets (Figure 5).

Table 3. Survival and new tissue regrowth of micro-cuttings cultured in different growth media.

Mediums	Potato ZHB			Ginger G-2		
	Survival Rate (%)	Regrowth (%)	Malformation	Survival Rate (%)	Regrowth (%)	Malformation
MS	100.0 ± 0.0 a	100.0 ± 0.0 a	−	100.0 ± 0.0 a	100.0 ± 0.0 a	−
1/4 MS	100.0 ± 0.0 a	100.0 ± 0.0 a	−	100.0 ± 0.0 a	100.0 ± 0.0 a	−
1/4 MS 60 B9	91.7 ± 8.3 ab	91.7 ± 8.3 ab	+	91.7 ± 8.3 a	91.7 ± 8.3 ab	+
1/4 MS 80 B9	91.7 ± 8.3 ab	91.7 ± 8.3 ab	+	100.0 ± 0.0 a	91.7 ± 8.3 ab	+
1/4 MS 100 B9	75.0 ± 14.4 ab	75.0 ± 14.4 abc	+	75.0 ± 14.4 ab	41.7 ± 30.5 abc	+
1/4 MS 120 B9	58.3 ± 8.3 cd	58.3 ± 8.3 abc	+	91.7 ± 8.3 a	91.7 ± 8.3 ab	+
1/4 MS 140 B9	83.3 ± 8.3 ab	83.3 ± 16.7 ab	+	83.3 ± 8.3 a	83.3 ± 8.3 ab	+
1/4 MS 25 CCC	100.0 ± 0.0 a	75.0 ± 25.0 abc	+	100.0 ± 0.0 a	91.7 ± 8.3 ab	−
1/4 MS 50 CCC	75.0 ± 0.0 ab	83.3 ± 16.7 ab	+	75.0 ± 14.4 ab	50.0 ± 0.0 abc	−
1/4 MS 75 CCC	100.0 ± 0.0 a	100.0 ± 0.0 a	−	50.0 ± 0.0 c	41.7 ± 8.3 abc	−
1/4 MS 100 CCC	100.0 ± 0.0 a	100.0 ± 0.0 a	−	100.0 ± 0.0 a	91.7 ± 8.3 ab	−
1/4 MS 1 PP333	66.7 ± 8.3 ab	62.5 ± 12.5 abc	+	91.7 ± 8.3 a	83.3 ± 8.3 ab	−
1/4 MS 2 PP333	66.7 ± 8.3 ab	75.0 ± 14.4 abc	+	91.7 ± 8.3 a	83.3 ± 8.3 ab	+
1/4 MS 3 PP333	83.3 ± 8.3 ab	83.3 ± 8.3 ab	+	83.3 ± 8.3 a	58.3 ± 16.7 abc	+
1/4 MS 4 PP333	58.3 ± 8.3 cd	58.3 ± 22.1 abc	+	91.7 ± 8.3 a	83.3 ± 8.3 ab	+
1/4 MS 1 ABA	41.7 ± 8.3 cd	41.7 ± 8.3 bc	+	33.3 ± 8.3 c	16.7 ± 8.3 c	+
1/4 MS 2 ABA	91.7 ± 8.3 ab	91.7 ± 8.3 ab	+	41.7 ± 8.3 c	25.0 ± 0.0 bc	+
1/4 MS 3 ABA	58.3 ± 8.3 cd	58.3 ± 30.1 abc	+	52.2 ± 4.2 c	50.0 ± 0.0 abc	+
1/4 MS 4 ABA	58.3 ± 8.3 cd	50.0 ± 14.4 abc	+	41.7 ± 8.3 c	25.0 ± 0.0 bc	+

1/4 MS 60 B9 refers to 1/4 MS medium supplemented with 60 mg/L B9, and so on. Data were surveyed 60 days post-culture. Data are presented as means ± SE. Letters indicate significant differences. Data were analyzed using one-way ANOVA at $p < 0.05$.

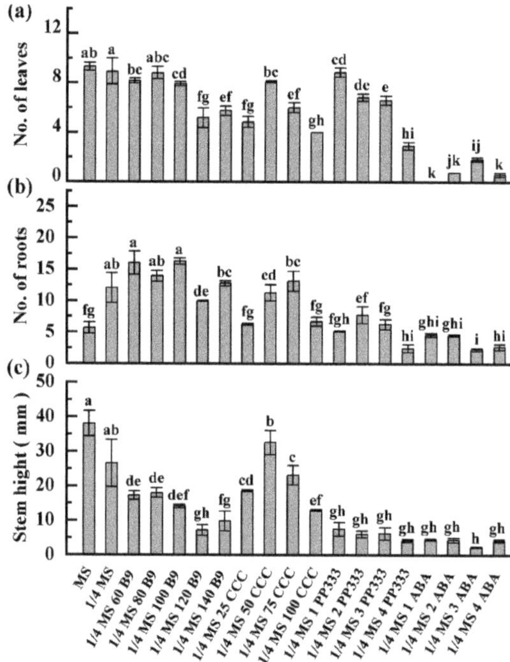

Figure 4. Growth status of ZHB plantlets incubated in slow-growth media. Data were surveyed 60 d post-culture. ¼ MS 60 B9 refers to ¼ MS medium supplemented with 60 mg/L B9, and so on. (**a**–**c**): Average plant height, number of new leaves, and number of new roots of potato ZHB, respectively. Data are presented as means ± SE. Significant differences were analyzed as one-way ANOVA at $p < 0.05$.

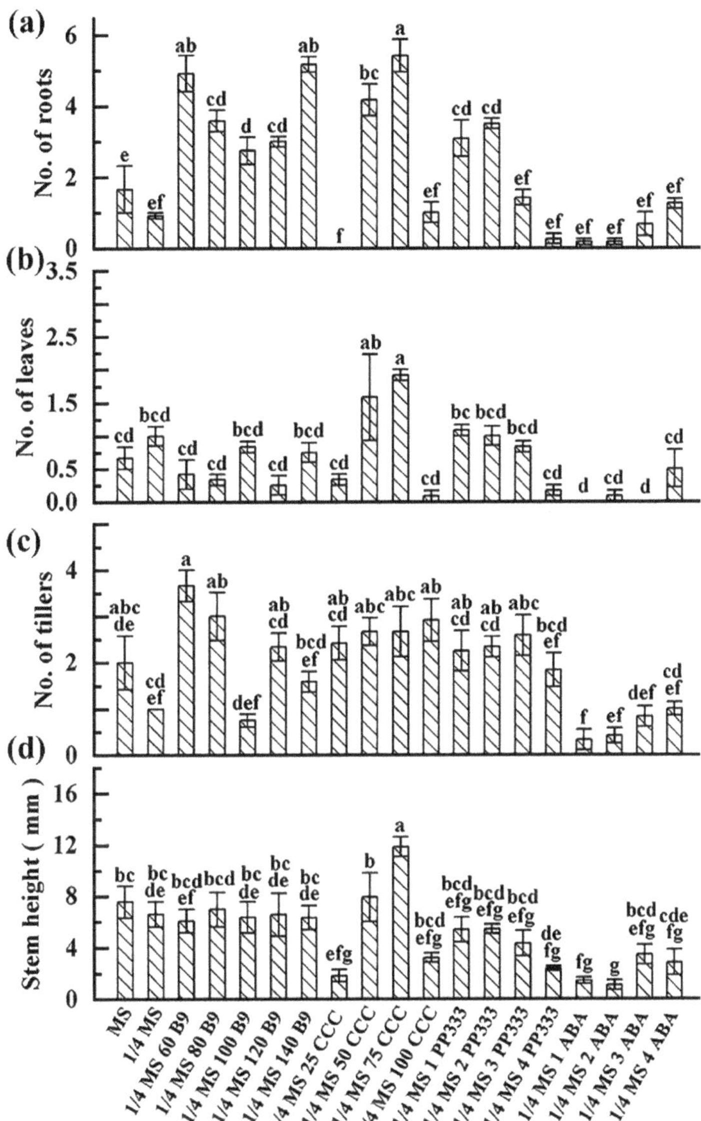

Figure 5. Growth statues of G-2 plantlets incubated in slow-growth media. Data were surveyed 60 d post-culture. ¼ MS 60 B9 refers to ¼ MS medium supplemented with 60 mg/L B9, and so on. (a–d): Average plant height, number of new leaves, number of new tillers, and number of new roots of ginger G-2, respectively. Data are presented as means ± SE. Significant differences were analyzed as one-way ANOVA at $p < 0.05$.

High survival, "true to type" morphology, and inhibition of vegetative growth ensure a stronger shipping resistance. Thus, 1/4 MS medium treated with 100 mg/L CCC led to 100% survival, no malformation, and restricted vegetative growth on both ZHB and G-2. Such plantlets were selected for the following experiments.

3.4. Regrowth of Packaged Micro-Cuttings

Taking ¼ MS 100 CCC as the culture medium, ZHB and G-2 micro-cuttings were encapsulated in microplates in vacuum-sealed packages. Sweet potato HXS were used to test the feasibility of this system on different plant germplasms. Rates of 100.0%, 75.3%, and 100% survival were identified for ZHB, G-2, and HXS, respectively, and their regeneration rates were 0% for the first two and 6.7% for HXS (Table 4). Although the survival was lower in G-2, the new tissue regrowth was significantly decreased in both ZHB and G-2 when compared to those incubated in MS medium in the same incubation environment (Table 2). The multiplication, which indicates the quantities of a shoot that a micro-cutting with a single bud would generate, was measured. After releasing them from the plate, an average of 4.4 nodes from ZHB cuttings were harvested after the first cycle of incubation (first 30 days), and significantly higher number of nodes were measured from the other 30-day cultured plantlets (6.9 nodes). Similarly, in the case of sweet potato HXS, an average of 2.4 and 3.6 nodes from the 1st and 2nd cycle of subcultured plantlets, respectively, were identified. Interestingly, for G-2, significantly more shoots were tillered from packed cuttings whereas 4.3 buds were found in the first cycle of culture; the multiplication was decreased to 2.5 on average after the 2nd cycle of incubation (Table 4).

Table 4. Plant survival, new tissue regrowth, and subculture of packaged micro-cuttings conserved at 10 °C in the dark with 1/4 MS 100 CCC.

Sample	Survival (%)	New Tissue Regrowth (%)	Multiplication	
			1st Cycle	2nd Cycle
Potato ZHB	100.0 ± 0.0	0.0 ± 0.0	4.4 ± 0.2 b	6.9 ± 0.1 a
Ginger G-2	75.3 ± 6.7	0.0 ± 0.0	4.3 ± 0.1 a	2.5 ± 0.1 b
Sweet potato HXS	100.0 ± 0.0	6.7 ± 0.7	2.4 ± 0.2 b	3.6 ± 0.4 a

Data were analyzed after 60 days of incubation and are presented as mean values ± SE. Letters indicate significant differences. Significant differences were analyzed by one-way ANOVA at $p < 0.05$.

Similar patterns of vegetative generation of preserved ZHB and HXS recovered from packed microplates were observed (Figure 6). Generally speaking, the stem height, number of leaves, number of roots, and length of roots were significantly short and few in the 1st cycle of cultured micro-cuttings; however, this inhibition was reversed by further subcuture. The vegetative growth of plantlets was restored to similar levels after another cycle of subculture compared to untreated plants, except for the stem height of sweet potato HXS, which was shorter than in the control after the 2nd cycle of subculture (Figure 6). The average stem height for ginger G-2, similar to ZHB and HXS, was significantly inhibited by the packaging and recovered following successive subculture. However, the growth and differentiation of leaves, roots, and tillers were significantly stimulated by the packaged conservation, and these changes were temporary as well; the growth of cuttings resumed to control levels following the 2nd cycle of subculture (Figure 6). No malformation or changes in morphology of plantlets were observed (Figure S1).

3.5. Genetic Stability

ISSR and SSR profiles from recovered ZHB and G-2 plantlets stored in different slow-growth media were compared to those obtained from their respective untreated and normally cultured mother plants. The primers that we used showed different abilities for detecting genetic variation. In both ZHB and G-2, out of 50 primers for each method tested, 5 primers produced strong, clear, and reproducible bands, which differed between the two species (Table 5; Figure 7). In ZHB, 11–14 bands were scorable; in 30 individual samples, SSR primers produced 1860 bands in total (62 bands * 30 samples), and no polymorphic bands were identified (data not shown). Using the ISSR marker, the primers yielded 1470 clean bands (49 bands * 30 samples). Three polymorphic bands were detected, which accounted for about 0.2% genetic variation. In G-2, using the SSR marker, the number of bands for each primer varied between 6 and 9, and a total of 1110 bands

(37 bands * 30 samples) were generated from all samples (data not shown). One specific band was detected, which accounted for less than 0.1% of genetic variation. Using SSR, five primers generated 47 bands in a single sample. However, we were unable to detect any signs of genetic variation (Table 5; Figure 7).

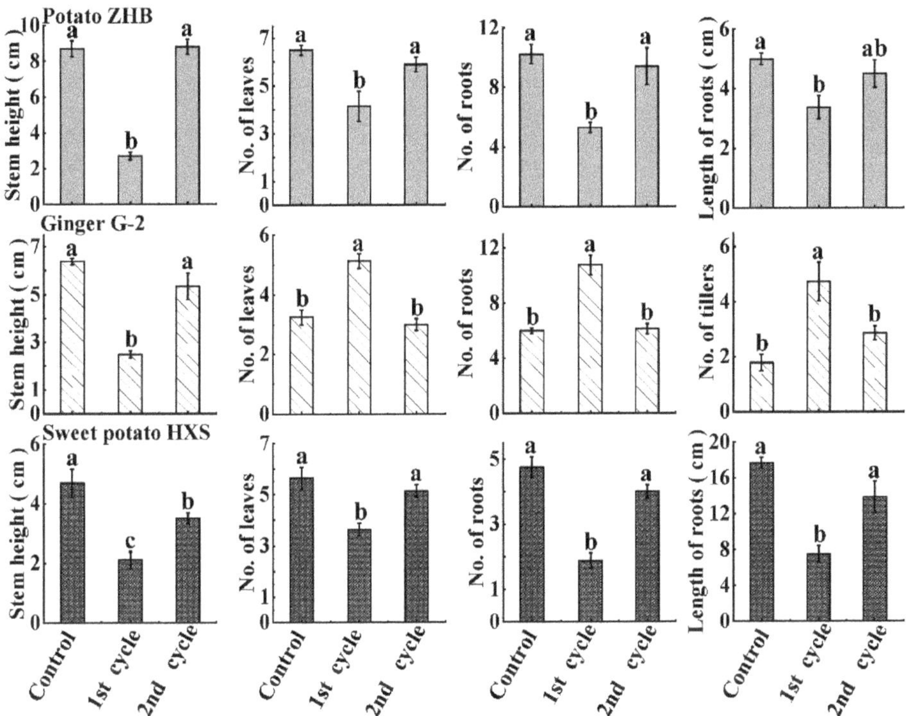

Figure 6. Growth of ZHB, G-2, and HXS after recovery in MS medium. Statistical analysis of growth indices. Data are presented as means ± SE. Letters indicate significant differences. Significant differences were analyzed via one-way ANOVA at $p < 0.05$.

Table 5. The numbers of amplified bands in plantlets regenerated after recovery in MS medium under normal subculture conditions by SSR and ISSR.

	Primer Name	No. of Bands	No. of Polymorphic Bands		Primer Name	No. of Bands	No. of Polymorphic Bands
				Potato ZHB			
	RSS2428	11	0		6	9	0
	RSS1457	12	0		8	12	0
SSR	RSS0881	14	0	ISSR	13	8	1
	RSS2112	13	0		807	9	2
	RSS75	12	0		868	11	0
	Total	62	0		Total	49	3
				Ginger G-2			
	RSS2898	9	0		6	8	0
	RSS2114	9	0		17	11	0
SSR	RSS0347	6	0	ISSR	19	9	0
	RSS2474	7	0		20	9	0
	RSS2428	6	1		807	10	0
	Total	37	1		Total	47	0

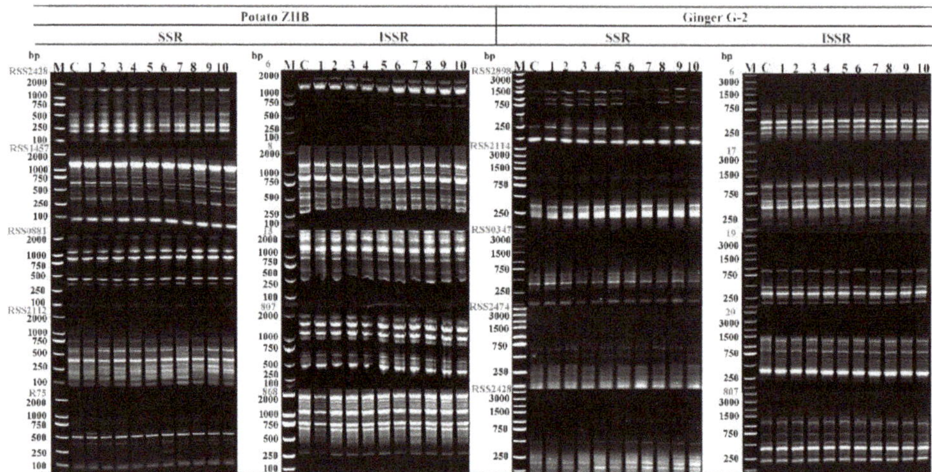

Figure 7. SSR and ISSR banding patterns in recovered potato ZHB and ginger G-2. Three repeats of 10 randomly selected samples were detected.

3.6. Transportation Feasibility

The transportation resistance of the present device was tested via express automobile transportation and railway transportation in pocket and suitcase, across thousands of kilometers. The packages remained complete and the micro-cuttings were tightly attached to the medium in each cup (Figure 8).

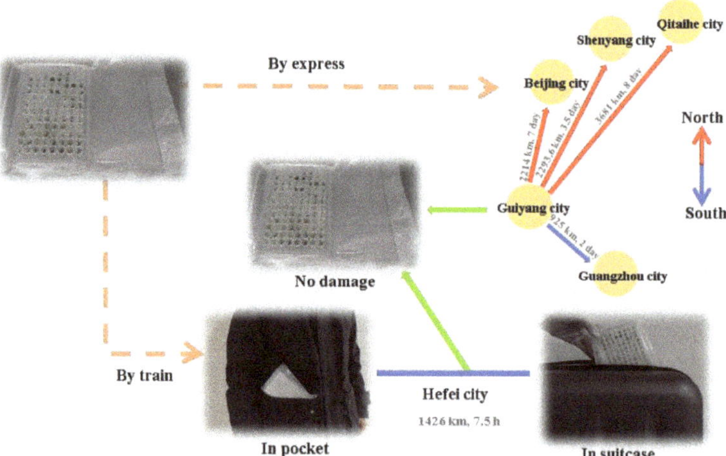

Figure 8. The transportation and shipping feasibility of micro-cuttings encapsulated in vacuum-packed microplates.

4. Discussion

In vitro plant germplasm resources are currently transported in tubes, jars, or plastic bags that are usually used for plant tissue culture. Different packages have drawbacks and technical constraints that limit their efficiency and decrease the security of the shipped biomaterials. Hence, it is essential to optimize the shipping systems in order to limit the factors that can lead to the loss of plant germplasm and bio-contamination. This can mainly be achieved by transportation-resistant packages and slow-growth systems.

The results demonstrate that it is possible to store potato ZHB and ginger G-2 germplasms in space-limited, vacuum-packed, slow-growth, dark conditions. The system was effective in terms of 60-day storage, which requires no additional effort and ensures the vitality of samples. Under normal culture conditions, the samples are not well-conserved in space-limited culture in microplates and vacuum-sealed packages. A mere 85.8% of micro-cuttings survived and the majority of them (62.5%) showed stem germination. In the case of G-2, merely 20% of the cuttings exhibited fresh and live tissues during their 30-day incubation period (Table 1). These cases suggest the disadvantages of long-distance shipping, which decreased the vitality of plant germplasm in vitro [14]. An optimized result was achieved with conserved and packed plates at 10 °C in the dark (Table 2). Lowering the culture temperature and shading from light are important ways to delay the growth of in vitro plants [26]. This strategy successfully resulted in the slow-growth conservation of globe artichoke in vitro [27,28]. It was emphasized that, compared to osmotic stress, storing in cold and dark conditions facilitated the medium-term conservation of globe artichoke [27]. According to De Lacerda et al. (2021) [17], rapidly growing plants *P. glomerata* and *L. filifolia* were maintained for up to 360 days with 100% or 50% survival in 5 mL of mineral oil at a temperature of 15 °C, while higher temperatures (20 °C and 25 °C) decreased plant survival. Light is essential for plant growth and development as well as improving cutting survival and regrowth of both ZHB and G-2 in the present study. Considering that increased germination rates were a disadvantage during conservation and shipping, darkness was selected to simulate the real transportation environment. Our findings suggest that the germplasm should be transported during late autumn or late spring but not in extremely cold winter, and the package should be shipped in an opaque express box under relatively ambient air temperature.

To further postpone the germination of sample cuttings during transportation and conservation, we tested the role of ionic concentration in the culture medium and growth retardants on plantlet growth. Subsequently, we used an efficient combination in the packaging system. Reducing the MS medium composition to 1/4 did not affect sample survival or regrowth (Table 3) but resulted in a relatively slower shoot growth (Figures 4 and 5). Our results were consistent with Catană (2010) [7], who reported that 1/4 or 1/10 MS medium for the *Caryophyllaceae* family facilitated the conservation with optimal parameters depending on the genotype. However, Arbeloaa et al. (2017) [29] reported totally different results involving 138 different fruit trees belonging to 18 species, with generally higher multiplication rates in 1/2 MS medium than in MS medium. Plant growth retardants, including B9, CCC, PP333, and ABA were usually used in plant slow-growth conservation in vitro [20,30–33]. Compared to B9, PP333, and ABA, 100 mg/L CCC in this study did not inhibit sample germination and regrowth; however, it ensured that all samples were alive and maintained normal morphology of plantlets (Table 3). In addition, compared to those incubated in 1/4 MS blank medium, the regrowth of plantlets was significantly inhibited when 100 mg/L CCC was combined (Figures 4 and 5). Furthermore, the application of 100 mg/L CCC supplemented with 1/4 MS medium in the vacuum-packed plates led to significantly lower new tissue regrowth when compared to MS medium, which indicates the advantages of this medium in delaying the germination of micro-cuttings, though it slightly decreased the survival of G-2 (Tables 2 and 4). Notably, compared to ZHB, G-2 was more sensitive to the incubation parameters, including temperature, illumination and plant growth regulators selected in this study. Obviously, the packing and conservation systems presented here are not optimal for G-2. Further studies are needed to develop a "special to genotype" system for different kinds of plant genotypes or a broader spectrum adaptive method.

Compared to unpacked micro-cutting controls, variations in morphology were observed after 60 days of storage in vacuum-sealed packages and without the addition of plant growth regulators. Two manners of regrowth of ZHB micro-cuttings were observed, including callus regeneration (though very few) or revival of normal stems. However, no callus was seen in G-2; instead of shoot elongation, which was observed on ZHB, tillering was the main manner of G-2 regrowth in wells of plates (Figure 3). Callus formation may increase the probability of somatic variation [34]. Evidence suggests that calluses carry high levels of genome-wide CHH methylation, particularly across heterochromatic regions [35]. The biotechnological interventions in vitro may cause somatic variation [35,36], suggesting the need for genetic stability assessment [19,37]. The genetic stability is usually estimated via morphological identification and molecular marker assessments [19,37–39]. In order to confirm the genetic stability of the present shipping system, we further compared the morphology and reproduction of recovered plantlets, which germinated in vacuum-packed microplates supplemented with 100 mg/L CCC. Molecular markers were analyzed as well.

Cuttings with callus germination or other malformed samples were not seen when the packing system was treated with 100 mg/L CCC after 60 days of conservation (data not shown). Although the analysis of the morphological data of the 1st cycle regenerated plants revealed significant morphological differences, the differences were restored to control levels after the second cycle of subculture (Figure 6). Interestingly, in contrast to ZHB and HXS plants, whose growth was significantly slower in the 1st cycle of subculture, G-2 tillers produced double shoots after release (Figure 6). More particularly, even without the supplementation of a plant growth regulator, G-2 generated tillers during 60-day packing and conservation instead of stem elongation (Figure 3). The number of tillers is used to evaluate the reproductive ability of ginger [40,41]. We found that space-limited incubation and vacuum packing rather than plant growth regulators may have stimulated tillering in the present study. Therefore, this method can be used commercially to accelerate the propagation of in vitro ginger, and the underlying mechanism requires further study.

Molecular markers are frequently used to assess the genetic stability of in-vitro-derived plantlets [19,42–46]. In the present study, SSR and ISSR were used to assess the genetic stability of artichoke plantlets regenerated from a slow-growth packing system used by Tavazza et al. (2015) [19]. Very limited polymorphism was detected in the present study; in potato, about 0.2% genetic variations were detected by ISSR, while for G-2, less than 0.1% genetic variations were identified by SSR but not by ISSR (Table 5). According to Liu and Yang (2012) [42] and Yin et al. (2013) [43], the somaclonal variation rate ranges from 0.37% to 4.2%, suggesting "true-to-type" plants. Our results show that both ZHB- and G-2-regenerated plants were genetically stable using the primers used, indicating that this system did not have a detectable impact on genetic stability in this experimental system.

Ultimately, shipping resistance of the present packing and conservation system was assessed using express automobile transportation or train transportation, which are relatively slow and bumpy for logistics. The destinations were spread across China's provinces and the temperatures varied from 7 to 22 °C at the time of delivery. The packed plant samples were shipped across thousands of kilometers and reached destinations in 2–8 days without package and sample damage. This indicates that the present methods had positive effects on sample integrity during express transportation under a wide range of temperatures. However, further assessment should be conducted by international sea freight lasting several months.

As far as we know there are very limited published reports describing a safe and highly efficient protocol for long-distance shipping and conservation of in vitro plants, validated by appropriate transportation tests. In conclusion, our results show that it is possible to store and transport in vitro cultures of potato, ginger, and sweet potato for a prolonged period of time by slowing down growth in a vacuum-sealed microplate. This protocol can aid transnational transportation or long-distance shipping of plant germplasm, which usually lasts months, without extreme high or low temperature variation or violent collision. The significantly reduced risk of germplasm loss and biosafety issues have been detected compared to conventional methods of shipping or custom inspection. Our results facilitate the distribution of germplasm to curators and the wider user community.

Supplementary Materials: The following supporting information can be downloaded at: https://www.mdpi.com/article/10.3390/horticulturae8070609/s1, Table S1: SSR and ISSR primers and primer sequences; Figure S1: Growth feature of ZHB, G-2, and HXS after recovery in MS medium.

Author Contributions: Project administration, J.L.; methodology, J.L., M.H. and T.H.; formal analysis, J.L. and X.X.; writing—original draft preparation, J.L.; writing—review and editing, W.Z. Investigation, H.T. All authors have read and agreed to the published version of the manuscript.

Funding: This research was funded by the "Research and application of rapid propagation technology of virus-free Guizhou yellow ginger" project, which received financial support from the Guizhou Modern Seed Industry Group Co., Ltd. and the "Talent introduction" [Guidarenjihezi (2019)73] program supported by Guizhou University.

Institutional Review Board Statement: Not applicable.

Informed Consent Statement: Not applicable.

Data Availability Statement: The data presented in this study are available within the article.

Conflicts of Interest: The authors have no conflict of interest to declare.

References

1. Priyanka, V.; Kumar, R.; Dhaliwal, I.; Kaushik, P. Germplasm conservation: Instrumental in agricultural biodiversity—A review. *Sustainability* **2021**, *13*, 6743. [CrossRef]
2. Chawla, A.; Kumar, A.; Warghat, A.; Singh, S.; Bhushan, S.; Sharma, R.K.; Bhattacharya, A.; Kumar, S. Approaches for conservation and improvement of Himalayan plant genetic resources. *Adv. Crop Improv. Tech.* **2020**, *18*, 297–317.
3. Staub, J.; Chatterton, J.; Bushman, S.; Johnson, D.; Jones, T.; Larson, S.; Robins, J.; Monaco, T. A history of plant improvement by the USDA-ARS forage and range research laboratory for rehabilitation of degraded western U.S. rangelands. *Rangelands* **2016**, *38*, 233–240. [CrossRef]
4. Schiva, T. Bamboo germplasm: Perspectives for ornamental purposes. *Acta Hortic.* **1999**, *486*, 255–260. [CrossRef]
5. Santos, I.R.I.; Salomão, A.N. In vitro germination of zygotic embryos excised from cryopreserved endocarps of queen palm (*Syagrus romanzoffiana* (Cham.) Glassman). *In Vitro Cell. Dev. Biol. Plant* **2017**, *53*, 418–424. [CrossRef]
6. Chalam, C.V. Elimination of plant viruses by certification and quarantine for ensuring biosecurity. *Appl. Plant Virol.* **2020**, *52*, 749–762.
7. Catană, R.; Mitoi, E.M.; Helepciuc, F.; Holobiuc, I. In vitro conservation under slow growth conditions of two rare plant species from *Caryophyllaceae* family. *Electron. J. Biol.* **2010**, *6*, 86–91.
8. Reed, B.M.; Sarasan, V.; Kane, M.; Bunn, E.; Valerie, C.P. Biodiversity conservation and conservation biotechnology tools. *In Vitro Cell. Dev. Biol. Plant* **2011**, *47*, 1–4. [CrossRef]
9. Waterworth, P.; Kahn, R.P. Thermotherapy and aseptic bud culture of sugarcane to facilitate the exchange of germplasm and passage through quarantine. *Plant Dis. Rep.* **1978**, *62*, 772–776.
10. Kumar, C.A.; Chandel, K.P.S. Strategy for the production of virus-free plant germplasm under in-vitro conservation and exchange—A brief review. *Indian J. Plant Genet. Resour. (IJPGR)* **1992**, *5*, 67–78.
11. Bekheet, S.A. In vitro culture techniques for conservation of date palm germplasmin in Arab countries. *Acta Hortic.* **2010**, *882*, 211–217. [CrossRef]
12. Negahdar, N.; Hashemabadi, D.; Kaviani, B. In vitro conservation and cryopreservation of *Buxus sempervirens* L., a critically endangered ornamental shrub. *Russ. J. Plant Physiol.* **2021**, *68*, 661–668. [CrossRef]
13. Filloux, D.; Girard, J.C. Indexing and elimination of viruses infecting yams (*Dioscorea* spp.) for the safe movement of germplasm. In Proceedings of the 14th Triennial Symposium of the ISTRC, Trivandrum, India, 20–26 November 2006; pp. 1–13.
14. Reed, B.M.; Paynter, C.; Bartlet, B. Shipping Procudures for Plant Tissue Culture. Available online: https://www.ars.usda.gov/research/publications/publication/?seqNo115=119618# (accessed on 7 January 2022).
15. El-Dawayati, M.M.; Zaid, Z.E.; Elsharabasy, S.F. Effect of conservation on steroids contents of callus explants of *Date palm* cv. sakkoti. *Aust. J. Basic Appl. Sci.* **2012**, *6*, 305–310.
16. Ramírez-Pérez, Y.; Tapia-Campos, E.; Cruz-Gutiérrez, E.J.; Barba-Gonzalez, R. Obtaining a protocol for slow growth for in vitro conservation of eustoma cultivars (*Gentianaceae*). *Acta Hortic.* **2020**, *1288*, 185–188. [CrossRef]
17. De Lacerda, L.F.; Gomes, H.T.; Bartos, P.M.C.; Vasconcelos, J.M.; Filho, S.C.V.; De Araújo Silva-Cardoso, I.M.; Scherwinski-Pereira, J.E. Growth, anatomy and histochemistry of fast growing species under in vitro conservation through mineral oil and low-temperature combination. *Plant Cell Tissue Organ Cult.* **2021**, *144*, 143–156. [CrossRef]
18. Oseni, O.M.A. Review on plant tissue culture, a technique for propagation and conservation of endangered plant species. *Int. J. Curr. Microbiol. App. Sci.* **2018**, *7*, 3778–3786. [CrossRef]
19. Tavazza, R.; Rey, N.A.; Papacchioli, V.; Pagnotta, M.A. A validated slow-growth in vitro conservation protocol for globe artichoke germplasm: A cost-effective tool to preserve from wild to elite genotypes. *Sci. Hortic.* **2015**, *197*, 135–143. [CrossRef]
20. Tahtamouni, R.; Shibli, R.; Al-Abdallat, A.; Al-Qudah, T. Turkish journal of agriculture and forestry analysis of growth, oil yield, and carvacrol in thymbra spicata l. after slow-growth conservation. *Turk. J. Agric. For.* **2016**, *40*, 213–221. [CrossRef]
21. Fernando, I.; Fei, J.; Stanley, R. Measurement and analysis of vibration and mechanical damage to bananas during longdistance interstatetransport by multitrailer road trains. *Postharvest Biol. Technol.* **2019**, *158*, 11097. [CrossRef]

22. Nikulina, E.O.; Ivanova, G.V.; Kolman, O.I.; Ivanova, A.N.; Perestoronin, D.Y. Research of the influence of vacuum packaging on the quality and safety of meat semi-finished products. In *IOP Conference Series: Earth and Environmental Science*; IOP Publishing: Bristol, UK, 2021; Volume 667, p. 032066.

23. Jiang, Q.J.; Jin, W.W.; Zhang, W.Y.; Zhang, Z.C.; You, L.F.; Bi, Y.Q.; Yuan, L.M. Analysis of vibration acceleration levels and quality deterioration of Chinese bayberry fruit in semi–vacuum package by express delivery. *J. Food Process. Eng.* **2021**, *44*, e13899. [CrossRef]

24. Anca, B.; Adriana, P.V.; Rodica, Z.; Luiza, M.; Mihaela, P. The results in the field of the in vitro conservation of the cultivars when using classic and modern conservation methods. *An. Univ. Oradea-Fasc. Biol.* **2007**, *14*, 23–26.

25. Murashige, T.; Skoog, F. A revised medium for rapid growth and bio assays with tobacco tissue cultures. *Physiol. Plant.* **1962**, *15*, 473–497. [CrossRef]

26. Monticelli, S.; Gentile, A.; Forni, C.; Caboni, E. Slow growth in â Pisanaâ apricot and in â Cerinaâ. apple, two Italian cultivars. *Acta Hortic.* **2020**, *1285*, 117–124. [CrossRef]

27. Bekheet, S.; Usama, I.A. In vitro conservation of globe artichoke (*Cynara scolymus* L.) Germplasm. *Int. J. Agric. Biol.* **2007**, *9*, 404–407.

28. Benelli, C.; Carlo, A.D.; Previati, A.; Roncasaglia, R. Recenti acquisizioni sulla conservazione in vitro in crescita rallentata. In Proceedings of the IX Giornate Scientifiche SOI Italus Hortus Conference, Florence, Italy, 10–12 March 2010; p. 91.

29. Arbeloaa, A.; Marí n, J.A.; Andreu, P.; Garcí a, E.; Lorente, P. In vitro conservation of fruit trees by slow growth storage. *Acta Hortic.* **2017**, *1155*, 13. [CrossRef]

30. Žiauka, J.; Kuusienė, S. Different inhibitors of the gibberellin biosynthesis pathway elicit varied responses during in vitro culture of aspen (*Populus tremula* L.). *Plant Cell Tissue Organ Cult.* **2010**, *102*, 221–228.

31. Ciobanu, I.B.; Constantinovici, D.; Crețu, L. Influence of genotype and cultivation conditions on vitroplantlets evolution of solanum *tuberosum* L. local varieties. *An. Ştiinţifice Ale Univ. Al. I. Cuza Iaşi.* **2011**, *2*, 13–20.

32. Indrayanti, R.; Putri, R.E.; Sedayu, A.; Adisyahputra. Effect of paclobutrazol for in vitro medium-term storage of banana variant *cv.* Kepok (*Musa acuminata* × *Balbisiana* Colla). In Proceedings of the 9th International Conference on Global Resource Conservation (ICGRC 2018), Malang, Indonesia, 7–8 March 2018; pp. 020001–020009.

33. Buldakov, S.A. Use of growth inhibitor chlormequat chloride in potato culture in vitro. In *E3S Web of Conferences*; EDP Sciences: Les Ulis, France, 2021; Volume 285, p. 03003.

34. Kumar, S.; Kumari, R.; Baheti, T.; Thakur, M.; Ghani, M. Plant regeneration from axillary bud, callus and somatic embryo in carnation (*Dianthus caryophyllus*) and assessment of genetic fidelity using RAPD–PCR analysis. *Indian J. Agric. Sci.* **2016**, *86*, 1482–1488.

35. Lizamore, D.; Bicknell, R.; Winefield, C. Elevated transcription of transposable elements is accompanied by het-siRNA-driven de novo DNA methylation in grapevine embryogenic callus. *BMC Genom.* **2021**, *22*, 676. [CrossRef]

36. Kaeppler, S.M.; Kaeppler, H.F.; Rhee, Y. Epigenetic aspects of somaclonal variation in plants. *Plant Mol. Biol.* **2000**, *43*, 179–188. [CrossRef]

37. Li, J.W.; Chen, H.Y.; Li, X.Y.; Zhang, Z.B.; Blystad, D.R.; Wang, Q.C. Cryopreservation and evaluations of vegetative growth, microtuber production and genetic stability in regenerants of purple-fleshed potato. *Plant Cell Tissue Organ Cult.* **2016**, *128*, 641–653. [CrossRef]

38. Mendes, M.; Verde, D.; Ramos, A.; Gesteira, A.; Souza, A. In vitro conservation of citrus rootstocks using paclobutrazol and analysis of plant viability and genetic stability. *Sci. Hortic.* **2021**, *286*, 110231. [CrossRef]

39. Parab, A.R.; Lynn, C.B.; Subramaniam, S. Assessment of genetic stability on in vitro and ex vitro plants of *Ficus carica* var. black jack using issr and damd markers. *Mol. Biol. Rep.* **2021**, *48*, 7223–7231. [CrossRef] [PubMed]

40. Kandiannan, K.; Prasath, D.; Sasikumar, B. Biennial harvest reduces rhizome multiplication rate and provides no yield advantage in ginger (*zingiber officinale* roscoe.). *J. Spices Aromat. Crops* **2016**, *25*, 79–83.

41. Marsh, L.; Hashem, F.; Smith, B. Organic ginger (*Zingiber officinale* Rosc.) development in a short temperate growing season: Effect of seedling transplant type and mycorrhiza application. *Am. J. Plant Sci.* **2021**, *12*, 14. [CrossRef]

42. Liu, X.M.; Yang, G.C. Adventitious shoot regeneration of oriental lily (*Lilium orientalis*) and genetic stability evaluation based on ISSR marker variation. *In Vitro Cell. Dev. Biol. Plant* **2012**, *48*, 172–179. [CrossRef]

43. Yin, Z.F.; Zhao, B.; Bi, W.L.; Chen, L.; Wang, Q.C. Direct shoot regeneration from basal leaf segments of *Lilium* and assessment of genetic stability in regenerants by ISSR and AFLP markers. *In Vitro Cell. Dev. Biol. Plant* **2013**, *49*, 333–342. [CrossRef]

44. Yin, Z.F.; Bi, W.L.; Chen, L.; Zhao, B.; Volk, G.M.; Wang, Q.C. An efficient, widely applicable cryopreservation of *Lilium* shoot tips by droplet vitrification. *Acta Physiol. Plant* **2014**, *36*, 1683–1692. [CrossRef]

45. Moniruzzaman, M.; Yaakob, Z.; Anuar, N. Factors affecting in vitro regeneration of *Ficus carica* L. and genetic fidelity studies using molecular marker. *J. Plant Biochem. Biotechnol.* **2020**, *30*, 304–316. [CrossRef]

46. Gautam, N.; Bhattacharya, A. Molecular marker based assessment of genetic homogeneity within the in vitro regenerated plants of *Crocus sativus* L.–a globally important high value spice crop. *S. Afr. J. Bot.* **2021**, *140*, 461–467. [CrossRef]

Article

In Vitro and *In Vivo* Performance of Plum (*Prunus domestica* L.) Pollen from the Anthers Stored at Distinct Temperatures for Different Periods

Milena Đorđević [1], Tatjana Vujović [1], Radosav Cerović [2], Ivana Glišić [1], Nebojša Milošević [1], Slađana Marić [1], Sanja Radičević [1], Milica Fotirić Akšić [3] and Mekjell Meland [4,*]

[1] Fruit Research Institute, Čačak, Kralja Petra I/9, 32000 Čačak, Serbia; mdjordjevic@institut-cacak.org (M.Đ.); tvujovic@institut-cacak.org (T.V.); iglisic@institut-cacak.org (I.G.); nmilosevic@institut-cacak.org (N.M.); smaric@institut-cacak.org (S.M.); sradicevic@institut-cacak.org (S.R.)
[2] Innovation Centre at Faculty of Technology and Metallurgy, University of Belgrade, Karnegijeva 4, 11120 Belgrade, Serbia; radosav.cerovic@gmail.com
[3] Faculty of Agriculture, University of Belgrade, Nemanjina 6, 11080 Belgrade, Serbia; fotiric@agrif.bg.ac.rs
[4] NIBIO Ullensvang, Norwegian Institute of Bioeconomy Research, Ullensvangvegen 1005, N-5781 Lofthus, Norway
* Correspondence: mekjell.meland@nibio.no

Citation: Đorđević, M.; Vujović, T.; Cerović, R.; Glišić, I.; Milošević, N.; Marić, S.; Radičević, S.; Fotirić Akšić, M.; Meland, M. *In Vitro* and *In Vivo* Performance of Plum (*Prunus domestica* L.) Pollen from the Anthers Stored at Distinct Temperatures for Different Periods. *Horticulturae* **2022**, *8*, 616. https://doi.org/10.3390/horticulturae8070616

Academic Editors: Jean Carlos Bettoni, Min-Rui Wang and Qiao-Chun Wang

Received: 13 June 2022
Accepted: 5 July 2022
Published: 7 July 2022

Publisher's Note: MDPI stays neutral with regard to jurisdictional claims in published maps and institutional affiliations.

Abstract: A study was conducted to investigate the effect of different storage periods and temperatures on pollen viability *in vitro* and *in vivo* in plum genotypes 'Valerija', 'Čačanska Lepotica' and 'Valjevka'. *In vitro* pollen viability was tested at day 0 (fresh dry pollen) and after 3, 6, 9 and 12 months of storage at four different temperatures (4, −20, −80 and −196 °C), and *in vivo* after 12 months of storage at distinct temperatures. *In vitro* germination and fluorescein diacetate (FDA) staining methods were used to test pollen viability, while aniline blue staining was used for observing *in vivo* pollen tube growth. Fresh pollen germination and viability ranged from 42.35 to 63.79% ('Valjevka' and 'Čačanska Lepotica', respectively) and 54.58 to 62.15%, ('Valjevka' and 'Valerija', respectively). With storage at 4 °C, pollen viability and germination decreased over the period, with the lowest value after 12 months of storage. Pollen germination and viability for the other storage temperatures (−20, −80 and −196 °C) were higher than 30% by the end of the 12 months. Pollination using pollen stored at 4 °C showed that pollen tube growth mostly ended in the lower part of the style. With the other storage temperatures, pollen tube growth was similar, ranging between 50 and 100% of the pistils with pollen tubes penetrated into the nucellus of the ovule in the genotype 'Čačanska Lepotica'. The results of these findings will have implications for plum pollen breeding and conservation.

Keywords: European plum; pollen germination *in vitro*; pollen viability; pollen storage; low and sub-zero temperatures

1. Introduction

Plums include a large group of closely related *Prunus* species of the *Rosaceae* family. Among them, the European plum (*Prunus domestica* L.) and the Japanese plum (*Prunus salicina* Lindl.) are the most extensively cultivated plum species worldwide [1]. According to FAO production data, in 2020, among temperate fruit species, the plum is in fourth place globally, after apple, pear and peach [2].

Today in Europe, most of the modern plum cultivars belong to *P. domestica*. A previous study [3] reported that the European plum, which includes many old English and Eastern European cultivars, showed the highest level of genetic diversity and on the basis of the fruit characteristics can be divided into several groups: plums, prunes, greengages or Reine Claudes and mirabelles [4]. Plums are mostly used for fresh consumption, but also for drying and processing in different forms [5]. Therefore, in Europe, different plum breeding programmes have been developed focused on defined breeding goals based on the market requirement. One of the leading plum breeding programmes is that of the Fruit Research Institute, Čačak, Serbia, which is principally aimed at solving the most significant problems in production [6]. Thanks to plum production in Serbia, during the XX century, the former Yugoslavia was one of the largest plum producers in the world [7].

In addition to the common objectives aimed for in plum breeding such as excellent internal and external quality, self-fertility, regular productivity and resistance to *Plum pox virus* (PPV) [8], in recent times, there is an increasing need in plum breeding to obtain genotypes well adapted to different climates and biotic stresses [9]. Conventional methods are still largely used in breeding programs and among them the most important method is hybridization [4–6,8]. Although the results of cross-pollination mostly depend on the female parents, great attention should be paid to the pollen quality. Commonly, pollen quality is assessed on the basis of pollen viability and vigour [10].

Therefore, when there is a need for a readily available supply of pollen in experimental studies, e.g., if there is asynchrony in flowering between parental cultivars or in the case of spatial and temporal isolation between parents, a long period of pollen storage may effectively be utilized [11]. Furthermore, pollen storage is also important in the conservation of genetic material [12]. In an extended period of storage, it is important to preserve the pollen without a significant loss of its viability. Pollen viability includes different features of pollen performance such as germinability, stainability and fertilization ability [13]. A considerable number of methods for testing pollen viability have been introduced so far— germination *in vivo, in vitro* and the histochemical approach. Even though *in vivo* methods give the best prediction of pollen behaviour, the seed set may not depend on fertilization alone [14]. Germinability tests *in vitro* have the potential to provide the best basis for fast predicting pollen performance [15]. In the histochemical approach, one of the viability testing methods is based on the fluorochromatic reaction, which tests for the presence of enzyme activity and the membrane integrity of the vegetative cell [16].

Factors affecting pollen viability during storage were the physiological stage of the flower, pollen age and moisture content [17]. Results indicated that when dispersing from the anther, developmentally immature pollen—binucleate (in *Rosaceae* and most other angiosperms) had reduced metabolic activity, low moisture content and can better tolerate desiccation compared to trinucleate pollen [18]. On the contrary, trinucleate pollen grains (in papaya, dianthus, beet, spinach, quinoa, cabbage, radish, flax, carrot, celery, wheat, corn, oat, barley, sorghum, range grasses) are very sensitive to dehydration due to less pronounced exine and decreased level of reserves after the second mitotic division [19].

So far, the study of the viability of stored pollen at different temperatures has been discussed by many authors for various fruit species such as almond [20], pear [21], mango [22], kiwifruit [23], sweet cherry [24] and apple [17]. There is no previous report on the effects of storage periods up to one year at low and sub-zero temperatures on plum pollen viability. The objective of this study was to assess the effect of different storage periods and temperatures on pollen viability *in vitro* and *in vivo* in three plum genotypes. Determination of the most suitable storage period and temperature for pollen preservation can contribute not only to improving breeding efficiency but also to biodiversity conservation.

2. Materials and Methods

2.1. Field Site and Plum Genotypes Used

The study was conducted in the plum orchard of the Fruit Research Institute, Čačak located in Serbia at 43°55′24″ N and 20°26′51″ E at an elevation of 497 metres. Three plum genotypes, developed at the Fruit Research Institute, used for investigation were chosen on the basis of the flowering time—'Valerija' (early), 'Čačanska Lepotica' (mid-early) and 'Valjevka' (mid-late) [25]. Branches of the chosen genotypes with flowers in the late balloon stage were taken on 28 and 29 May 2017 and collected pollen was used for storage experiments as well as for pollination of the selected female parent. The average air temperature and humidity on May 28 were 9.26 °C and 47.02%, while on the other day, they were 12. 38 °C and 46.99%, respectively (data were provided by the local Experimental Meteorological Station located near the experimental orchard). Trees of the selected genotypes were healthy and free from diseases and pests. A study on pollen behaviour *in vivo* (in the pistils) was performed after one year of storage at different temperatures. The genotype 'Čačanska Lepotica' was used as the female parent.

2.2. Anthers' Collection and Storage

In the laboratory, anthers from the flowers in the late balloon phase (BBCH61) of the chosen genotypes were collected into paper dishes [26]. Anthers were left to desiccate at room temperature until their dehiscence and release of pollen grains (up to 48 h). In order to determine moisture content (MC), anthers with pollen grains were oven-dried to a constant weight (Memmert GmbH + Co.KG, Büchenbach, Germany) for 48 h at 65 °C and 1 h at 105 °C. Moisture content was determined from the constant dry weight of anthers and weight measured immediately after dehiscence. Data presented in Table 1 are the mean values of three independent moisture measurements in each genotype. After desiccation, the anthers were placed in marked vials and stored in darkness under the following storage conditions: at 4 °C in a refrigerator; at −20 °C (Artiko PR700, Esbjerg, Denmark); at −80 °C (Artiko ULUF GG, Esbjerg, Denmark) in a freezer; and at −196 °C (Cryo Diffusion B2020, Lery, France) by the direct plunging of closed cryotubes with desiccated pollen in liquid nitrogen (LN). The viability of the pollen was estimated every 3 months during the storage period, namely at day 0 (fresh dry pollen) and after 3, 6, 9 and 12 months (until the full blooming time of the female parent). Each of the storages included four tubes with pollen per each genotype (one tube for each storage period). After defrosting, pollen stored at −20 °C was incubated for a short period at room temperature, while after storage at −80 °C and in LN, fast thawing was done in a water bath at 38 °C for 2 min and 5 min, respectively.

2.3. In Vitro Pollen Germination

A nutrition medium for pollen grains containing 12% sucrose and 1% agar was poured into Petri dishes (10 mL per dish). Fresh pollen taken immediately after desiccation as well as pollen stored at different temperatures were dusted onto the nutrition medium with a fine brush. The incubation period was 24 h at 23 °C. *In vitro* germination was observed under a light microscope (Olympus BX61, Tokyo, Japan) with a 20× ocular by counting germinated pollen grains in three different ocular fields, with an average of about 80 pollen grains in each. A Pollen grain was considered as germinated if the length of the pollen tubes exceeded their diameter [27]. Germinability of pollen was tested for each genotype, storage period and temperature.

2.4. Pollen Viability Test

To determine pollen viability the FDA test was used [15]. Fresh staining solution was poured into vials, after which, pollen was stirred into the solution USING a stainless steel needle. Vials were shaken and left in the dark for 20 min. Then, 1–2 drops of the solution were put into microscope slides, covered with a cover slip and immediately observed under UV light with an Olympus BX61 fluorescence microscope. Pollen grains which showed bright green fluorescence were considered viable, while those with light exine fluorescence and slight adsorption were considered non-viable grains. For each genotype, storage period and temperature, a similar number of pollen grains were observed (on average 50 in three different ocular fields).

2.5. Stigmatic Germinability and Fertilization Ability of Stored Pollen

For testing pollen germinability and fertilization ability after one year of storage at different temperatures, the genotype 'Čačanska Lepotica' was used as the mother parent. At the late balloon phase, flowers of this genotype were emasculated and isolated with paper bags. At the anthesis (BBCH65), pollination was carried out using pollen defrosted as previously described (Section 2.2) and performed by finger (two touches of stigma). Branches were then re-bagged and marked. Sampling of 30 pistils in FPA solution was carried out on the 3rd, 6th and 10th day after anthesis (DAA) for each tested storagetemperature. FPA solution was prepared using 70% ethyl alcohol, propionic acid and formaldehyde, 90:5:5 by volume. Samples were kept at 4 °C until further observation.

Pistils were washed in water for a short time and softened in 8N NaOH for 24 h. Test tubes were occasionally shaken to gently mix the pistils. After softening, pistils were stained with 0.1% aniline blue to view the pollen tubes according to the protocol described by Preil [28] and Kho and Baër [29]. On the microscope slides, the ovaries were separated from the pistils and opened along the suture. The pistils were divided into two parts with needles, covered with cover slips and gently squashed. One or two drops of glycerol were put on two sides of the cover slips to avoid the drying of the sample. The samples were observed under UV light with an Olympus BX61 fluorescence microscope.

For each genotype, the dynamics of pollen tube growth through the different parts of the style and ovary were presented as a percentage of the longest pollen tube penetrating a certain part of the pistils.

2.6. Statistical Analysis

Arcsine root square transformation was used for data in percentage and ANOVA analysis was performed within Statistica 8.0 (StatSoft, Inc., Tulsa, OK, USA). Results were analysed by two-way analysis of variance and the means were separated using the LSD test at $p \leq 0.05$. Data were shown as means $\pm$ SD.

3. Results

3.1. Anthers Moisture Content

The results indicate that in all tested genotypes, the percentage of moisture in anthers after drying was above 6%, with the lowest value in 'Valerija' and the highest in 'Čačanska Lepotica' (Table 1).

Table 1. Anthers moisture content measured immediately after dehiscence.

Genotype	Moisture Content (%)
'Čačanska Lepotica'	6.78 ± 0.31
'Valerija'	6.15 ± 0.18
'Valjevka'	6.72 ± 0.65

Values in the table show mean $\pm$ SD.

3.2. Pollen Germinability and Viability

Germination and viability of fresh pollen for all tested plum genotypes was over 40% (Tables 2 and 3). Pollen germination was rated from 42.35% in 'Valjevka' to 63.79% in 'Čačanska Lepotica'. The viability percentage ranged from 54.58% to 62.15% ('Valjevka' and 'Valerija', respectively). For fresh pollen, the difference in genotype germinability and viability was found to be significant. Only with pollen of 'Čačanska Lepotica', was slightly lower viability observed compared to germinability, while in the other two tested genotypes, the opposite situation was found.

Table 2. *In vitro* pollen germination stored at different temperatures for different periods.

Genotype	Temperature	Fresh Pollen/Day 0 of the Storage	3 Months of Storage	6 Months of Storage	9 Months of Storage	12 Months of Storage	ANOVA
'Čačanska Lepotica'	4 °C		57.73 ± 1.06 a/BCD	55.38 ± 0.70 a/DEF	48.56 ± 0.64 b/F	3.17 ± 1.97 d/G	genotype *, temperature *, genotype × temperature *
	−20 °C	63.79 ± 0.87 a/A	59.60 ± 0.98 a/ABCD	57.61 ± 1.19 a/CD	56.81 ± 0.45 a/CDE	52.41 ± 0.91 a/EFG	
	−80 °C		59.83 ± 0.71 a/ABC	56.61 ± 0.98 a/CDE	55.73 ± 0.68 a/CDEF	55.27 ± 0.54 a/DEF	
	−196 °C		62.04 ± 0.93 a/AB	58.94 ± 0.57 a/BCD	57.68 ± 1.38 a/CD	51.77 ± 0.89 a/FG	
'Valerija'	4 °C		33.93 ± 0.38 b/E	25.05 ± 0.78 d/F	18.40 ± 0.19 f/G	0.00 ± 0.00 f/H	genotype *, temperature *, genotype × temperature *
	−20 °C	44.73 ± 0.94 b/A	41.32 ± 1.08 b/BCD	38.33 ± 0.19 bc/CD	37.71 ± 0.03 cd/CDE	37.41 ± 1.43 bc/DE	
	−80 °C		40.30 ± 1.08 b/BCD	38.24 ± 0.68 bc/CD	37.74 ± 0.07 cd/CDE	36.77 ± 0.27 bc/DE	
	−196 °C		43.87 ± 1.14 b/AB	41.03 ± 2.57 b/ABC	40.87 ± 0.56 c/ABCD	40.77 ± 0.36 b/ABCD	
'Valjevka'	4 °C		30.65 ± 1.18 c/C	21.43 ± 1.08 e/D	13.66 ± 0.36 g/D	1.68 ± 1.28 e/E	genotype *, temperature *, genotype × temperature *
	−20 °C	42.35 ± 0.92 c/A	40.30 ± 1.03 c/BC	34.63 ± 1.55 c/BC	33.49 ± 0.54 e/BC	35.27 ± 0.83 c/BC	
	−80 °C		37.80 ± 0.76 c/AB	36.80 ± 1.57 bc/B	34.57 ± 1.37 de/BC	35.58 ± 0.52 c/B	
	−196 °C		41.92 ± 0.94 c/A	41.75 ± 0.76 b/A	36.45 ± 1.02 cd/AB	33.81 ± 1.21 c/BC	
		genotype *, temperature ns, genotype × temperature ns	genotype *, temperature *, genotype × temperature ns	genotype *, temperature *, genotype × temperature *	genotype *, temperature *, genotype × temperature *	genotype *, temperature *, genotype × temperature *	ANOVA

Values in the table show mean ± SD, ns—non-significant; * significant at $p \leq 0.05$. Means followed the same lowercase in columns (comparison between all genotypes and all temperatures within one storage period) and capital in the rows (comparison between all temperatures and all storage periods within one genotype) are not significantly different.

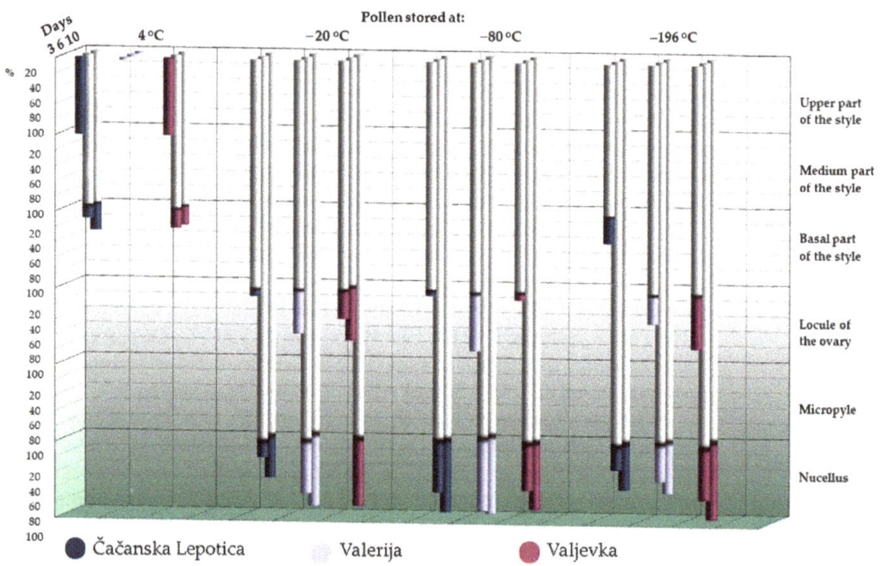

Figure 2. Dynamics of pollen tube growth in the pistils of the 'Čačanska Lepotica' after 12 months of pollen storage at different temperatures.

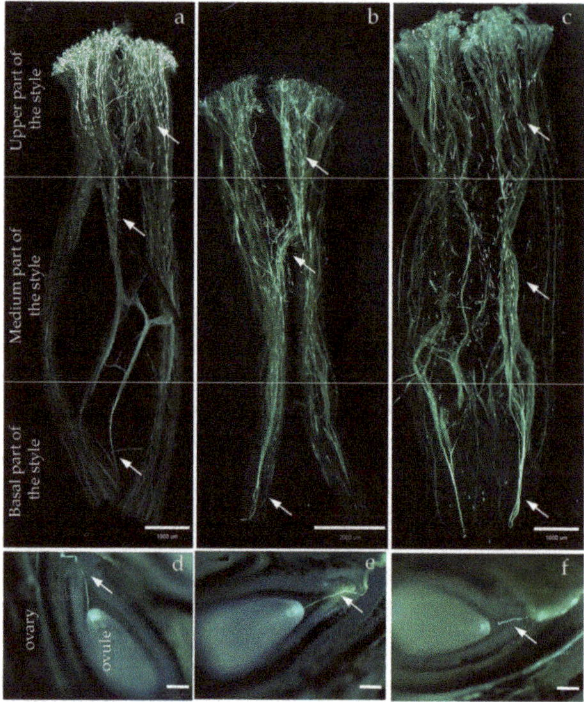

Figure 3. Growth of pollen tubes into certain parts of the pistils of 'Čačanska Lepotica' from the pollen of: (**a**,**d**) 'Čačanska Lepotica' stored at −20 °C; (**b**,**e**) 'Valerija' stored at −80 °C; (**c**,**f**) 'Valjevka' stored at −196 °C. Scale bars: (**a**,**c**)—1000 μm; (**b**)—2000 μm; (**d**,**e**,**f**)—200 μm. Arrows—indicate pollen tubes.

The highest average number of pollen tubes in the upper part of the style was observed in all tested genotypes for pollen stored at −20 °C ('Čačanska Lepotica', 51.53; 'Valjevka', 47.66; 'Valerija', 41.87) (Figure 4). The analysis of variance for the average number of pollen tubes in the upper part of the style indicates that storage temperature and interaction between genotype and temperature had a significant effect. In the ovary, quite surprising results were obtained for the average number of pollen tubes. Namely, the highest average number of pollen tubes was observed with the pollen of 'Valerija' stored at −20 °C (5.73) and 'Valjevka' stored at −196 °C (5.16). As compared to these genotypes, a markedly lower average number of pollen tubes in the locule of the ovary was determined using the pollen of 'Čačanska Lepotica' stored at different temperatures. The analysis of variance showed that the effect of genotype, storage temperature and their interaction on the average number of pollen tubes in the locule of the ovary was significant.

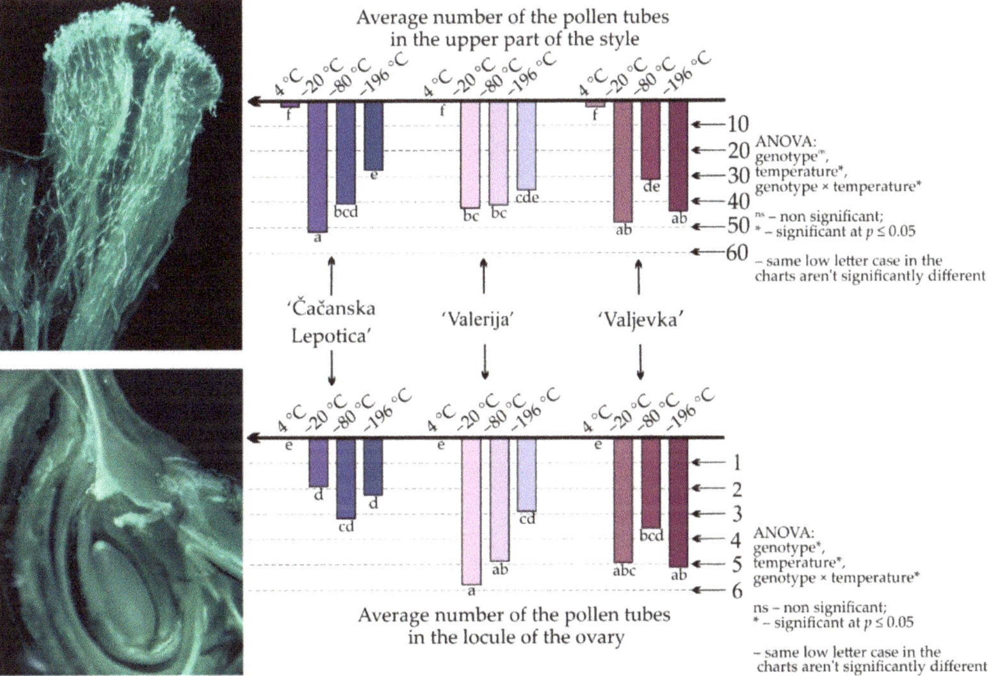

Figure 4. Average number of pollen tubes in certain parts of the pistil of 'Čačanska Lepotica'.

4. Discussion

4.1. Moisture Content of Anthers

Most of the mature pollen grains, when released from the anther, are metabolically inactive and desiccated, with water content ranging from 15 to 35% [30]. Pollen with high initial water content has been observed to be sensitive to stress and is typically short-lived during storage periods [31]. About 70% of plant species, including *Prunus* species, have binucleate pollen classified as tolerant to drying [18]. According to some authors, the longevity of stored pollen may be related to the presence of proteins and starches contained within [32]. Results for successful long-term conservation of desiccation-tolerant pollen indicate that moisture content should be lowered to between 5% and 10% on a fresh weight basis to avoid ice crystals forming during the freezing process [33]. In this current study, the moisture content of anthers after desiccation was less than 7% which can be considered as a low moisture content, which prevents damage during storage at low temperatures and thawing at room temperature.

4.2. Pollen Viability after Different Storage Periods and Temperatures

The obtained results for the germination of freshly collected pollen of genotype 'Čačanska Lepotica' are in agreement with the results from other authors for this genotype [34,35], but this value is higher than reported in the previous study [36], probably due to the fact that in the earlier work the constant temperature during pollen germination was lower. Pollen germination of the other two genotypes 'Valerija' and 'Valjevka' was above 40%, and according to Wertheim [37], these are considered genotypes with good pollen germination rates. Furthermore, other factors, responsible for variation in *in vitro* germination should be taken into account, such as the physiological condition of plants as well as environmental conditions, the time of pollen maturation and shedding that can cause strong variations in pollen germination [38]. Results of some studies [39] indicate that the concentration of some carbohydrates such as sucrose and starch increases in pollen due to the reduction of the metabolism under heat stress. Interaction between exogenous and endogenous factors during pollen grain development may affect its capacity to germinate [35].

With regard to FDA, with fresh pollen of genotype 'Čačanska Lepotica' a slightly lower percentage of pollen was stained in comparison with the percentage of pollen which germinated on the medium, while in the other two genotypes, better value was obtained with FDA.

Generally, the viability of pollen stored for different periods and at different temperatures was lower than for fresh pollen. Similar results were observed with fresh almond [20], mango [22] and apple [17] pollen. After 9 months of storage, pollen viability decreased slightly (but still above 30% in all genotypes), indicating that a temperature of 4 °C is suitable for short-term storage of these plum genotypes. In all genotypes, from 9 months onwards the viability of pollen stored at 4 °C significantly decreased having its lowest value by the end of the storage period. Complete loss of pollen viability, or its gradually decrease, was reported in other *Prunus* species such as almond [20] and sweet cherry [14] at 4 °C after 2 or 12 months of storage, respectively.

The results of this study show that pollen germinability and viability can be preserved for a one-year period by storing at sub-zero temperatures (−20, −80 and −196 °C). Generally, a slight and almost linear decrease in pollen viability was observed in all genotypes at all storage temperatures during the 12-month period. Pollen germination after 12 months of storage at sub-zero temperatures declined by 8.83% to 20.17%, while with the FDA test, this decline was 10% to 33.49%. Pollen germination percentages were similar for pollen stored at −20 and −80 °C, while the greatest differences in pollen viability were observed for those kept at −196 °C. The germination value of pollen stored for 12 months at this temperature was highest in 'Valerija', while in 'Čačanska Lepotica' and 'Valjevka', significantly lower values of this parameter were observed.

In genotypes 'Valerija' and 'Valjevka', higher percentages of pollen grains were stained by FDA in comparison with the percentage of pollen germinated on the medium containing 12% of sucrose. The results obtained in this study are in agreement with previous results for sweet cherry [40] and apple [17], where FDA staining tended to overestimate the ability of pollen to germinate. With the genotype 'Čačanska Lepotica', the opposite situation was observed. A lower value after FDA staining was obtained at all temperatures for all storage periods, which could indicate either a low level of enzyme esterase or its low activity in the pollen grain.

4.3. Pollen Germination In Vivo

Successful pollination and double fertilization are two essential processes in seed production. The progamic phase consists of a number of successive steps started after the pollen lands on the stigma: its adhesion, hydration, germination and production of a pollen tube [41]. This phase is crucial since the process of pollen acceptance/rejection occurs during this period [42,43]. *In vivo* pollen tube growth has been demonstrated as an effective method for assessing pollen behaviour [44]. A strong correlation between the number of pollen tubes in the upper third of the style and pollen germination *in vitro* was reported in sweet [45] and sour cherry [46], which decreases with the increased distance from the stigma. On the other hand, in plum, [35] no such correlation was found, which was explained by different conditions for pollen tube growth.

Pollen grains of 'Čačanska Lepotica' and 'Valjevka' stored for 12 months at 4 °C germinated on the stigma surface. However, pollen tube growth *in vivo* on the 10th day after pollination mostly ended in the lower parts of the style. Bearing in mind that the average number of pollen tubes in the upper part of the style was lower than 5, we could say that the pollen functionality of these genotypes was lost. Namely, pollen grain can germinate, as the germination test has shown, but could not achieve fertilization. In different plant species, during low-temperature storage, free radicals accumulate due to disorders of the oxidative system which leads to the low energy of pollen and ultimately to the inhibition of pollen germination [47,48]. Furthermore, it is found that during pollen storage, the content of glucose and proline increased when pollen germination decreased. However, storage of pollen at ultra-low temperatures inhibits metabolism, reduces enzyme activity and slows down the respiration rate at which pollen viability declines [49].

The most important findings in this study are that the pollen of all the tested genotypes stored at −20, −80 and −196 °C for 12 months, germinated on the stigma, pollen tubes grew through the style of 'Čačanska Lepotica' and penetrated through the micropyle into the nucellus. Except for the pollen of 'Valjevka' stored at −20 °C, where on the 6th day after pollination the pollen tube entered into the locule of the ovary, tubes of the pollen of the other genotypes stored at different temperatures were observed to enter into the nucellus of the ovary. On the 10th day after pollination, in all tested genotypes, the percentages of pollen tubes entering the nucellus were even higher.

For the genotype 'Čačanska Lepotica', the best dynamic of pollen tube growth was observed with pollen stored at −80 °C. This was followed by the highest average number of pollen tubes entering the locule of the ovary. The dynamics of pollen tube growth for pollen stored at −20 °C and −196 °C are very similar, with nearly the same average number of pollen tubes entering the locule of the ovary. The dynamics of pollen tube growth for the genotype 'Valerija' with pollen stored at −20 °C and −80 °C were similar. The highest average numbers of pollen tubes in the upper part of the style and in the locule of the ovary were observed with pollen stored at these temperatures. As regards 'Valjevka', the best pollen tube dynamics, as well as average pollen tube number in a certain part of the pistils, were observed with pollen stored at −196 °C.

Recent studies investigating the effect of pollinizers on pollen tube growth in the European plum revealed that the best dynamics are achieved in cross-pollination variants [35,36,50]. The results obtained in this work are in accordance with these findings. Good dynamics of pollen tube growth observed in self-pollination of 'Čačanska Lepotica' using pollen stored at different temperatures confirms the self-fertility of this genotype as was previously proved [51,52].

5. Conclusions

To the best of our knowledge, this is the first report on the effect of one-year pollen storage at different temperatures on the behaviour of plum pollen *in vitro* and *in vivo*. This study provides useful information about the possibility of pollen storage for up to one year at different temperatures in plum genotypes 'Čačanska Lepotica', 'Valerija' and 'Valjevka'. These findings are of great importance to plum breeding programmes where there is spatial and temporal isolation between parental genotypes. Furthermore, there is no need for the growth of pollen parent and stored pollen could be a good source of germplasm in different exchange programmes. Based on the observed results, storage of pollen at 4 °C was acceptable for storage up to 3 months in all tested genotypes. Storage of pollen at −20, −80 and −196 °C for a one-year period was possible in all analysed genotypes, although the viability of pollen stored at these temperatures slightly decreased in comparison with fresh pollen. The dynamics of pollen tube growth of pollen from anthers stored at sub-zero temperatures, which did not differ from the dynamics previously determined using fresh pollen [50,52], is a good indicator that such pollen can be safely used for breeding purposes.

Supplementary Materials: The following supporting information can be downloaded at: https://www.mdpi.com/article/10.3390/horticulturae8070616/s1, Figure S1: Pollen germination *in vitro*: (a) 'Čačanska Lepotica', fresh pollen; (b) 'Valerija', fresh pollen; (c) 'Valjevka', 3 months at 4 °C; (d) 'Valerija', 6 months at 4 °C; (e) 'Čačanska Lepotica', 9 months at −20 °C; (f) 'Čačanska Lepotica', 9 months at −80 °C; (g) 'Valerija', 12 months at 4 °C; (h) 'Čačanska Lepotica', 12 months at −196 °C. Scale bar = 200 μm.

Author Contributions: Conceptualisation, methodology and formal analysis, M.Đ. and T.V.; investigation, M.Đ., T.V., I.G., N.M., S.M. and S.R., writing-draft preparation, review, and edition, M.Đ., T.V., R.C., M.F.A. and M.M.; project administration, M.M. All authors have read and agreed to the published version of the manuscript.

Funding: This research was funded by the Ministry of Education, Science and Technological Development of the Republic of Serbia (grant number 451-03-68/2022-14/200215, 451-03-9/2021-14/200287 and 451-03-68/2022-14/200116) and by The Research Council of Norway (project No. 280376).

Institutional Review Board Statement: Not applicable.

Informed Consent Statement: Not applicable.

Data Availability Statement: All data are presented in this manuscript.

Conflicts of Interest: The authors declare no conflict of interest.

References

1. Jayasankar, S.; Dowling, C.; Selvaraj, D.K. Plums and Related Fruits. In *Encyclopedia of Food and Health*; Caballero, B., Finglas, P., Tondrá, F., Eds.; Academic Press: Oxford, UK, 2016; pp. 401–405.
2. FAOStat 2021. Available online: https://www.fao.org/faostat/en/#data/QCL (accessed on 19 April 2022).
3. Zhebentyayeva, T.; Shankar, V.; Scorza, R.; Callahan, A.; Ravelonandro, M.; Castro, S.; DeJong, T.; Saski, C.A.; Dardick, C. Genetic characterization of worldwide *Prunus domestica* (plum) germplasm using sequence-based genotyping. *Hortic. Res.* **2019**, *6*, 12. [CrossRef] [PubMed]
4. Hartmann, W.; Neumüller, M. Plum breeding. In *Breeding Plantation Tree Crops Temperate Species*; Mohan Jain, S., Priyadarshan, P.M., Eds.; Springer: New York, NY, USA, 2009; pp. 161–231. [CrossRef]
5. Butac, M. Plum breeding. In *Prunus*; Küden, A., Ali, A., Eds.; IntechOpen: London, UK, 2020; pp. 1–24. [CrossRef]
6. Milošević, N.; Glišić, I.; Đorđević, M.; Radičević, S.; Jevremović, D. An overview of plum Breeding at Fruit Research Institute, Čačak. *Acta Hortic.* **2021**, *1322*, 7–11. [CrossRef]
7. Halapija Kazija, D.; Jelačić, T.; Vujević, P.; Milinović, B.; Čiček, D.; Biško, A.; Pejić, I.; Šimon, S.; Žulj Mihaljević, M.; Pecina, M.; et al. Plum germplasm in Croatia and neighboring countries assessed by microsatellites and DUS descriptors. *Tree Genet. Genomes* **2014**, *10*, 761–778. [CrossRef]
8. Milošević, T.; Milošević, N. Plum (*Prunus* spp.) Breeding. In *Advances in Plant Breeding Strategies: Fruits*; Al-Khayri, J.M., Jain, S.M., Johnson, D.V., Eds.; Springer: Cham, Switzerland, 2018; pp. 165–215. Available online: https://lib.ugent.be/catalog/ebk01:4100000005323086 (accessed on 10 May 2022).
9. Neumüller, M. Fundamental and applied aspects of plum (*Prunus domestica*) breeding. *Fruit Veg. Cereal Sci. Biotechnol.* **2011**, *5*, 139–156.
10. Aldahadhal, A.; Samarah, N.; Bataineh, A. Effect of storage temperature and duration on pollen viability and *in vitro* germination of seven pistachio cultivars. *J. Appl. Hortic.* **2020**, *22*, 184–188. [CrossRef]
11. Akihama, T.; Omura, M.; Kozaki, I. Long-term storage of fruit tree pollen and its application in breeding. *JARQ* **1979**, *14*, 53–56.
12. Hanna, W.; Towill, L. Long-term pollen storage. In *Plant Breeding Reviews*; Janic, J., Ed.; John Wiley and Sons: Chichester, UK, 1995; pp. 197–207.
13. Dafni, A.; Firmage, D.H. Pollen viability and longevity: Practical, ecological and evolutionary implications. *Plant Syst. Evol.* **2020**, *222*, 113–132. [CrossRef]
14. Özcan, A. Effect of low-temperature storage on sweet cherry (*Prunus avium* L.) pollen quality. *HortScience* **2020**, *55*, 258–260. [CrossRef]
15. Heslop-Harrison, J.; Heslop-Harrison, Y. Evaluation of pollen viability by enzymatically induced fluorescence; intracellular hydrolysis of fluorescein diacetate. *Stain Technol.* **1970**, *45*, 115–120. [CrossRef]
16. Masum, A.; Pounders, C.; Blythe, E.; Wang, X. Longevity of crapemyrtle pollen stored at different temperatures. *Sci. Hortic.* **2012**, *139*, 53–57. [CrossRef]

17. Čalić, D.; Milojević, J.; Belić, M.; Miletić, R.; Zdravkovic Korać, S. Impact of storage temperature on pollen viability and germinability of four Serbian autochthon apple cultivars. *Front. Plant Sci.* **2021**, *12*, 709231. [CrossRef]

18. Franchi, G.G.; Piotto, B.; Nepi, M.; Baskin, C.C.; Baskin, J.M.; Pacini, E. Pollen and seed desiccation tolerance in relation to degree of developmental arrest, dispersal and survival. *J. Exp. Bot.* **2011**, *62*, 5267–5281. [CrossRef]

19. Frankel, R.; Galun, E. *Pollination Mechanisms, Reproduction and Plant Breeding. Monographs on Theoretical and Applied Genetics*; Springer: Berlin/Heidelberg, Germany, 1977; pp. 1–284. [CrossRef]

20. Martínez-Gómez, P.; Gradziel, T.M.; Ortega, E.; Dicenta, F. Low temperature storage of almond pollen. *HortScience* **2002**, *37*, 691–692. [CrossRef]

21. Bhat, Z.A.; Dhillon, W.S.; Shafi, R.H.S.; Rather, J.A.; Mir, A.H.; Shafi, W.; Rashid, R.; Bhat, J.A.; Rather, T.R.; Wani, T.A. Influence of storage temperature on viability and *in vitro* germination capacity of pear (*Pyrus* spp.) pollen. *J. Agric. Sci.* **2012**, *4*, 128–135. [CrossRef]

22. Dutta, S.K.; Srivastav, M.; Chaudhary, R.; Lal, K.; Patil, P.; Singh, S.K.; Sing, A.K. Low temperature storage of mango (*Mangifera inidica* L.) pollen. *Sci. Hortic.* **2013**, *161*, 193–197. [CrossRef]

23. Naik, S.; Rana, P.; Rana, V. Pollen storage and use for enhancing fruit production in kiwifruit (*Actinidia deliciosa* A. Chev.). *J. Appl. Hortic.* **2013**, *15*, 128–132. [CrossRef]

24. Alburquerque, N.; Garcia Montiel, F.; Burgos, L. Influence of storage temperature on the viability of sweet cherry pollen. *Span. J. Agric. Res.* **2007**, *5*, 86–90. [CrossRef]

25. Lukić, M.; Pešaković, M.; Marić, S.; Glišić, I.; Milošević, N.; Radičević, S.; Leposavić, A.; Đorđević, M.; Miletić, R.; Karaklajić Stajić, Ž.; et al. *Fruit Cultivars Developed at the Fruit Research Institute, Čačak (1946–2016)*; Fruit Research Institute: Čačak, Serbia, 2016; pp. 1–180.

26. Meier, U. Growth stages of mono- and dicotyledonous plants. In *Federal Biological Research Centre for Agriculture and Forestry*, 2nd ed.; Meier, U., Ed.; BBCH Monograph; Federal Biological Research Centre for Agriculture and Forestry: Berlin/Brunswick, Germany, 2001; pp. 1–158.

27. Galleta, G.J. Pollen and seed management. In *Methods in Fruit Breeding*; Moore, J.N., Janick, J., Eds.; Purdue University Press: West Lafayette, IN, USA, 1983; pp. 23–47.

28. Preil, W. Observing the growth of pollen tubes in pistil and ovarian tissue by means of fluorescence microscopy. *Zeiss Inf.* **1970**, *75*, 24–25.

29. Kho, Y.O.; Baër, J. Fluorescence microscopy in botanical research. *Zeiss Inf.* **1971**, *76*, 54–57.

30. Heslop-Harrison, J. An interpretation of the hydrodynamics of pollen. *Am. J. Bot.* **1979**, *66*, 737–743. [CrossRef]

31. Georgieva, I.D.; Kruleva, M.M. Cytochemical investigation of long-term stored maize pollen. *Euphytica* **1993**, *72*, 87–94. [CrossRef]

32. Li, Y.M.; Chen, L.B. Vigor change of several grasses pollen stored in different temperature and humidity conditions. *Plant Physiol. Commun.* **1998**, *34*, 35–37.

33. Dinato, N.B.; Imaculada Santos, I.R.; Zanotto Vigna, B.B.; de Paula, A.F.; Fávero, A.P. Pollen cryopreservation for plant breeding and genetic resources conservation. *CryoLetters* **2020**, *41*, 115–127. [PubMed]

34. Milatović, D.; Nikolić, D.; Radović, M. Influence of temperature on pollen germination and pollen tube growth of plum cultivars. In Proceedings of the Sixth International Scientific Agricultural Symposium "Agrosym 2015", Jahorina, Bosnia and Herzegovina, 15–18 October 2015; pp. 378–382.

35. Cerović, R.; Fotirić Akšić, M.; Đorđević, M.; Meland, M. The effect of the pollinizers on pollen tube growth and fruit set of European plum (*Prunus domestica* L.) in a Nordic climate. *Sci. Hortic.* **2021**, *288*, 110390. [CrossRef]

36. Glišić, I.; Milatović, D.; Cerović, R.; Radičević, S.; Đorđević, M.; Milošević, N. Examination of suitability of the cultivar 'Čačanska Lepotica' as a pollenizer for promising plum genotypes developed at FRI, Čačak (Serbia). *Acta Hortic.* **2020**, *1289*, 213–220. [CrossRef]

37. Wertheim, S.J. Methods for cross pollination and flowering assessment and their interpretation. *Acta Hortic.* **1996**, *423*, 237–241. [CrossRef]

38. Impre, D.; Reitz, J.; Köpnick, C.; Rolletschek, H.; Börner, A.; Senula, A.; Nagel, M. Assessment of pollen viability for weat. *Front. Plant Sci.* **2020**, *10*, 1588. [CrossRef]

39. Aloni, B.; Peet, M.; Pharr, M.; Karni, L. The effect of high temperature and high atmospheric CO_2 on carbohydrate changes in bell pepper (*Capsicum annum*) pollen in relation to its germination. *Physiol. Plant.* **2001**, *112*, 505–512. [CrossRef]

40. Cerovic, R.; Micic, N.; Djuric, G.; Nikolic, M. Determination of pollen viability in sweet cherry. *Acta Hortic.* **1998**, *468*, 559–566. [CrossRef]

41. Zheng, Y.-Y.; Lin, X.-J.; Liang, H.-M.; Wang, F.-F.; Chen, L.-Y. The long journey of pollen tube in the pistil. *Int. J. Mol. Sci.* **2018**, *19*, 3529. [CrossRef]

42. Montal, R.; Cuenca, J.; Vives, M.C.; Navarro, L.; Ollitrault, P.; Aleza, P. Influence of temperature on the progamic phase in Citrus. *Environ. Exp. Bot.* **2019**, *166*, 103806. [CrossRef]

43. Qu, H.; Guan, Y.; Wang, Y.; Zhang, S. PLC-mediated signaling pathway in pollen tubes regulates the gametophytic self-incompatibility of *Pyrus* species. *Front. Plant Sci.* **2017**, *8*, 1164. [CrossRef]

44. Radičević, S.; Cerović, R.; Nikolić, D.; Đorđević, M. The effect of genotype and temperature on pollen tube growth and fertilization in sweet cherry (*Prunus avium* L.). *Euphytica* **2016**, *209*, 121–136. [CrossRef]

45. Cerović, R.; Ružić, Đ. Pollen tube growth in sour cherry (*Prunus domestica* L.) at different temperatures. *J. Hortic. Sci. Biotechnol.* **1992**, *67*, 333–340. [CrossRef]
46. Sanzol, J.; Hererro, M. The "effective pollination period" in fruit tree. *Sci. Hortic.* **2001**, *90*, 1–17. [CrossRef]
47. Muhlemann, J.K.; Younts, T.L.B.; Muday, G.K. Flavonols control pollen tube growth and integrity by regulating ROS homeostasis during high-temperature stress. *Proc. Natl. Acad. Sci. USA* **2018**, *115*, 11188–11197. [CrossRef]
48. Dai, H.F.; Jiang, B.; Zhao, J.S.; Li, J.C.; Sun, Q.M. Metabolomics and transcriptomics analysis of pollen germination response to low-temperature in pitaya (*Hylocereus polyrhizus*). *Front. Plant Sci.* **2022**, *13*, 866588. [CrossRef]
49. Du, G.; Xu, J.; Gao, C.; Lu, J.; Li, Q.; Du, J.; Lv, M.; Sun, X. Effect of low storage temperature on pollen viability of fifteen herbaceous peonies. *Biotechnol. Rep.* **2019**, *21*, e00309. [CrossRef]
50. Đorđević, M.; Cerović, R.; Radičević, S.; Nikolić, D.; Milošević, N.; Glišić, I.; Marić, S.; Lukić, M. Pollen tube growth and embryo sac development in 'Pozna Plava' plum cultivar related to fruit set. *Erwerbs-Obstbau* **2019**, *61*, 313–322. [CrossRef]
51. Ogašanović, D. The most suitable pollinators for new plum cultivars. *J. Yugosl. Pomol.* **1985**, *19*, 109–115.
52. Kuzmanović, M.; Cerović, R.; Radičević, S. Study of the progamic phase of pollination in the plum cultivar Čačanska Lepotica. *J. Pomol.* **2007**, *41*, 89–93. Available online: https://www.institut-cacak.org/vocarstvo.html (accessed on 10 May 2022).

horticulturae

Article

Evaluation of Critical Points for Effective Cryopreservation of Four Different *Citrus* spp. Germplasm

Damla Ekin Ozkaya [1,2], Fernanda Vidigal Duarte Souza [3] and Ergun Kaya [1,*]

[1] Molecular Biology and Genetics Department, Faculty of Science, Mugla Sitki Kocman University, Kotekli 48000, Mugla, Turkey
[2] Pathology Laboratory Techniques Program, School of Advanced Vocational Studies, Dogus University, Umraniye 34775, Istanbul, Turkey
[3] Embrapa Cassava & Fruits, Caixa Postal 007, Cruz das Almas 44380-000, BA, Brazil
* Correspondence: ergunkaya@mu.edu.tr

Abstract: The different pre- and post-treatments are critical in cryopreservation procedures and affect the shoot tip regrowth after freezing. In the present study, the long-term storage of four citrus cultivars [Bodrum Mandarin (*Citrus deliciosa* Ten.); Klin Mandarin (*Citrus nobilis* Lauriro); White grapefruit and Red grapefruit (*Citrus paradisi* L.)] were carried out by droplet vitrification methods, and the critical points for effective cryopreservation of these species were determined. In this study, we investigated the effect of explant size, cold hardening treatments, sucrose concentrations, and media combinations on shoot regrowth after cryopreservation. The highest shoot tip regrowth, ranging from 13.3 to 33.3%, was achieved when they were obtained from 0.3 to 0.7 mm explants excised from cold hardened seedlings at 4 °C for three days that were then precultured in a medium containing 0.25 M of sucrose and treated with PVS2 at 0 °C for 45 min. In addition, it has been determined that a regeneration medium containing boric acid (H_3BO_3) or ferric ethylenediaminetetraacetate (FeEDDHA) increased the regeneration up to 33.3% after cryopreservation.

Keywords: cold hardening; cryostorage; droplet vitrification; liquid nitrogen; sucrose preculture

Citation: Ozkaya, D.E.; Souza, F.V.D.; Kaya, E. Evaluation of Critical Points for Effective Cryopreservation of Four Different *Citrus* spp. Germplasm. *Horticulturae* **2022**, *8*, 995. https://doi.org/10.3390/horticulturae8110995

Academic Editors: Jean Carlos Bettoni, Min-Rui Wang and Qiao-Chun Wang

Received: 6 September 2022
Accepted: 23 October 2022
Published: 26 October 2022

Publisher's Note: MDPI stays neutral with regard to jurisdictional claims in published maps and institutional affiliations.

1. Introduction

Cryopreservation is the most suitable preservation process for living cells, tissues, and organs in liquid nitrogen (LN, −196 °C) for long periods of time as the mitotic and metabolic activities are reduced at a basal level [1]. It allows for the use of different parts of plants, such as the shoot tips, seeds, nodal and dormant buds, pollen grains, somatic and zygotic embryos, calli, and cell suspensions. The genetic stability also maintains for many years during cold storage [2,3]. However, during and after cryopreservation, the cells, tissues, and organs can suffer from freezing damage and even lose their viability, resulting in failed cryopreservation. Therefore, the selection and pretreatment of excised explants are often needed as the first step of cryopreservation [4]. In addition, fatal ice crystals can form inside the cells, as freezing occurs during the liquid nitrogen treatment [5,6]. The reason for this is the amount of water in the cell. During the frozen storage of plant materials, the water content of the cell should be reduced, and ice crystal formation should be prevented. Plant cells cannot survive LN treatment without the use of cryoprotective solutions [7–9]. Therefore, in order to prevent the cells and shoot tips from freezing and to provide vitrification with liquid nitrogen, dehydration should be performed with cryoprotective solutions [10].

Cryoprotectants are generally divided into penetrating cryoprotectors and non-penetrating cryoprotectors [11]. Penetrating solutions pass through the cell membrane to maintain extracellular and intracellular balance, and non-penetrating solutions accumulate in the extracellular solution without passing through the cell membrane [11]. Penetrating cryoprotectants contain dimethyl sulfoxide (DMSO), glycerol, and amino acids such as proline, while non-penetrating cryoprotectants contain sugars and alcohols [12]. When these solutions were examined, it was determined that DMSO and glycerol were the most efficient cryoprotectants when used appropriately [13]. DMSO is often preferred because of its rapid penetration into cells, but it also has a toxic effect. Therefore, the PVS with an optimized concentration of DMSO is often required in vitrification-based methods [14].

Among various PVSs, PVS2 proved applicable to the shoot tip cryopreservation of a wide range of plant genera. However, it also poses osmotic stress and chemical toxicity to plant tissues, resulting in excessive damage to the cells and poor shoot regrowth after cryopreservation [15,16]. Therefore, the sucrose preculture and loading (with 2 M glycerol + 0.4 M sucrose) are often applied to induce osmotolerance prior to PVS2 treatment [16]. In addition, the duration of PVS2 exposure and temperature should be optimized for the establishment of an optimized vitrification-based protocol [17]. It was observed that shoot tips were formed at different rates in the samples treated with PVS2 at different times [18].

Citrus germplasms, which are taken under protection in their natural habitats and collection gardens, are suppressed by biotic stresses such as insects, nematodes, viruses, bacteria, and fungi, and abiotic stresses such as extreme heat and cold, soil salinity, and acidity [19]. The application of in vitro techniques on these species has some limitations as they have phenolic compounds, similar to some other woody species. Therefore, it is of great value to apply cryopreservation strategies for the safe conservation of *Citrus* germplasm [20–22]. In the literature, there are some protocols applied to the cryopreservation of citrus species. In a study applying shoot tip cryopreservation, nine different citrus species from greenhouse stock plants in Fort Collins (CO, USA) were cryopreserved using the droplet vitrification technique [23]. The same research group further established a droplet vitrification technique for cryopreservation of three *Citrus* accessions [24]. In both of these studies, the thawed shoot tips after liquid nitrogen were recovered using the micrografting method [22,23]. Droplet vitrification combines the use of aluminum foils that facilitate ultra-rapid freezing and thawing with the PVS vitrification and has been widely applied in shoot tip cryobanking [24–26]. Compared to other vitrification techniques, higher cooling and warming rates are observed in the droplet vitrification technique since the samples are placed on aluminum foil with very high thermal conductivity [27,28]. The use of aluminum foil allows explants to reach ultra-low temperatures quickly during contact with liquid nitrogen. Thus, the chance of reaching the vitrified state of the cytoplasm during ultra-fast freezing is increased [29,30].

In this context, the present study aimed to investigate the critical points such as shoot tip size, cold hardening, sucrose preculture, PVS2 treatment time, and culture media on four different citrus cultivars [Bodrum Mandarin (*Citrus deliciosa* Ten.); Klin Mandarin (*Citrus nobilis* Lauriro); cultivars of white grapefruit and red grapefruit (*Citrus paradisi* L.)]. The direct post-thaw culture was tested without micrografting to facilitate the easier operation of droplet vitrification.

2. Materials and Methods

2.1. Plant Material, Surface Sterilization and In Vitro Propagation of Citrus Micro-Shoots

The seeds of *Citrus deliciosa* Ten. cv. "Bodrum Mandarin", *C. nobilis* Lauriro cv. "Klin Mandarin", *C. paradisi* L. cv. "white grapefruit" and cv. "red grapefruit" were obtained from Mugla Metropolitan Municipality, Agricultural Services Department, Mugla Local Seed Bank collection. Surface sterilization was performed via the protocol of Ozudogru et al. [31]. Each of these four *Citrus* spp. cultivars were used to evaluate the effects of shoot tip size, cold hardening, and sucrose preculturing on shoot tip cryopreservation with at least three replications per treatment. The seeds were treated with 70% ethyl alcohol for 5 min, 10% H_2O_2, and twice with 20% commercial bleach for 10 min with active chlorine, and then they were washed with sterile distilled water until completely rinsed. After drying the seeds for 10 min in a laminar flow cabinet, they were transferred to solid Woody Plant Medium (WPM, Duchefa Biochemie) nutrient medium [32] supplemented with 20 g.L^{-1}

and 7 g.L^{-1} agar (pH, 5.8) without any growth regulators and incubated in a growth room under standard conditions (16 h light/8 h dark conditions on 50 μmol^{-1}m^{-2}s^{-1} with white cool fluorescent light, 25 ± 2 °C). The four weeks old micro-shoots (Figure 1a) obtained from the germinated seeds were subcultured on WPM medium supplement with 10 g.L^{-1} charcoal, 1 mg.L^{-1} 6-Benzylaminopurine (BAP), 7 g.L^{-1} agar (pH, 5.8) in Magenta™ vessel GA-7 (Sigma-Aldrich) to obtain a sufficient number of micro-shoots (16 explants were cultured per vessel) for cryopreservation applications.

Figure 1. The micro-shoots and shoot tips of *Citrus* spp. used in experiments. The micro-shoots of *C. deliciosa* cv. "Bodrum Mandarin" derived from in vitro clonal propagiton (**a**), the shoot tips in different sizes derived from micro-shoots of *C. nobilis* cv. "Klin Mandarin" (**b**,**c**), bars 1 mm.

In the preliminary trials, two different nutrient media, Murashige and Skoog (MS) nutrient medium [33] or WPM, were tested in combinations with or without charcoal for shoot tip regeneration. In these studies, the regeneration percentage, the average shoots formed from each shoot tip, and the shoot length were scored. The shoot-forming capacity index (SFC) was calculated via these obtained data [SFC = (average no of shoots per regenerating explant) × (% of regenerating explant)/100] [34]. In the light of this preliminary study, a WPM medium supplement with 10 g.L^{-1} charcoal, 1 mg.L^{-1} BAP, 7 g.L^{-1} agar (pH, 5.8), yielded the best stem-forming capacity index compared to other media combinations, and was decided as the regeneration medium for all of the shoot tips used in this study. In addition to this treatment, some combinations of additional chemical compounds were tested for the increased compacity of shoot tip regrowth.

2.2. Cold Hardening

The micro-shoots that belonged to four different *Citrus* spp. cultivars (16 micro-shoots were cultured per vessel) were reproduced in four-week subculture periods. They were covered with aluminum foil in GA-7 and cold-hardened at 4 °C in the dark following different incubation durations: 24 h, 3 days, and 7 days (Figure 2). To evaluate the effect of cold-hardening on shoot-tip viability, 15 shoot tips were excised from cold-hardened micro-shoots after the cold exposure. The cold-hardened shoot tips were then directly transferred to the regeneration medium and cultured for four weeks under the standard conditions of the control group.

2.3. Isolation of Shoot Tips

For all pre- and post-cryopreservation treatments, two sizes of shoot tips belonging to four different *Citrus* spp. cultivars were used as control or liquid nitrogen groups. The shoot tips were cut between 0.3 and 0.7 mm (Figure 1b) or larger than 0.7 mm in size (Figure 1c). Then, 15 shoot tips of each size were then precultured with enriched sucrose levels and cryopreserved following a droplet-vitrification procedure.

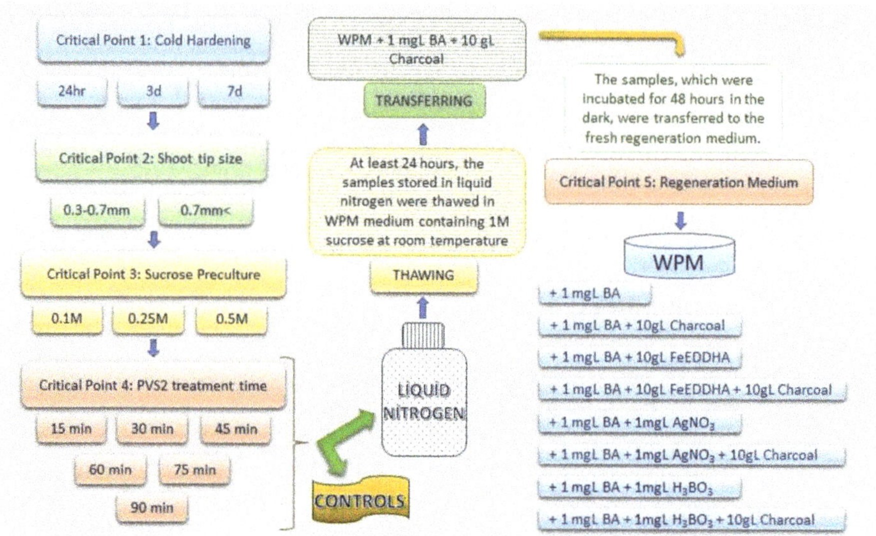

Figure 2. Schematic representation of the stages that are crucial for regeneration before and after cryopreservation of four different *Citrus* spp. cultivars via droplet vitrification.

2.4. Sucrose Preculture

The shoot tips excised from the cold-hardened micro-shoots were transferred separately to WPM media containing 0.1 M, 0.25 M, or 0.5 M sucrose, 7 g.L^{-1} agar (pH, 5.8), and they were incubated at growth room for 24 h. In order to evaluate the effect of each different concentration of sucrose on shoot tip regeneration, 15 shoot tips treated with WPM media containing 0.1 M, 0.25 M, or 0.5 M sucrose, 7 g.L^{-1} agar (pH, 5.8) for incubation of 24 h in a growth room.

2.5. Application of Droplet Vitrification Technique

For cryopreservation, a droplet vitrification technique, which is a more effective, easily applicable, and inexpensive method for many species, was applied [29]. After cold-hardening and sucrose preculture treatment, the shoot tips were transferred to aluminum foil strips (~0.5 × 2 cm in size) containing 3 µL PVS2 for each drop [35]. A total of five drops were dropped on the aluminum foil strip, and each shoot tip was placed on each drop (Figure 3a). Each shoot tip was treated with PVS2 for 15, 30, 45, 60, 75, or 90 min on ice (to prevent cell damage due to the rapid infiltration of osmotic agents). After treatment with PVS2, aluminum foils with shoot tips were plunged into liquid nitrogen by transferring them into cryovials in liquid nitrogen (Figure 3b). The control samples were treated with PVS2 for 15, 30, 45, 60, 75, or 90 min but not immersed in liquid nitrogen.

2.6. Thawing and Post-Thaw Recovery

The samples of each cultivar were sucrose-precultured by removing the shoot tips in two different sizes after the cold preculture step. Afterward, the vitrification solution application (at the different treatment times mentioned above) and immersion in liquid nitrogen and thawing steps were applied to the shoot tips. The samples stored in liquid nitrogen for at least 24 h were thawed by direct transfer to a liquid WPM nutrient medium containing 1 M sucrose (pH, 5.8) at room temperature. For this process, aluminum foils containing shoot tips, which are treated with liquid nitrogen and the control group (not treated with liquid nitrogen) were removed from the cryovials and immersed in the sucrose liquid nutrient medium in Petri dishes. The shoot tips were washed in the same nutrient medium for 15 min and the PVS2 solution was diluted from the cells and tissues. Afterwards, the samples were transferred on solid WPM medium supplement with 10 g.L^{-1} charcoal, 1 mg.L^{-1} BAP and 7 g.L^{-1} agar (pH, 5.8) for 24 h (Figure 3c), and after 24 h, the samples were transferred to different nutrient media [WPM medium supplemented with 1 mg.L^{-1} BAP; 1 mg.L^{-1} BAP and 10 g.L^{-1} charcoal; 1 mg.L^{-1} BAP and 10 g.L^{-1} FeEDDHA; 1 mg.L^{-1} BAP, 10 g.L^{-1} FeEDDHA and 10 g.L^{-1} charcoal; 1 mg.L^{-1} BAP and 10 g.L^{-1} AgNO$_3$; 1 mg.L^{-1} BAP, 10 g.L^{-1} AgNO$_3$ and 10 g.L^{-1} charcoal; 1 mg.L^{-1} BAP and 10 g.L^{-1} H$_3$BO$_3$; 1 mg.L^{-1} BAP, 10 g.L^{-1} H$_3$BO^3 and 10 g.L^{-1} charcoal, and each media also contains 7 g.L^{-1} agar (pH, 5.8)] to observe their regeneration during the final recovery (Figure 2).

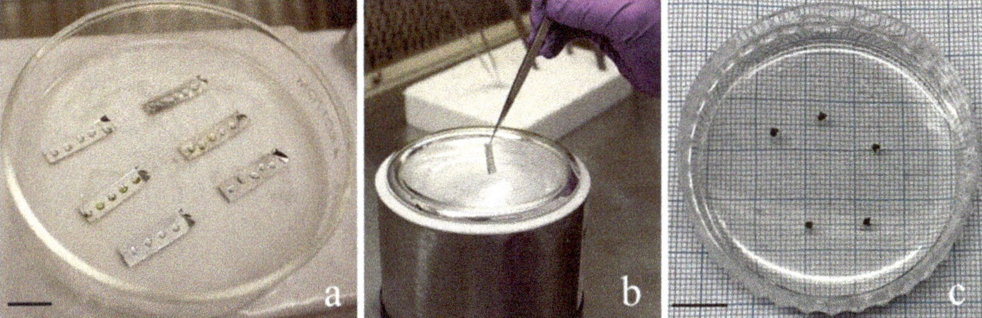

Figure 3. (**a**) the demonstration of droplet-vitrification performed in this study. (**b**) The ultra-rapid freezing performed with aluminum foil as the carrier of shoot tips. (**c**) Post-thaw cultured shoot tips after cryopreservation.

2.7. Data Analyses

A high percentage of regeneration, over 90%, was obtained from each control group sample (Tables 1–4) of all critical point treatments except the sucrose preculture treatment on WPM medium supplemented with 0.5 M sucrose and different time PVS2 treatments. In this context, the study is based on the results obtained after liquid nitrogen treatments, and each critical point was studied in conjunction with the other. After thawing, shoot tips incubated in the dark for 48 h in regeneration nutrient medium (described before) were transferred to different nutrient media and incubated for 4 weeks under the above-mentioned standard conditions (Figure 2). After four weeks of incubation, at least one leaf formation from the shoot tips of incubation was determined as the shoot tip regeneration. In total, five different critical steps were evaluated in the present study. Four of these were the application parameters before cryopreservation, and one of them was applied after cryopreservation. A total of 15 shoot-tips were used for each parameter, which was performed in three replicates. Including the controls, 2 different sizes of shoot tips, 3 different sucrose concentrations containing sucrose preculture, 6 different times of cryoprotectant application, and finally, after 24 h of incubation after thawing, the shoot tips were transferred to 8 different nutrient media (described in Figure 2). In all these studies, 864 different parameters were tried in tree replicates, and a total of 6720 shoot tips were used for each cultivar, including controls. The percentages of regeneration after cryopreservation were calculated for each cultivar, and IBM® SPSS Statistics 24.0 was used for the statistical analysis of data.

Table 1. The shoot tip regeneration, the number of shoots, and the shoot forming capacity (SFC) of four different *Citrus* spp. cultivars tested in different media combinations.

	MS + 1 mg.L^{-1} BAP			MS + 1 mg.L^{-1} BAP + 10 g.L^{-1} Charcoal			WPM + 1 mg.L^{-1} BAP			WPM + 1 mg.L^{-1} BA + 10 g.L^{-1} Charcoal		
	Regeneration (%)	Avarage Nuber of Shoots	Shoot Forming Capacity Index	Regeneration (%)	Avarage Nuber of Shoots	Shoot Forming Capacity Index	Regeneration (%)	Avarage Nuber of Shoots	Shoot Forming Capacity Index	Regeneration (%)	Avarage Nuber of Shoots	Shoot Forming Capacity Index
Bodrum Mandarin	100 ± 0.0 a *	1.4 ± 0.16 D	1.4	100 ± 0.0 a	2.0 ± 0.13 C	2.0	100 ± 0.0 a	3.1 ± 0.06 B	3.1	100 ± 0.0 a	3.7 ± 0.04 A	3.7
Klin Mandarin	93.3 ± 2.11 b	1.3 ± 0.11 D	1.2	100 ± 0.0 a	1.8 ± 0.08 C	1.8	100 ± 0.0 a	2.8 ± 0.09 B	2.8	100 ± 0.0 a	5.3 ± 0.07 A	5.3
White grapefruit	96.7 ± 1.7 b	1.1 ± 0.06 D	1.06	100 ± 0.0 a	2.8 ± 0.13 B	2.8	100 ± 0.0 a	2.3 ± 0.04 C	2.3	100 ± 0.0 a	3.1 ± 0.06 A	3.1
Red grapefruit	100 ± 0.0 a	2.0 ± 0.12 D	2.0	100 ± 0.0 a	2.9 ± 0.11 C	2.9	100 ± 0.0 a	3.1 ± 0.03 B	3.1	100 ± 0.0 a	3.4 ± 0.06 A	3.4

* Values followed by the same letter within each variable and cultivar area are not significantly different ($p < 0.05$), according to the LSD test. The upper- and lowercase letters next to the values indicate statistical homology between the values. The regeneration percentage is grouped among themselves (lowercase), the number of stems formed per shoot tip (with capital letters). Each *Citrus* spp. cultivar was statistically evaluated within itself via the test of homogeneity of variance.

Table 2. The shoot tip regeneration of four different *Citrus* spp. cultivars cold hardened in different treatment times. The shoot tips treated with PVS2 for 45 min were incubated in regeneration medium for the first 24 h, and then they transferred to WPM medium supplemented with 1 mg.L^{-1} BAP, 10 mg.L^{-1} activated charcoal, and 1 mg.L^{-1} H$_3$BO$_3$ medium for four weeks under standard culture conditions.

	The Shoot Tips No Cold Hardened		The Shoot Tips Cold Hardened for 24 h		The Shoot Tips Cold Hardened for 3 days		The Shoot Tips Cold Hardened for 7 Days	
	Regeneration (%) (Control)	Regeneration (%) (After Cryostorage)	Regeneration (%) (Control)	Regeneration (%) (after Cryostorage)	Regeneration (%) (Control)	Regeneration (%) (after Cryostorage)	Regeneration (%) (Control)	Regeneration (%) (after Cryostorage)
Bodrum Mandarin	100 ± 0.0 a *	0	100 ± 0.0 a	13.3 ± 1.26 d	96.7 ± 0.67 b	33.3 ± 0.60 c	0	0
Klin Mandarin	100 ± 0.0 a	0	100 ± 0.0 a	8.3 ± 0.99d	100 ± 0.0 a	26.7 ± 0.58b	16.7 ± 0.54c	0
White grapefruit	100 ± 0.0 a	0	100 ± 0.0 a	0	100 ± 0.0 a	13.3 ± 0.57 b	3.3 ± 052 c	0
Red grapefruit	100 ± 0.0 a	0	100 ± 0.0 a	1.7 ± 0.38 d	93.3 ± 0.61 b	26.7 ± 1.11 c	0	0

* Values followed by the same letter within each variable and cultivar area are not significantly different ($p < 0.05$), according to the LSD test. The lowercase letters next to the values indicate statistical homology between the values. The regeneration percentage is grouped among themselves. Each *Citrus* spp. cultivar was statistically evaluated within itself via the test of homogeneity of variance.

Table 3. The regeneration of two differently sized shoot tips of four different *Citrus* spp. Cultivars. The shoot tips treated with PVS2 for 45 min were incubated in regeneration medium for the first 24 h, and then they transferred to WPM medium supplemented with 1 mg.L^{-1} BAP, 10 mg.L^{-1} activated charcoal, and 1 mg.L^{-1} H$_3$BO$_3$ medium for four weeks under standard culture conditions.

	The Shoot Tips 0.3–0.7 mm in Size		The Shoot Tips More Than 0.7 mm in Size	
	Regeneration (%) (Control)	Regeneration (%) (after Cryostorage)	Regeneration (%) (Control)	Regeneration (%) (after Cryostorage)
Bodrum Mandarin	100 ± 0.0 a *	33.3 ± 1.26 b	100 ± 0.0 a	0
Klin Mandarin	100 ± 0.0 a	26.7 ± 1.14 b	100 ± 0.0 a	0
White grapefruit	100 ± 0.0 a	13.3 ± 1.05 b	100 ± 0.0 a	0
Red grapefruit	100 ± 0.0 a	26.7 ± 1.97 b	100 ± 0.0 a	0

* Values followed by the same letter within each variable and cultivar area are not significantly different ($p < 0.05$), according to the LSD test. The lowercase letters next to the values indicate statistical homology between the values. The regeneration percentage is grouped among themselves. Each *Citrus* spp. cultivar was statistically evaluated within itself via the test of homogeneity of variance.

Table 4. The shoot tip regeneration of four different *Citrus* spp. cultivars tested in different sucrose preculture media for 24 h. The shoot tips treated with PVS2 for 45 min were incubated in regeneration medium for the first 24 h, and then they transferred to WPM medium supplemented with 1 mg.L^{-1} BAP, 10 mg.L^{-1} activated charcoal, and 1 mg.L^{-1} H$_3$BO$_3$ medium for four weeks under standard culture conditions.

	The Shoot Tips No Sucrose Precultured		The Shoot Tips Sucrose Precultured on WPM Medium Containing 0.1 M Sucrose		The Shoot Tips Sucrose Precultured on WPM Medium Containing 0.25 M Sucrose		The Shoot Tips Sucrose Precultured on WPM Medium Containing 0.5 M Sucrose	
	Regeneration (%) (Control)	Regeneration (%) (after Cryostorage)	Regeneration (%) (Control)	Regeneration (%) (after Cryostorage)	Regeneration (%) (Control)	Regeneration (%) (after Cryostorage)	Regeneration (%) (Control)	Regeneration (%) (after Cryostorage)
Bodrum Mandarin	100 ± 0.0 a *	0	100 ± 0.0 a	11.7 ± 1.02 d	100 ± 0.0 a	33.3 ± 0.6 c	53.3 ± 0.55 b	0
Klin Mandarin	100 ± 0.0 a	0	100 ± 0.0 a	16.7 ± 0.52 d	100 ± 0.0 a	26.7 ± 0.58 c	66.7 ± 0.54 b	8.3 ± 0.28 e
White grapefruit	100 ± 0.0 a	0	100 ± 0.0 a	0	100 ± 0.0 a	13.3 ± 0.57 c	63.3 ± 0.53 b	6.7 ± 0.55 d
Red grapefruit	100 ± 0.0 a	0	100 ± 0.0 a	13.3 ± 0.48 d	100 ± 0.0 a	26.7 ± 1.12 c	56.7 ± 1.03 b	0

* Values followed by the same letter within each variable and cultivar area are not significantly different ($p < 0.05$), according to the LSD test. The lowercase letters next to the values indicate statistical homology between the values. The regeneration percentage is grouped among themselves. Each *Citrus* spp. cultivar was statistically evaluated within itself via the test of the homogeneity of variance.

3. Results

3.1. The Determination of the Medium for Shoot Tip Regrowth

Pre-trials of the present study were conducted to optimize the optimal regeneration medium for the shoot tips of four different *Citrus* spp. cultivars. Two different nutrient media supplemented with 1 mg.L^{-1} BAP were tested with or without charcoal, and WPM supplemented with charcoal yielded the best results for the shoot tip regeneration. Very high regeneration percentages ranging between 93.3 and 100 were obtained in all of the tested media. However, in the tested media, the highest number of stems per shoot ranged between 3.1 and 5.3, was obtained from WPM supplemented with 1 mg.L^{-1} and 10 g.L^{-1} Charcoal, pH 5.8 (Table 1). For this reason, this medium was used as the regeneration medium in further trials based on the calculated Shoot Forming Capacity (SFC) index [34].

3.2. The Evaluation of Cold-Hardening

Considering each treatment as an integrated (Figure 4), it was observed that shoot tip explants cut from shoots acclimatized to cold for 3 days had a significantly positive effect on regeneration after cryopreservation and the best regeneration percentages were 33.3%, 26.7%, 13.3%, and 26.7, respectively, in all four citrus cultivars (Table 2). It was observed that the regeneration of the non-acclimatized samples was significantly reduced, and they even died after cryopreservation, evidencing that this step is critical for the process.

Figure 4. The shoot tip regrowth of *Citrus* spp. after optimized cryopreservation. After three days of cold-hardening application, shoot tip explants cut in 0.3–0.7 mm size were taken to the sucrose preculture stage on solid WPM nutrient medium containing 0.25 M sucrose. Then, they were kept in droplets containing 3 μL PVS2 for 45 min, and they were incubated on solid WPM nutrient medium containing 1 mg.L^{-1} BAP and 10 mg.L^{-1} charcoal for 24 h. Then, they were transferred to WPM solid nutrient medium supplemented with 1 mg.L^{-1} BAP, 1 mg.L^{-1} H$_3$BO$_3$, and 10 mg.L^{-1} charcoal (**a**) cv. Bodrum Mandarin; (**b**) cv. Klin Mandarin; (**c**) white grapefruit and (**d**) red grapefruit after four weeks incubation; (**e**) cv. Bodrum Mandarin; (**f**) cv. Klin Mandarin; (**g**) red grapefruit after eight weeks incubation, bars 1 cm.

Local necrosis formations and etiolated shoots were observed in the one-week-cold-hardened micro-shoots of all *Citrus* spp. cultivars. For this reason, the healthy shoot tips could not be cut from them. On the other hand, in the control groups of cold hardening, all of the shoot tips of each cultivar cut from 24 h- or three-days cold hardened micro-shoots obtained 100% regeneration. However, in the post-cryo applications, the regeneration was just only observed in three-day cold-hardened shoot tips (Figure 4), and the viability was not observed in shoot tips obtained from 24 h-cold hardened micro-shoots (Table 2).

3.3. The Effect of Shoot Tip Size on Cryopreservation

In the control groups, 100% shoot regrowth was obtained from all four tested *Citrus* spp. After cryopreservation, no viability was obtained in the shoot tips larger than 0.7 mm. On the other hand, the regenerations were obtained from all four different *Citrus* spp. cultivars from the applications where 0.3–0.7 mm cut shoot tips were used (Table 3).

3.4. The Evaluation of Sucrose Preculture

As for the optimized shoot tip preculture with elevated sucrose levels, the shoot tips were transferred to solid WPM medium containing sucrose at three different concentrations (0.1 M, 0.25 M, and 0.5 M) and incubated for 24 h at the standard growth room conditions described above and then they subjected to other cryogenic applications. For the shoot tips tested in the control group, 100% regeneration was obtained after preculture with 0.1 and 0.25 M sucrose for 24 h. However, a significant decrease was observed in the shoot tip regrowth of those precultured with 0.5 M sucrose. Of these three parameters, the samples precultured in a nutrient medium containing 0.25 M sucrose provided the best results in recovery after cryopreservation (Table 4).

3.5. The Evaluation of Cryoprotectant Treatment Time

The PVS2 treatment was also optimized in this study with exposure durations ranging from 0 to 90 min. In the control groups without freeze–thaw cycles, high shoot regrowth levels (83.3–100%) were obtained. However, when the samples were cryopreserved after various PVS2 exposures, the highest shoot regrowth levels ranging from 13.3% to 33.3% were obtained after the 45 min of PVS2 treatment for all four tested genotypes (Table 5, Figures 5 and 6).

Table 5. The shoot tip regeneration of four different *Citrus* spp. cultivars tested in different PVS2 treatment times. After PVS2 treatment, the shoot tips were incubated in regeneration medium for the first 24 h, and then they transferred to WPM medium supplemented with 1 mg.L^{-1} BAP, 10 mg.L^{-1} activated charcoal, and 1 mg.L^{-1} H$_3$BO$_3$ medium for four weeks under standard culture conditions.

PVS2 Treatment Time	Regeneration (%)	Bodrum Mandarin	Klin Mandarin	White Grapefruit	Red Grapfruit
0 min	Control	100 ± 0.0 a *	96.7 ± 0.71 b	93.3 ± 0.51 c	100 ± 0.0 a
	Cryostoraged	0	0	0	0
15 min	Control	100 ± 0.0 a	90.0 ± 0.0 d	100 ± 0.0 a	100 ± 0.0 a
	Cryostoraged	0	0	0	0
30 min	Control	96.7 ± 1.67 b	100 ± 0.0 a	96.7 ± 0.49 b	100 ± 0.0 a
	Cryostoraged	0	3.3 ± 0.56 f	0	0
45 min	Control	93.3 ± 1.42 c	96.6 ± 0.65 b	100 ± 0.0 a	86.7 ± 1.07 d
	Cryostoraged	33.3 ± 1.67 d	26.7 ± 1.09 e	13.3 ± 0.88 f	26.7 ± 0.75 e
60 min	Control	100 ± 0.0 a	90.0 ± 1.89 d	86.7 ± 1.26 d	90.0 ± 0.0 c
	Cryostoraged	6.7 ± 0.79 e	0	6.7 ± 0.86 g	0
75 min	Control	83.3 ± 3.02 d	93.3 ± 1.23 c	93.3 ± 0.94 c	86.7 ± 0.39 d
	Cryostoraged	0	0	0	0
90 min	Control	83.3 ± 2.11 d	96.7 ± 0.59 b	83.3 ± 0.92 e	93.3 ± 0.78 b
	Cryostoraged	3.3 ± 0.74 f	0	0	6.7 ± 0.73 f

* Values followed by the same letter within each variable and cultivar area are not significantly different ($p < 0.05$), according to the LSD test. The lowercase letters next to the values indicate statistical homology between the values. The regeneration percentage is grouped among themselves. Each *Citrus* spp. cultivar was statistically evaluated within itself via the test of homogeneity of variance.

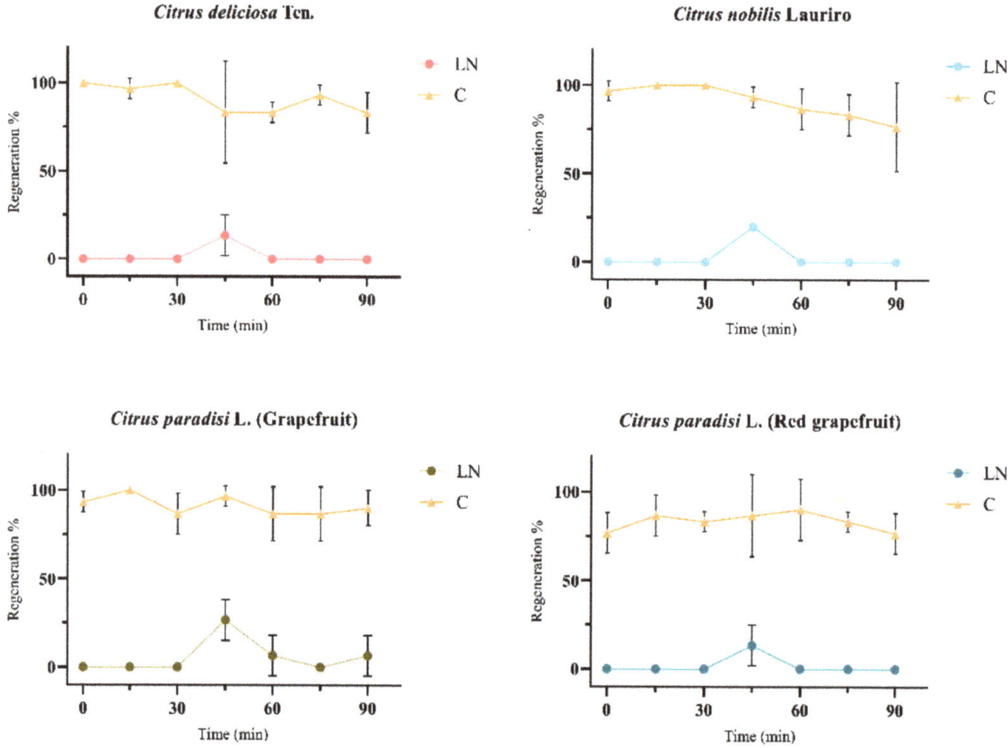

Figure 5. The shoot-tips regenerations of liquid nitrogen (LN) and control (C) groups of four different *Citrus* spp. cultivars after cryogenic applications on solid WPM medium supplemented with 1 mg.L^{-1} BAP, 10 mg.L^{-1} activated charcoal, and 1 mg.L^{-1} FeEDDHA. The graphs were formed according to the shoot tip regeneration data, which obtained the best regeneration in the optimized cryogenic applications and then treated with PVS2 at different times. In addition, shoot tips untreated with PVS2 (0 min) were also given as the control group of PVS2 treated with different time in graph. The graphs represent the mean ± SD ($p < 0.05$).

3.6. The Evaluation of the Optimized Post Thaw Culture after Cryopreservation

In this study, cryopreserved shoot tips were first post-thaw cultured on solid WPM medium supplemented with 1 mg.L^{-1} BAP and 10 mg.L^{-1} charcoal before transferring to eight different nutrient mediums for final recovery (Figure 2). For the four tested cultivars, as the optimized explant size, preculture, PVS2 exposure, etc., have already been presented in the results, the only solid WPM nutrient medium containing 10 mg.L^{-1} activated charcoal and 1 mg.L^{-1} BAP supplemented with 1 mg.L^{-1} H$_3$BO$_3$ (Figure 5) or 1 mg.L^{-1} FeEDDHA (Figure 6) led to shoot regrowth after cryopreservation. Noticeably, the highest shoot regrowth levels, ranging from 13.3% to 33.3%, were obtained for all the tested cultivars after the post-thaw recovery with 1 mg.L^{-1} H$_3$BO$_3$ (Figure 6).

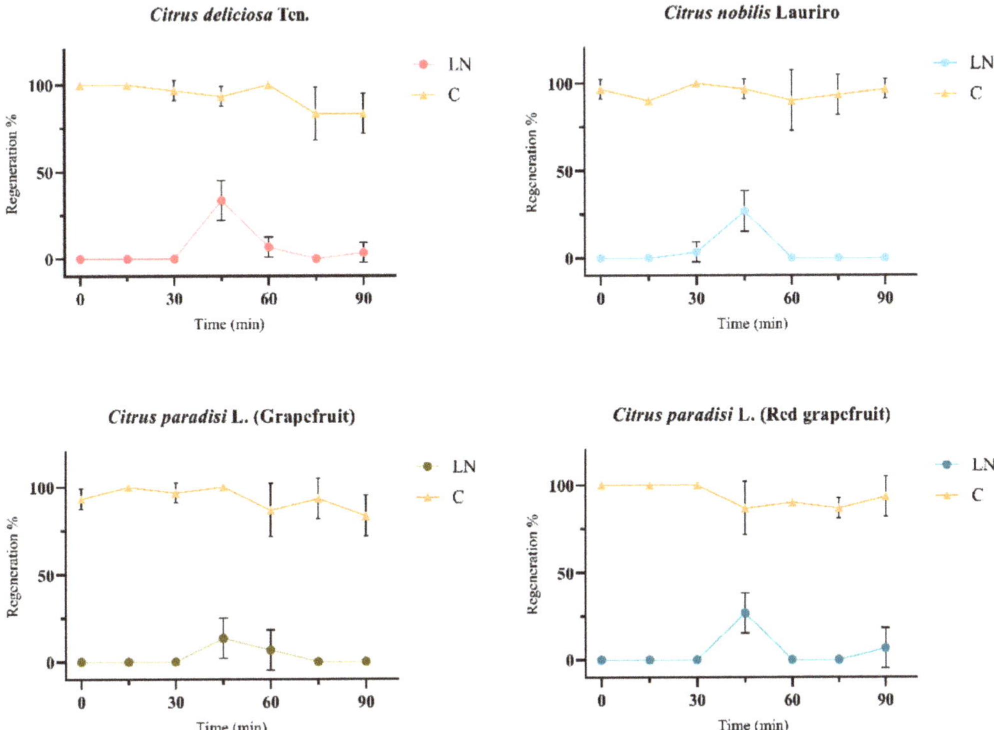

Figure 6. The shoot-tips regenerations of liquid nitrogen (LN) and control (C) groups of four different *Citrus* spp. cultivars after cryogenic applications on solid WPM medium supplemented with 1 mg.L^{-1} BAP, 10 mg.L^{-1} activated charcoal, and 1 mg.L^{-1} H$_3$BO$_3$. The graphs were formed according to the shoot tip regeneration data, which obtained the best regeneration in the optimized cryogenic applications and then treated with PVS2 at different times. In addition, shoot tips untreated with PVS2 (0 min) were also given as the control group of PVS2 treated with different time in graph. The graphs represent the mean ± SD ($p < 0.05$).

4. Discussion

In this study, five critical points following a droplet-vitrification protocol were evaluated for cryopreservation of four *Citrus* spp. Each critical point was evaluated in different parameters, and the optimum combination was obtained.

First, in the current work, the positive effects of cold-hardening for three days of cold-hardening proved necessary to obtain shoot regrowth after cryopreservation. It has been proven by studies that some genes for the adaptation of plants to low temperatures are expressed during cold adaptation [36]. It is known that proline synthesis is induced during cold hardening, especially in plants, and the accumulation of this amino acid in tissues increases the plant's resistance to both osmotic stress and low temperatures [37]. In a study on the cryopreservation of the date plant, meristems were used, and it was proven that cold preculture increased the accumulation of proline in the tissues [38]. Similarly, in studies on the cryopreservation of blackberry [39], apple [40], and heaven bamboo [41] plants, it has been proven that cold preculture provides very successful results in recovery after cryopreservation. In our study, cold hardening led to successful shoot recovery after cryopreservation.

The size of the explant used in cryopreservation is another factor affecting the cryopreservation success, and the shoot tip size should be large enough for the tissue to regenerate during the recovery after cryopreservation but small enough to prevent excessive fatal ice crystallization due to the higher water content of the vacuole in the mature tissues during the treatment with liquid nitrogen [42,43]. In this study, successful shoot tip regrowth was only obtained with the use of small shoot tips (0.3–0.7 mm). The small shoot tips consist of more cells that could survive after cryogenic treatments due to the reduced number and size of the vacuoles [44,45]. In another study for cryopreservation of *Vitis* spp., four different shoot tips, 0.5, 1.0, 1.5, and 2.0 mm, in size, were used, and the best regeneration was obtained from meristems 1 mm in size. An earlier study made for citrus cryopreservation also found that explants cut to 1 mm in size presented better results [22,46]. Nevertheless, some other factors, such as the anatomical and morphological characteristics of the in vitro shoots may also affect the optimized explant size for shoot tip cryopreservation.

Preculturing the shoot tips with a high concentration of sucrose is important to ensure shoot tip recovery after cryopreservation. In this study, 0.25 sucrose treatment for 24 h resulted in improved shoot regrowth after cryopreservation. During the preculture, sugar accumulation in the extracellular compartments will ensure the transfer of the existing water in the vacuoles of the cells to the intercellular compartments [47–49]. The critical point to be considered here is to optimize the sugar concentration to be used for sucrose preculture so that it is both high enough to ensure maximum removal of water in the vacuole and low enough to not damage tissues and cells while removing water [20,50]. Similar to our study, the sucrose preculture at 0.25 M to 1.0 M for 24 **h** has proved effective in the successful cryopreservation of many species such as heavenly bamboo [38], eucalyptus [3], sweet orange [51], and pineapple [52].

In vitrification-based methods for shoot tip cryopreservation, the exposure time to the PVS is a critical point for recovery after cryopreservation [53]. For example, insufficient PVS2 dehydration can cause freezing injuries due to ice crystallization, while excessive PVS exposure would lead to greater osmotic stress and toxicity to the tissues [54]. In the present study, PVS2 exposure for 45 min proved optimal for obtaining shoot regrowth after cryopreservation for all of the tested cultivars. We also applied PVS2 exposure on ice as it could result in the slow penetration of cryoprotectants into the tissues for alleviated osmotic stress and toxicity. Similarly, a PVS2 exposure on ice for 30 min was applied in a droplet-vitrification protocol for the shoot tip cryopreservation of *Citrus* spp. [22].

In the present study, various chemical additives were tested in the recovery medium for improved shoot tip regrowth after cryopreservation. In the present study, it was observed that the medium containing activated charcoal yielded healthier shoots compared to those without (Table 1, Figure 4). Activated charcoal provides the retention of various chemicals, especially the phenolic components present in the environment. Thus it can prevent shoot tips from the excessive damage caused by pre- and post-cryopreservation applications from inhibiting shoot development by keeping these chemicals [28]. They act against this stress and limit the growth of the plant by passing into the nutrient medium [54]. Moreover, in the present study, successful shoot regrowth was obtained only after the addition of H_3BO_3 or FeEDDHA in the post-thaw recovery, whereas shoot tip micrografting was applied to assist in the regrowth of *Citrus* spp. Shoot tips cryopreservation [22]. H_3BO_3 is important in the recovery after liquid nitrogen may be due to its supporting effect on the cell wall [55]. Therefore, its deficiency can be a limiting factor for growth and development [55,56].In our study, the positive response of H_3BO_3 in, thus, the use of H_3BO_3 in cryogenic applications may have ensured the preservation of cell integrity due to the support it provides the cell wall. FeEDDHA, which also led to shoot recovery after cryopreservation, is effective in metabolic activities due to its chelating feature and the chlorophyll ratio in plants. Therefore, it is important to shoot growth [57]. In this context, the current study may suggest that FeEDDHA may have positive effects on post-cryo shoot development, and this is the first use of these substances in shoot tip cryopreservation.

There are two publications by the same researcher in the literature on citrus shoot tip cryopreservation. In one of the studies, shoot tips cryopreserved with PVS2 vitrification [22], and in the second study, shoot tips cryopreserved with droplet vitrification technique [23];

similar to our study, the micrografting method was used for the recovery of shoot tips after liquid nitrogen treatment. In our study, the shoot tips were transferred to different nutrient media with different contents for recovery after thawing (Figure 2). Our study, which is basically based on the same principle, aimed to determine the most suitable recovery medium by transferring the shoot tips to different nutrient media so that they can overcome these post-cryo stresses. However, the method we applied was relatively less time-consuming and easier to manipulate than the micrografting method. Thus, it was implemented more effectively during the study. In addition, with this application, the healthier growth of shoot tips was ensured in the recovery after cryopreservation.

5. Conclusions

The determination of the physical, molecular, and biochemical changes associated with successful regeneration after cryopreservation is a critical point for developing cryopreservation protocols [58–61]. In this study, an optimized explant size, preculture, and PVS2 exposure, and the addition of H_3BO_3 and FeEDDHA in the post-thaw recovery, are critical factors affecting the direct shoot tip regrowth of *Citrus* spp. after cryopreservation. We believe that the combined optimization of these critically important treatments in difficult-to-cryopreservation species, such as *Citrus* spp., would yield positive results in the cryopreserve of similar species or difficult woody species in the future. In this context, the present study will be a very useful resource for scientists working in similar fields for their future work.

Author Contributions: Conceptualization, E.K. and F.V.D.S.; methodology, D.E.O., E.K. and F.V.D.S.; validation, D.E.O., E.K. and F.V.D.S.; investigation, D.E.O. and E.K.; writing—original draft preparation, D.E.O., E.K. and F.V.D.S.; writing—review and editing, D.E.O., E.K. and F.V.D.S.; supervision, E.K. and F.V.D.S. All authors have read and agreed to the published version of the manuscript.

Funding: This study was supported by Mugla Sitki Kocman University, Scientific Research Projects Coordination Unit (Mugla, Turkey, MSKU-BAP, Project Number: 19/076/08/1/2).

Data Availability Statement: Not applicable.

Acknowledgments: The authors thank the Muğla Local Seed Bank staff (Muğla Metropolitan Municipality, Agricultural Services Department) for providing the *Citrus* spp. seeds and to Naeem Abdul Ghafoor for helping the graph construction from the statistical data in the study.

Conflicts of Interest: The Authors declare to have no conflicts of interests.

References

1. Thormann, I.; Dulloo, M.; Engels, J. Techniques for *ex situ* plant conservation. In *Plant Conservation Genetics*; Henry, R.J., Ed.; The Haworth Press: New York, NY, USA, 2006; pp. 7–36.
2. Kaya, E.; Alves, A.; Rodrigues, L.; Jenderek, M.; Hernandez-Ellis, M.; Ozudogru, A.; Ellis, D. Cryopreservation of *Eucalyptus* genetic resources. *CryoLetters* **2013**, *34*, 608–618.
3. Kaya, E.; Souza, F.V.D. Comparison of two PVS2-based procedures for cryopreservation of commercial sugarcane (*Saccharum* spp.) germplasm and confirmation of genetic stability after cryopreservation using ISSR markers. *Vitr. Cell Dev. Biol. Plant.* **2017**, *53*, 410–417. [CrossRef]
4. Kalaiselvi, R.; Rajasekar, M.; Gomathi, S. Cryopreservation of plant materials-a review. *Int. J. Chem. Stud.* **2017**, *5*, 560–564.
5. Sakai, A.; Yoshida, S. Survival of plant tissue at super-low temperature VI. Effects of cooling and rewarming rates on survival. *Plant Physiol.* **1967**, *42*, 1695–1701. [CrossRef] [PubMed]
6. Sakai, A. Cryopreservation of Germplasm of Woody Plants. In *Cryopreservation of Plant Germplasm I*; Bajaj, Y.P.S., Ed.; Springer: Berlin/Heidelberg, Germany, 1995; Volume 32, pp. 53–69. [CrossRef]
7. Gnanapragasam, S.; Vasil, I.K. Ultrastructural changes in suspension culture cells of Panicum maximum during cryopreservation. *Plant Cell Rep.* **1992**, *11*, 169–174. [CrossRef]
8. Volk, G.M.; Walters, C. Plant vitrification solution 2 lowers water content and alters freezing behavior in shoot tips during cryoprotection. *Cryobiology* **2006**, *52*, 48–61. [CrossRef]
9. Panis, B. Sixty years of plant cryopreservation: From freezing hardy mulberry twigs to establishing reference crop collections for future generations. *Acta Hortic.* **2019**, *1234*, 1–8. [CrossRef]
10. Sakai, A. Potentially valuable cryogenie procedures for cryopreservation of cultured plant meristems. In *Conservation of Plant Genetic Resources In Vitro*; Razdan, M.K., Cocking, E.C., Eds.; Science Publishers Inc.: Enfield, CT, USA, 1997; Volume 1, pp. 53–66.

11. Reed, B.M. Cryopreservation—Practical Considerations. In *Plant Cryopreservation: A Practical Guide*; Reed, B.M., Ed.; Springer: New York, NY, USA, 2008; pp. 3–13. [CrossRef]

12. Panis, B.; Lambardi, M. Status of cryopreservation technologies in plants (crops and forest trees). In *The Role of Biotechnology in Exploring and Protecting Agricultural Genetic Resources*, 2nd ed.; Ruane, J., Sonnino, A., Eds.; FAO: Rome, Italy, 2006; pp. 61–78.

13. Nag, K.K.; Street, H.E. Freeze preservation of cultured plant cells: I. the pretreatment phase. *Physiol. Plant.* **1975**, *34*, 254–260. [CrossRef]

14. Zamecnik, J.; Faltus, M.; Bilavcik, A. Vitrification Solutions for Plant Cryopreservation: Modification and Properties. *Plants* **2021**, *10*, 2623. [CrossRef]

15. Kim, H.H.; No, N.Y.; Shin, D.J.; Ko, H.C.; Kang, J.H.; Cho, E.G.; Engelmann, F. Development of alternative plant vitrification solutions to be used in droplet-vitrification procedures. *Acta Hortic.* **2011**, *908*, 181–186. [CrossRef]

16. Ping, K.S.; Poobathy, R.; Zakaria, R.; Subramaniam, S. Development of a PVS2 Droplet-vitrification Cryopreservation Technique for Aranda Broga Blue Orchid Protocorm-like Bodies (PLBs). *CryoLetters* **2017**, *38*, 290–298.

17. Reed, B.M. *Shoot Tips Cryopreservation Manual*; USDA-ARS National Clonal Germplasm Repository: Corvallis, OR, USA, 2004; pp. 1–39. Available online: https://www.ars.usda.gov/ARSUserFiles/4630/cryopreservation/20.%20Cryopreservation%20Manual%20(2004).pdf (accessed on 25 October 2022).

18. Matsumoto, T.; Sakai, A.; Yamada, K. Cryopreservation of *in vitro*-grown apical meristems of wasabi (*Wasabia japonica*) by vitrification and subsequent high plant regeneration. *Plant Cell Rep.* **1994**, *13*, 442–446. [CrossRef]

19. Roose, M.L.; Gmitter, F.G.; Lee, R.F.; Hummer, K.E. Conservation of citrus germplasm: An international survey. *Acta Hortic* **2015**, *1101*, 33–38. [CrossRef]

20. Gonzalez-Arnao, M.T.; Engelmann, F. Cryopreservation of plant germplasm using the encapsulation-dehydration technique: Review and case study on sugarcane. *CryoLetters* **2006**, *27*, 155–168.

21. Volk, G.M.; Bonnart, R.; Krueger, R.; Lee, R. Cryopreservation of citrus shoot tips using micrografting for recovery. *CryoLetters* **2012**, *33*, 418–426.

22. Kaya, E.; Souza, F.; Yilmaz-Gokdogan, E.; Ceylan, M.; Jenderek, M. Cryopreservation of *Citrus* Seed via Dehydration Followedby Immersion in Liquid Nitrogen. *Turk. J. Biol.* **2017**, *41*, 242–248. [CrossRef]

23. Volk, G.M.; Bonnart, R.; Shepherd, A.; Yin, Z.; Lee, R.; Polek, M.L.; Krueger, R. Citrus cryopreservation: Viability of diverse taxa and histological observations. *Plant Cell Tissue Organ Cult.* **2017**, *128*, 327–334. [CrossRef]

24. Wang, M.R.; Lambardi, M.; Engelmann, F.; Pathirana, R.; Panis, B.; Volk, G.M.; Wang, Q.C. Advances in cryopreservation of *in vitro*-derived propagules: Technologies and explant sources. *Plant Cell Tiss Organ Cult.* **2021**, *144*, 7–20. [CrossRef]

25. Bettoni, J.C.; Bonnart, R.; Volk, G.M. Challenges in implementing plant shoot tip cryopreservation technologies. *Plant Cell Tiss Organ Cult.* **2021**, *144*, 21–34. [CrossRef]

26. Kim, H.H.; Lee, Y.G.; Park, S.U.; Lee, S.C.; Baek, H.J.; Cho, E.G.; Engelmann, F. Development of Alternative Loading Solutions in Droplet-Vitrification Procedures. *CryoLetters* **2009**, *30*, 291–299.

27. Cruz-Cruz, C.A.; González-Arnao, M.T.; Engelmann, F. Biotechnology and conservation of plant biodiversity. *Resources* **2013**, *2*, 73–95. [CrossRef]

28. Kulus, D.; Zalewska, M. Cryopreservation as a tool used in long-term storage of ornamental species—A review. *Sci. Hortic.* **2014**, *168*, 88–107. [CrossRef]

29. Panis, B.; Piette, B.; Swennen, R. Droplet vitrification of apical meristems: A cryopreservation protocol applicable to all Musaceae. *Plant Sci.* **2005**, *168*, 45–55. [CrossRef]

30. Benelli, C.; De Carlo, A.; Engelmann, F. Recent advances in the cryopreservation of shoot-derived germplasm of economically important fruit trees of *Actinidia, Diospyros, Malus, Olea, Prunus, Pyrus* and *Vitis. Biotechnol. Adv.* **2013**, *31*, 175–185. [CrossRef] [PubMed]

31. Ozudogru, E.A.; Kirdok, E.; Kaya, E.; Capuana, M.; Benelli, C.; Engelmann, F. Cryopreservation of Redwood (*Sequoia sempervirens* (D. Don.) Endl.) in vitro Buds Using Vitrification-Based Techniques. *CryoLetters* **2011**, *32*, 99–110.

32. Mc Cown, B.H.; Lloyd, G. Woody Plant Medium (WPM)—A Mineral Nutrient Formulation for Microculture of Woody Plant Species. *Hort. Sci.* **1981**, *16*, 453.

33. Murashige, T.; Skoog, F. A Revised Medium for Rapid Growth and Bio Assays with Tobacco Tissue Cultures. *Plant Physiol.* **1962**, *15*, 473–497. [CrossRef]

34. Lambardi, M.; Sharma, K.K.; Thorpe, T.A. Optimization of in vitro bud induction and plantlet formation from mature embryos of Aleppo pine (*Pinus halepensis* Mill.). *Vitr. Cell Dev. Biol. Plant.* **1993**, *29*, 189–199. [CrossRef]

35. Sakai, A.; Kobayashi, S.; Oiyama, I. Cryopreservation of nucellar cells of navel orange (*Citrus sinensis* Osb. var. *brasiliensis* Tanaka) by vitrification. *Plant Cell Rep.* **1990**, *9*, 30–33. [CrossRef]

36. Carpentier, S.C.; Vertommen, A.; Swennen, R.; Fortes, C.W.; Souza, M.T.; Panis, B. Sugar-Mediated Acclimation: The Importance Of Sucrose Metabolism in Meristems. *J. Proteome Res.* **2010**, *9*, 5038–5046. [CrossRef]

37. Pociecha, E.; Plazek, A.; Janowiak, F.; Zwierzykowski, Z. ABA level, proline and phenolic concentration, and PAL activity induced during cold acclimation in androgenic Festulolium forms with contrasting resistance to frost and pink snow mould (*Microdochium nivale*). *Physiol. Mol. Plant Pathol.* **2009**, *73*, 126–132. [CrossRef]

38. Fki, L.; Bouaziz, N.; Chkir, O.; Benjemaa-Masmoudi, R.; Rival, A.; Swennen, R.; Drira, N.; Panis, B. Cold hardening and sucrose treatment improve cryopreservation of date palm meristems. *Biol. Plant* **2012**, *57*, 375–379. [CrossRef]

39. Reed, B.M. Cold Hardening vs ABA as a Pretreatment For Meristem Cryopreservation. *Hort. Sci.* **1990**, *25*, 1086. [CrossRef]
40. Kushnarenko, S.V.; Romadanova, N.V.; Reed, B.M. Cold Acclimation Improves Regrowth of Cryopreserved Apple Shoot Tips. *CryoLetters* **2009**, *30*, 47–54.
41. Ozudogru, A.; da Silva, D.P.C.; Kaya, E.; Dradi, G.; Paiva, R.; Lambardi, M. In vitro Conservation and Cryopreservation of *Nandina domestica* an Outdoor Ornamental Shrub. *Not. Bot. Horti Agrobot. Cluj-Napoca* **2013**, *41*, 638–645. [CrossRef]
42. Takada, S.; Tasaka, M. Embryonic Shoot Apical Meristem Formation in Higher Plants. *J. Plant Res.* **2002**, *115*, 411–417. [CrossRef]
43. Benson, E.E. Cryopreservation Of Phytodiversity: A Critical Appraisal of Theory & Practice. *CRC Crit. Rev. Plant Sci.* **2008**, *27*, 141–219. [CrossRef]
44. Helliot, B.; Swenen, R.; Poumay, Y.; Frison, E.; Lepoivre, P.; Panis, B. Ultrastructural Changes Associated with Cryopreservation of Banana (*Musa* spp.) Highly Proliferating Meristems. *Plant Cell Rep.* **2003**, *21*, 690–698. [CrossRef]
45. Wang, Q.C.; Valkonen, J.P.T. Efficient elimination of sweetpotato little leaf phytoplasma from sweetpotato by cryotherapy of shoot tips. *Plant Pathol.* **2008**, *57*, 338–347. [CrossRef]
46. Bettoni, J.C.; Bonnart, R.; Shepherd, A.; Kretzschmar, A.A.; Volk, G.M. Modifications to a Vitis shoot tip cryopreservation procedure: Effect of shoot tip size and use of cryoplates. *CryoLetters* **2019**, *40*, 103–112.
47. Panis, B.; Strosse, H.; Van Den Hende, S.; Swennen, R. Sucrose preculture to simplify cryopreservation of banana meristem cultures. *CryoLetters* **2002**, *23*, 375–384.
48. Pinker, I.; Halmagyi, A.; Olbricht, K. Effects of sucrose preculture on cryopreservation by droplet-vitrification of strawberry cultivars and morphological stability of cryopreserved plants. *CryoLetters* **2009**, *30*, 202–211.
49. Folgado, R.; Panis, B.; Sergeant, K.; Renaut, J.; Swennen, R.; Hausman, J.F. Unravelling the effect of sucrose and cold pretreatment on cryopreservation of potato through sugar analysis and proteomics. *Cryobiology* **2015**, *71*, 432–441. [CrossRef] [PubMed]
50. Panis, B.; Totte, N.; Van Nimmen, K.; Withers, L.A.; Swennen, R. Cryopreservation of banana (*Musa* spp.) meristem cultures after preculture on sucrose. *Plant Sci.* **1996**, *121*, 95–106. [CrossRef]
51. Souza, F.V.D.; Kaya, E.; de Jesus Vieira, L.; da Silva Souza, A.; da Silva Carvalho, M.D.J.; Santos, E.B.; Alves, A.A.C.; Ellis, D. Cryopreservation of Hamilin sweet orange [(*Citrus sinensis* (L.) Osbeck)] embryogenic calli using a modified aluminum cryo-plate technique. *Sci. Hortic.* **2017**, *224*, 302–305. [CrossRef]
52. Souza, F.V.D.; de Souza, E.H.; Kaya, E.; de Jesus Vieira, L.; da Silva, R.L. Cryopreservation of Pineapple Shoot Tips by the Droplet Vitrification Technique. *Methods Mol. Biol.* **2018**, *815*, 269–277. [CrossRef]
53. Sakai, A.; Engelmann, F. Vitrification, encapsulation-vitrification and droplet-vitrification: A review. *CryoLetters* **2007**, *28*, 151–172.
54. Thomas, T.D. The role of activated charcoal in plant tissue culture. *Biotechnol. Adv.* **2008**, *26*, 618–631. [CrossRef]
55. Matoh, T. Boron in plant cell walls. *Plant Soil* **1997**, *193*, 59–70. [CrossRef]
56. Matoh, T.; Kobayashi, M. Boron Function in Plant Cell Walls. In *Boron in Plant and Animal Nutrition*; Goldbach, H.E., Brown, P.H., Rerkasem, B., Thellier, M., Wimmer, M.A., Bell, R.W., Eds.; Springer: Boston, MA, USA, 2002; pp. 143–155.
57. Van der Salm, T.P.; Van der Toorn, C.J.; Hänisch ten Cate, C.H.; Dubois, L.A.; De Vries, D.P.; Dons, H.J. Importance of the iron chelate formula for micropropagation of Rosa hybrida L.'Moneyway'. *Plant Cell Tissue Organ Cult.* **1994**, *37*, 73–77. [CrossRef]
58. Kaviani, B. Conservation of plant genetic resources by cryopreservation. *Aust. J. Crop. Sci.* **2011**, *5*, 778–800.
59. Kaya, E.; Souza, F.V.D.; Almeida dos Santos-Serejo, J.; Galatali, S. Influence of dehydration on cryopreservation of *Musa* spp. germplasm. *Acta Bot Croat.* **2020**, *79*, 99–104. [CrossRef]
60. Reed, B.M.; Kovalchuk, I.; Kushnarenko, S.; Meier-Dinkel, A.; Schoenweiss, K.; Pluta, S.; Straczynska, K.; Benson, E.E. Evaluation of critical points in technology transfer of cryopreservation protocols to international plant conservation laboratories. *CryoLetters* **2004**, *25*, 341–352.
61. Panis, B.; Nagel, M.; Van den houwe, I. Challenges and Prospects for the Conservation of Crop Genetic Resources in Field Genebanks, in In Vitro Collections and/or in Liquid Nitrogen. *Plants* **2020**, *9*, 1634. [CrossRef]

Article

Morpho-Anatomical and Physiological Assessments of Cryo-Derived Pineapple Plants (*Ananas comosus* var. *comosus*) after Acclimatization

Ariel Villalobos-Olivera [1], José Carlos Lorenzo-Feijoo [2], Nicolás Quintana-Bernabé [1], Michel Leiva-Mora [3], Jean Carlos Bettoni [4] and Marcos Edel Martínez-Montero [1,*]

[1] Facultad de Ciencias Agrícolas, Universidad de Ciego de Ávila Máximo Gómez Báez, Ciego de Ávila 65200, Cuba; villalobos.olivera@gmail.com (A.V.-O.); nquintana@unica.cu (N.Q.-B.)
[2] Laboratory for Plant Breeding and Conservation of Genetic Resources, Bioplant Centre, University of Ciego de Ávila, Ciego de Ávila 65200, Cuba; jclorenzo@bioplantas.cu
[3] Departamento de Agronomía, Facultad de Ciencias Agropecuarias, Universidad Técnica de Ambato (UTA-DIDE), Campus Querochaca, Cantón Cevallos Vía a Quero, Sector El Tambo-La Universidad, Cevallos 1801334, Tungurahua, Ecuador; m.leiva@uta.edu.ec
[4] Independent Researcher, 35 Brasil Correia Street, Videira 89560510, SC, Brazil; jcbettoni@gmail.com
* Correspondence: cubaplantas@gmail.com

Abstract: Studies on the morpho-physiology of cryo-derived pineapple plants after acclimatization have been quite limited. Therefore, in the present study, the morpho-anatomical and physiological characteristics of cryo-derived *Ananas comosus* var. *comosus* 'MD-2' plants after acclimatization were investigated. Plants obtained from cryopreserved and non-cryopreserved shoot tips, as well as in vitro stock cultures (control), showed similar morphological development (viz. plant height, number of leaves, D leaf length, D leaf width, D leaf area, diameter of stem base, number of roots, plant fresh weight and plant dry weight) to conventionally micropropagated and non-cryopreserved plants. The pineapple plantlets developed efficient anatomical leaf structures that allowed them to adapt to the transition process from in vitro to ex vitro. In all groups of plants, the content of water and chlorophylls (a, a + b, a/b) decreased during the first 15 days of acclimatization and then remained constant until the end of the evaluation. The mesophilic succulence index increased to its maximum value after 15 days, then decreased and remained constant up to 45 days. Although physiological indicators fluctuated during the 45 days of acclimatization, no differences were observed in any of the indicators evaluated when plantlets obtained from cryopreserved shoot tips were compared with controls. The results of the plants from cryopreserved shoot tips show that they switched from C3 to Crassulacean acid metabolism, which denoted metabolic stability during acclimatization.

Keywords: crassulacean acid metabolism; cryopreservation; metabolic stability; vegetative growth

Citation: Villalobos-Olivera, A.; Lorenzo-Feijoo, J.C.; Quintana-Bernabé, N.; Leiva-Mora, M.; Bettoni, J.C.; Martínez-Montero, M.E. Morpho-Anatomical and Physiological Assessments of Cryo-Derived Pineapple Plants (*Ananas comosus* var. *comosus*) after Acclimatization. *Horticulturae* **2023**, *9*, 841. https://doi.org/10.3390/horticulturae9070841

Academic Editor: Haiying Liang

Received: 21 May 2023
Revised: 17 July 2023
Accepted: 20 July 2023
Published: 24 July 2023

1. Introduction

The conservation of biological material through cryopreservation is now considered to be the safest and most cost-effective strategy for the long-term storage of plant genetic resources [1–4]. In the case of important edible horticultural crops like pineapples (*Ananas comosus* var. *comosus*), which are vegetatively propagated and where crosses between varieties produce botanical seeds that are highly heterozygous, their seeds are of limited interest for the conservation of specific gene combinations [5]. The cryopreservation of shoot tips is the most relevant strategy for the long-term conservation of the pineapple crop. This is because true-to-type, virus-free plants can be regenerated directly from cryopreserved shoot tips [2–4]. Once cryopreserved, shoot tips are stored in a state where cellular divisions and metabolic processes are halted, and theoretically, plant materials can be preserved without genetic alteration for an indefinite period of time [6].

Cryopreservation methods have already been developed for pineapple shoot tips [5,7–13]. The vitrification-based cryopreservation method is the most widely applied for cryopreserving pineapple shoot tips [5,10–13]. In this method, precultured shoot tips are exposed to a highly concentrated plant vitrification solution (PVS) before being directly immersed in liquid nitrogen (LN) [2,12,13]. To date, the droplet-vitrification technique has been shown to be the most effective cryopreservation method across diverse genotypes. Droplet-vitrification uses ultra-fast cooling and warming rates of shoot tips, an important requirement for successful cryopreservation protocols [14,15]. Souza et al. [10,11] described a successful droplet-vitrification protocol that was applied to 16 genotypes of *Ananas*, both wild and cultivated, belonging to four botanical varieties. In this procedure, shoot tips were precultured for 48 h in preculture medium containing 0.3 M sucrose and then transferred to aluminum foil in 4–5 µL of plant vitrification solution 2 (PVS2) and treated at 0 °C for 45 min before being directly submerged in LN. This procedure resulted in shoot tip regrowth ranging from 44% to 86% across the 16 pineapple genotypes [10].

A. comosus var. *comosus* 'MD-2' is a hybrid cultivar of interest to the agricultural industry due to its desirable characteristics, commercial potential and ability to meet the demands of the global market [16]. In this sense, the Bioplantas Center (www.bioplantas.cu; accessed on 10 July 2023) in Cuba initiated the development of a technological innovation project in 2010 to generalize the micropropagation of the 'MD-2' pineapple, promote the creation of donor plant banks in different regions of the country and introduce cryopreservation protocols into the practice.

Micropropagation involves growing plants in vitro under conditions that restrict gas exchange, maintain high humidity and low light and use a sucrose-based medium, which may interfere with photosynthesis [17,18]. In addition, the different steps during cryopreservation protocols may impose chemical or physical stresses to the plant tissues which would cause somaclonal, genetic and epigenetic variations in regenerates in addition to poor shoot tip regrowth due to reactive oxygen species (ROS) formation [19–21]. Treatments such as shoot tip excision, preculture, cryoprotection, dehydration, the freeze–thaw cycle and acclimatization have all been known to impose ROS-induced oxidative stress [21–23]. The overproduction of ROS is highly reactive and toxic and can cause lipid peroxidation, protein denaturation, alterations in nucleic acids and membrane disruptions that may lead to programmed cell death [24,25]. Therefore, it is necessary to not only ensure shoot tip viability after cryopreservation, but also the true-to-type of the regenerants [18–21].

Several research papers have mentioned success in acclimating cryo-derived pineapple plants [10,26–28]. However, there have been only a few studies on assessments of morpho-anatomical characteristics in cryo-derived pineapple plants during/after acclimatization [26,28]. Furthermore, these studies do not describe the metabolic changes in cryo-derived plants during the transition from the in vitro to the ex vitro environment. It is known that pineapple plants undergo a shift from C3 to Crassulacean acid metabolism (CAM) when adapting to ex vitro conditions [29,30]. Therefore, this study investigates in detail, for the first time, the morpho-anatomical and physiological characteristics of cryo-derived pineapple plants during acclimatization. The results reported here support the use of the droplet-vitrification cryopreservation procedure described by Souza et al. [10] and modified by Villalobos et al. [12] for the establishment of cryopreserved pineapple gene bank collections.

2. Materials and Methods

Plant material and stock cultures

Tissue-cultured plants of *A. comosus* var. *comosus* 'MD-2' were obtained from the Bioplantas Center of the Universidad de Ciego de Ávila Máximo Gómez Báez (www.unica.cu; accessed on 6 July 2023) and used in this study. In vitro stock cultures were propagated and maintained in active growth on Murashige and Skoog (MS) medium [31] supplemented with 1.0 mg·L^{-1} thiamine, 0.1 mg·L^{-1} myo-inositol, 0.3 mg·L^{-1} 1-naphthaleneacetic acid (NAA), 30 g·L^{-1} sucrose, 2.1 mg·L^{-1} 6-benzylaminopurine (BAP) and 6.5 g·L^{-1} agar at pH 5.7 [32]. Cultures were placed in glass vessels (55 × 95 mm) and incubated in a growth chamber at 25 ± 2 °C with a 16 h photoperiod with a photosynthetic flux density of 75 ± 3 μmol·s^{-1}·m^{-2}. Subculture in fresh culture medium was performed every 45 days. After 45 days of subculture and before cryopreservation, shoots were preconditioned for 45 days in a liquid MS culture medium containing 100 mg·L^{-1} myoinositol, 1.0 mg·L^{-1} thiamine and 30 g·L^{-1} sucrose at a pH of 5.7, as defined by Villalobos-Olivera et al. [12].

Shoot tip cryopreservation

To cryopreserve pineapple shoot tips, the droplet-vitrification method was performed as described by Souza et al. [10], with some modifications according to the studies of Martínez-Montero et al. [5,9]. Specifically, plant vitrification solution 3 (PVS3; solution consisting of MS + 50% (*w/v*) sucrose + 50% (*w/v*) glycerol) [33] was used instead of PVS2 (solution consisting of MS + 30% (*w/v*) glycerol, 15% (*w/v*) ethylene glycol + 15% (*w/v*) dimethylsulfoxide + 0.4 M sucrose) [34]. Shoot tips (1 mm length, 1 mm width) with three- or four-leaf primordia were excised from 45-day-old in vitro stock cultures and incubated on MS culture medium supplemented with 100 mg·L^{-1} myoinositol, 1.0 mg·L^{-1} thiamine, 10 g·L^{-1} sucrose and 6.5 g·L^{-1} agar at pH 5.7 for 24 h at 25 ± 2 °C under dark conditions.

Pretreatment with glycerol and sucrose: Next, 10 shoot tips with an average fresh weight of 5 mg were placed in each polypropylene cryovial with a volume of 2.0 mL. The fresh weight of each shoot tip was estimated using the rule of three, based on the proportionality of a shoot tip's weight to the weight of 20 shoot tips measured using an analytical balance (Sartorius Entris 64-1S, Germany). Immediately thereafter, the cryovials were filled with liquid MS culture medium containing 2 M glycerol and 0.4 M sucrose for 20 min at 25 ± 2 °C.

Treatment with PVS3 vitrification solution: The cryovials containing the shoot tips were poured onto the surface of 9 cm diameter Petri dishes, which had filter paper moistened with 5 mL of PVS3 solution precooled to 0 °C. The Petri dishes were placed on the surface of an ice bath to dehydrate the shoot tips for 60 min. Individual shoot tips were then placed on aluminum foil strips (7 mm × 20 mm × 50 μm) in droplets of 5 μL PVS3, with five shoot tips per slide.

Immersion in LN: After the end of the PVS3 dehydration time, the aluminum foil strips containing the shoot tips were plunged in LN. After LN exposure for a few seconds, the foils with shoot tips were transferred into 2 mL cryovials filled with LN and maintained in LN for 10 h.

Thawing: the aluminum foil strips with shoot tips were warmed quickly by inverting the aluminum foil strips into unloading solution (MS culture medium + 1 M sucrose) at 25 ± 2 °C for 5 min.

In vitro recovery after cryopreservation: Shoot tips were placed in 55 × 95 mm glass vessels with solid MS culture medium containing 1.0 mg·L^{-1} NAA, 1.0 mg·L^{-1} BAP, 10 g·L^{-1} sucrose and 6.5 g·L^{-1} agar at pH 5.7 and cultured for 7 weeks at 25 ± 2 °C in darkness. They were then grown in the same conditions as the in vitro stock cultures.

Acclimatization conditions

Plants with a height greater than 5 cm, 5 to 8 functional leaves and a fresh weight greater than 4.5 g were considered as optimal to be transferred to the ex vitro phase. Plants were dipped in the preventive fungicide Previcur Energy® (Bayer Cropscience) at a concentration of 1 mL·L^{-1} for 5 min. To determine the effectiveness of acclimatization in pineapple plants from cryopreserved apices, two different conditions were established. The first condition consisted of plastic trays with four 0.5 cm Ø drainage holes, containing 90 cm^3 of red ferric soil and filter cake (1:1). The second condition involved black polyethylene bags with a substrate volume of 400 cm^3. Both conditions were performed according to Pino et al. [35]. The substrate mixture consisted of typical red ferralitic soil and filter cake (from sugarcane) (1:1) (*v/v*), which had been previously sieved. Table 1 shows the chemical properties of the substrate used in the acclimatization phase.

Table 1. Chemical properties of the substrate used for ex vitro acclimatization.

Components	CaO	K$_2$O	P$_2$O$_5$	OM	EC	pH
	mg·L^{-1}			%	mS·cm^{-1}	
Soil + filter cake	211.32	108.86	1107.9	31.4	1.07	7.10

Abbreviations: CaO, calcium oxide; K$_2$O, potassium oxide; P$_2$O$_5$, phosphorus pentoxide; OM, organic matter; EC, electrical conductivity.

Irrigation was performed daily using microsprinklers twice a day, at 9:00 a.m. and 2:00 p.m., for 10 min. The microsprinklers used were from an atomizer located at a height of 1.80 m above the ground, a feature that allows water to reach the plants in the form of a mist to maintain moisture and temperature [35].

During the first 45 days, the plants were cultivated in a greenhouse with a ceiling that allowed the passage of 25% of sunlight, environmental conditions of 80 ± 3% of relative humidity and 26.5 °C (using a digital thermo-hygrometer TECPEL®, model DTM-303, Spain), with natural light and photoperiod and a photosynthetic flux density of 250 ± 30 µmol·m^{-2}·s^{-1} (using a digital Lux/Light Meter, model FT-710 Faithfull, China) and atmospheric conditions with a CO$_2$ concentration between 375 and 400 µmol·mol^{-1} according to Pino et al. [30].

During this stage, the experimental treatments were established as follows: control plants that were micropropagated in a conventional manner (Microp), plants derived from shoot tips that were cryoprotected but not cryopreserved (non-cryopreserved) and plants derived from cryopreserved shoot tips (Cryo).

Morpho-anatomical and physiological assessments

After 45 days of acclimatization, the morpho-anatomical and physiological indicators of 50 plants from non-cryopreserved, cryopreserved and micropropagated plants were determined. Plant height (cm), number of leaves, length and width of leaf D (cm) (the leaf of the best physiological characteristics), leaf area D (cm^2), diameter of stem base of the plant (cm), number of roots and plant fresh weight and dry weight (g) were evaluated.

For anatomical analysis, the procedures were performed as specified by Ebel et al. [36]. For leaf D, sections were made in the middle part to perform the anatomical analysis. The sections were fixed in 70% alcohol, formaldehyde and acetic acid (90:5:5) for 12 h and then freehand cross sections of 15–25 µm thickness were made with a razor blade. These were stained with safranin for 3 min and then with toluidine blue for 5 min. Visualization was performed using an inverted biological microscope (model NIB-100, China) in conjunction with a digital camera (model HDCE-50B, China).

For the physiological indicators, the mesophilic succulence index (MSI) was determined using the ratio between the aquifer and the photosynthetic tissue of the plants according to Rodríguez-Escriba et al. [29]. This assessment was performed on days 0, 15, 30 and 45 of acclimatization using the following equation: MSI = WC [Cfls (a + b)] − 1.

Chlorophyll contents (Cfls) a, b, a + b and a/b ratio were determined on slices from the central area of leaf D (diameter of 0.78 cm), as suggested by Rodríguez-Escriba et al. [29]. The water content (WC) of the slices was determined by the difference between the fresh weight and the dry weight of the samples after 72 h of drying in an oven at 60 °C.

Gas exchange rate: Fifty plants, both non-cryopreserved and cryopreserved, were sampled, and transpiration rate and CO$_2$ assimilation were determined at 12:00 p.m. and 12:00 a.m. as described by Rodríguez-Escriba et al. [29]. To determine water use efficiency, CO$_2$ assimilation was divided by transpiration rate.

Organic acid: Fifty samples of 1 g were taken from the center of the largest plant leaves of cryopreserved and non-cryopreserved plants at 12:00 p.m. and 12:00 a.m. of each experimental treatment. For these samples, 10 mL of 50% (*v*/*v*) ethanol was added and incubated in a thermal bath at 90 °C for 20 min. Then, the liquid phase was separated from the solid phase and acid–base evaluation was performed using 0.79 g·L^{-1} NaOH and 1 mg·L^{-1} phenolphthalein as indicators, as described by Rodríguez-Escriba et al. [29]. Organic acid content was expressed in µmol H$^+$·g^{-1} leaf fresh weight.

Statistical analysis

For data analysis, IBM SPSS version 22 was used. First, data from each treatment in each experiment were shown to meet the assumptions of normal distribution and homogeneity of variances; second, the Kolmogorov–Smirnov and Levene tests were both used with $p \leq 0.05$. Parametric tests were performed (one-way ANOVA and bifactorial). In addition, Tukey's HSD test with $p \leq 0.05$ was performed for the ANOVA, which showed significant differences. First, for this analysis, data were transformed in percentages according to the function $y' = 2 \arcsin (y/100)^{1/2}$.

3. Results

In general, the regeneration of shoot tips after cryoprotection treatment and cryopreservation exceeded 98%. Furthermore, all of the 'MD-2' pineapple plantlets developed from shoot tips in three variants (i.e., plants regenerated from cryopreserved shoot tips, conventionally micropropagated plants and non-cryopreserved plants) were successfully rooted in vitro and survived the acclimatization process.

3.1. Morphological Variables

Table 2 shows a morphological comparison of the 'MD-2' pineapple plants regenerated from micropropagation or cryopreservation via droplet-vitrification (before or after LN immersion) and acclimatized (45 days) under different conditions (plastic trays or black polyethylene bags). Firstly, it is noteworthy that all measured variables, including plant height, number of leaves, leaf dimensions (length, width, area), stem base diameter, number of roots and fresh and dry weights of plants, were comparable between cryopreserved and non-cryopreserved plants, as well as in vitro stock cultures. Secondly, the plants acclimatized in black polyethylene bags exhibited the highest results in terms of morphological variables, except for leaf length and stem base diameter, where no statistical differences were observed. Therefore, for the subsequent research, the acclimatization process using black polyethylene bags was selected.

Table 2. Morphological comparison of 'MD-2' pineapple plants regenerated from micropropagation or cryopreservation via droplet-vitrification before or after liquid nitrogen (LN) immersion, and acclimatized (45 days) in black polyethylene bags or plastic trays.

Variables	Plastic Trays			Black Polyethylene Bags			Average	SE
	Microp	Non-Cryo	Cryo	Microp	Non-Cryo	Cryo		
Plant height (cm)	10.48 b	10.52 b	10.53 b	11.98 a	11.97 a	11.99 a	11.24	±0.04
Number of leaves	8.26 b	8.24 b	8.21 b	9.20 a	9.23 a	9.21 a	8.65	±0.04
D leaf length (cm)	9.11 a	9.16 a	9.16 a	9.14 a	9.13 a	9.12 a	9.14	±0.05
D leaf width (cm)	1.55 b	1.42 b	1.54 b	1.60 a	1.63 a	1.62 a	1.56	±0.08
D leaf area (cm^2)	6.88 b	6.82 b	6.75 b	7.02 a	7.01 a	7.03 a	6.94	±0.07
Diameter of stem base (cm)	1.32 a	1.29 a	1.28 a	1.36 a	1.35 a	1.36 a	1.32	±0.04
Number of roots	10.88 b	11.02 b	10.96 b	12.24 a	12.17 a	12.31 a	11.59	±0.14
Plant fresh weight (g)	10.48 b	10.52 b	10.46 b	10.66 a	10.62 a	10.66 a	10.48	±0.06
Plant dry weight (g)	1.83 b	1.82 b	1.80 b	1.86 a	1.86 a	1.86 a	1.83	±0.03

Abbreviations: SE: standard error of the mean. Microp: plants that were micropropagated in a conventional manner. Non-cryo: plants derived from shoot tips that were cryoprotected but not cryopreserved. Cryo: plants derived from cryopreserved shoot tips. Values labeled with the same letters in each column were not significantly different at $p < 0.05$ using Tukey's mean separation test. Each data point represents the mean for n = 50.

3.2. Morpho-Anatomical Characteristics

Figure 1 shows that no visual differences were observed between different treatments after acclimatization. After 45 days of ex vitro acclimatization, the pineapple plants in all three groups had similar root and leaf systems (Figure 1A,D,G) and displayed functional leaves and roots (Figure 1A–C). The leaves were thick and erect, with a distinct adaxial epidermis, aquifer parenchyma, chlorophyll parenchyma, abaxial epidermis and scales (Figure 1B,E,H). Extra axillary fibers and vascular rods were also clearly visible (Figure 1C,F,I).

Figure 1. Morpho-anatomical characteristics of pineapple plants of 'MD-2' micropropagated control (**A–C**), regenerated from cryoprotected (**D–F**) and cryoprotected and cryopreserved (**G–I**) 45 days after acclimatization. (**A,D,G**) Plants after 45 days of acclimatization, (**B,E,H**) leaf D of plants at the end of the acclimatization period and (**C,F,I**) cross section of leaves D of plants at the end of the acclimatization period. h1: in vitro leaves, h2: ex vitro leaves, epad: adaxial epidermis, pac: aquifer parenchyma, fe: extraaxillary fibers, av: vascular bar, pcl: chlorophyll parenchyma, epab: abaxial epidermis, esc: scales. The bar represents the 1 cm dimension for (**A,D,G,B,E,H**) and the 100 μm dimension for (**C,F,I**).

3.3. Physiological Indicators

Although the variations in physiological indicators (Table 3) occurred along the 45 days of acclimatization, no differences were observed in any indicator evaluated when comparing plantlets recovered from cryopreserved shoot tips with the controls (micropropagated and non-cryopreserved), without LN exposure. Water content, chlorophyll a, chlorophylls a + b and a/b ratio decreased in the first 15 days and then started to increase after 30 days and reached their maximum value after 45 days. The chlorophyll b increased after 15 days and remained constant until the end of the assessment. The mesophilic succulence index reached its maximum value on day 15, and then decreased and remained constant until the end of the assessment.

Table 3. Ex vitro acclimatization effects on D leaf water content, chlorophyll a, chlorophyll b, chlorophyll a + b, chlorophyll a/b ratio and mesophilic succulence index of pineapple plantlets.

Evaluation Time (Days)	D Leaf Water Content (g H_2O cm^{-2})			Chlorophyll a (μg Chlorophylls cm^{-2} of D Leaf)			Chlorophyll b (μg Chlorophylls cm^{-2} of D Leaf)			Chlorophyll (a + b) (μg Chlorophylls cm^{-2} of D Leaf)			Chlorophylls a/b			Mesophilic SUCULENCE Index (g H_2O mg^{-1} Chlorophylls)		
	Microp	Non-Cryo	Cryo	Microp	Non-Cryo	Cryo	Microp	Non-Cryo	Cryo	Microp	Non-Cryo	Cryo	Microp	Non-Cryo	Cryo	Microp	Non-Cryo	Cryo
0	0.053 b	0.052 b	0.052 b	30.97 b	30.38 b	30.92 b	16.23 b	16.38 b	16.15 b	47.27 c	47.12 c	47.21 c	1.89 a	1.87 a	1.90 a	0.88 c	0.90 c	0.89 c
15	0.045 c	0.044 c	0.044 c	20.77 c	21.02 c	20.98 c	19.14 ab	19.27 ab	19.12 ab	39.91 b	39.78 b	39.88 b	1.08 b	1.07 b	1.06 b	1.88 a	1.89 a	1.87 a
30	0.048 b	0.049 b	0.047 b	21.14 c	21.12 c	21.19 c	20.18 b	19.88 b	19.95 b	40.99 b	40.20 b	40.52 b	1.09 b	1.10 b	1.08 b	1.54 b	1.52 b	1.53 b
45	0.081 a	0.080 a	0.080 a	33.47 a	33.56 a	33.54 a	24.27 a	24.14 a	24.19 a	58.05 a	58.02 a	57.98 a	1.88 a	1.89 a	1.87 a	1.44 b	1.42 b	1.42 b
SE	±0.012			±0.12			±0.15			±0.78			±0.02			±0.032		

Abbreviations: SE: standard error of the mean. Microp: plants that were micropropagated in a conventional manner. Non-cryo: plants derived from shoot tips that were cryoprotected but not cryopreserved. Values labeled with the same letters in each column were not significantly different at $p < 0.05$ using Tukey's mean separation test. Each data point represents the mean for n = 50.

3.4. Gas Exchange Rate and Organic Acid Levels

The gas exchange values found here confirm the natural expression of CAM metabolism in the plants of the three groups (Table 4). The gas exchange rate and organic acid levels were similar in plants from micropropagated, non-cryopreserved and cryopreserved shoot tips, and there was no difference between treatments. The highest gas exchange rate occurred at 12:00 a.m., while values were below the average at 12:00 p.m. The highest concentration of organic acids was also found at 12:00 a.m. (about 52 μmol H$^+$ g^{-1} fresh weight).

Table 4. Evaluation of gas exchange rate and organic acid levels in leaf D of ex vitro acclimatized pineapple plants after 45 days.

Indicator	Microp		Non-Cryo		Cryo		
	12:00 a.m.	12:00 p.m.	12:00 a.m.	12:00 p.m.	12:00 a.m.	12:00 p.m.	SE
D leaf transpiration rate (μmol H$_2$O m^{-2} s^{-1})	1.97 a	0.02 b	1.96 a	0.03 b	1.95 a	0.04 b	$\pm$0.01
D leaf stomatal conductance (μmol H$_2$O m^{-2}s^{-1})	0.10 b	58.22 a	0.26 b	58.57 a	0.25 b	58.56 a	$\pm$0.25
D leaf CO$_2$ assimilation (μmol CO$_2$ m^{-2} s^{-1})	8.26 a	0.01 b	8.20 a	0.07 b	8.21 a	0.06 b	$\pm$0.06
D leaf CO$_2$ assimilation percentage (%)	99.87 a	0.12 b	99.15 a	0.84 b	99.27 a	0.72 b	$\pm$0.71
D leaf water use efficiency (μmol CO$_2$ μmol^{-1} H$_2$O)	4.19 a	0.14 b	4.18 a	0.13 b	4.20 a	0.15 b	$\pm$0.01
D leaf organic acid levels (μmol H$^+$ g^{-1} fresh weight)	52.58 a	6.22 b	52.46 a	6.34 b	52.47 a	6.33 b	$\pm$0.12

Abbreviations: SE: standard error of the mean. Microp: plants that were micropropagated in a conventional manner. Non-cryo: plants derived from shoot tips that were cryoprotected but not cryopreserved Cryo: plants derived from cryopreserved shoot tips. Values labeled with different letters within each column differed significantly at $p < 0.05$ according to Tukey's mean separation test. Each data point represents the mean for n = 50.

4. Discussion

In this study, we report for the first time in Cuba the successful processes of in vitro propagation, cryopreservation and acclimatization of certified 'MD2' pineapples. This procedure provides an important tool to support in vitro gene banks and the long-term conservation of pineapple genetic resources. The availability of a cryopreservation method for pineapples may also facilitate the use of cryotherapy methods to eradicate viruses [37–40].

We found that cryopreservation using droplet-vitrification did not negatively affect the morphological characteristics of 'MD-2' pineapple plants, as they were comparable to those of non-cryopreserved plants and in vitro stock cultures (Table 2). Similar results of morphological characterization were also observed in the preliminary research of our group in two other pineapple cultivars 'Red Spanish Florencia' and 'Hybrid 54' (Smooth Cayenne/Red Spanish) after shoot tip cryopreservation and 45 days of acclimatization [27]. Similar morphological indicators were also observed in potatoes [2], shallots [41], apples [42], artichokes [43] and grapevines [44] between the cryopreserved plants and the control.

During the first few days after transplanting micropropagated plants under acclimatization conditions, relatively slow plant growth across all treatments was observed. This is likely due to the altered growth conditions and the shift in photosynthetic capacity [45], which is a common phenomenon observed in many plant species, including pineapple [30,35]. According to Villalobos et al. [18] and González et al. [46], pineapple plants grown in vitro develop a leaf system with autotrophic morphological structures during the initial days of ex vitro acclimatization. This helps plants to restore their photosynthetic machinery, allowing them to adapt to the low relative humidity conditions of ex vitro environments [29].

Previous studies by Hu et al. [47], Souza et al. [10] and Villalobos-Olivera et al. [12] described in vitro pineapple plants regenerated from shoot tip cryopreservation, but the adaptation of in vitro to ex vitro acclimatization conditions was not addressed. The success of in vitro propagation systems depends on effective acclimatization and this step is also critical for the establishment of seedlings under field conditions [48–50].

The acclimatization of pineapple plants in black polyethylene bags (size 400 cm^3) produced the best morphological results when compared to the use of plastic trays (size 90 cm^3), which is a novel finding (Table 2). The size of containers used for acclimatization presents a challenge in balancing optimal growing conditions and economic profitability in large-scale propagation [51]. While a reduced container size can provide significant economic benefits by lowering maintenance costs per unit of area, it can also negatively impact root development and photosynthetic activity, thereby affecting plantlet growth [52]. Furthermore, several studies have shown that small container size can restrict root growth and nutrient uptake, leading to reduced carbohydrate accumulation in leaves and an imbalanced source/sink ratio, which in turn can result in reduced shoot growth [53,54]. Kim et al. [55] demonstrated that the biggest-sized container was favorable for plant growth, as it showed higher plant height, leaf number, leaf area, fresh weight and dry weight for the growth of tissue-culture-propagated apple rootstock plants. Therefore, further studies on acclimatization based on industry requirements are needed.

The characteristics of morpho-anatomical development (Figure 1) correspond to those described by Pino et al. [35] and González et al. [46] during 45 days of ex vitro acclimatization of micropropagated pineapple plants. This reflects the adaptation process of leaf cells to the different environmental conditions, in which pineapple plantlets developed more functional leaf structures to adapt them to the transition process from in vitro to ex vitro [50,56].

Pineapple leaves formed in vitro have a short, club-shaped stem which narrow and strongly curved leaves grow around [30]. These leaves serve as a source of carbonaceous substances to meet metabolic needs and maintain plant adaptation during transitional stress in the first week under ex vitro conditions [30,45,46]. The newly developed leaves that adapted to ex vitro conditions (Figure 1B,E,H) exhibited a growth pattern with a more vertical orientation and increased width, which is indicative of greater metabolic efficiency and physiological adaptation to the specific environmental conditions [44]. These leaves exhibited a perfectly defined anatomical structure (Figure 1C,F,I), similar to that previously observed by Villalobos-Olivera et al. [28] and González et al. [46] in acclimatized pineapple plants from in vitro systems. These leaves exhibited the aquifer parenchyma, a specialized structure of plants with CAM metabolism that allows for higher efficiency in water use [36,57,58].

In the first phase of ex vitro acclimatization, pineapple plants switch from a C3 metabolism to a CAM metabolism to adapt to the increased temperature and light intensity [30,58–60]. The decrease in water content, chlorophyll a, chlorophylls a + b and a/b ratio in the first 15 days and the increase after 15 days could be related to the expression of the CAM metabolism (Table 3). The decrease in the water content of the leaves could be related to the loss of water vapor due to the inability of the stomata to function [29,58]. The increase in water content after 30 days could be related to further stomatal regulation [61]. This regulation aims to prevent water loss in the form of water vapor during respiration [29]. In addition, the plants of all groups develop aquifer tissue to store water.

The decrease in chlorophyll a in the first 15 days in plants might be related to tissue degradation during transition stress in the first week under ex vitro conditions due to increased temperature and light intensity [29]. Chlorophylls are sensitive indicators of the metabolic state of plants, and the decrease in their content reflects physiological disturbances or situations of biotic or abiotic stress [13]. Chlorophyll b increased consistently until the end of the evaluation. Working on the seedlings of four tropical woody species (Bignoniaceae) under stress conditions (low light intensity), Kitajima and Hogan [62] observed that while the chlorophyll a content decreased, the chlorophyll b content increased. This result is due to an adaptive response of the plant, in which chlorophyll b captures photons and transfers them to chlorophyll a to compensate for its function [63]. These observations could explain the decrease in chlorophyll a and increase in chlorophyll b in pineapple plants from cryopreserved shoot tips and controls (micropropagated and non-cryopreserved), as well as the results of Villalobos et al. [18]. Moreover, the fact that the chlorophyll a/b ratio remained at values above 1 can be understood as a typical photosynthetic plasticity response [29].

As for the mesophilic succulence index, the results found here are in agreement with those reported by Rodríguez-Escriba et al. [29] in pineapple plants under water stress situations. These authors reported that the increase in mesophilic succulence index was due to higher metabolic functionality. This increase was related to a decrease in chlorophylls and water content during the first 15 days under water stress conditions. The increase in the index indicates a better relationship between the hydric tissue and photosynthesis [64,65]. The plants in this study have values greater than one after 15 days of acclimatization, indicating the expression of CAM metabolism.

Gas exchange performed by plants under the three conditions after 45 days of acclimatization is characteristic of higher photosynthetic efficiency (Table 4). Pineapple plants exhibit C3 metabolism under optimal growth conditions and show CAM metabolism as an adaptation mechanism to biotic and abiotic stresses [66,67]. The CAM plants perform gas exchange at night [60,64,68]. The studied plants perform transpiration, stomatal conductance, CO_2 assimilation and water use efficiency at night time (12:00 a.m.), which is characteristic of CAM metabolism.

The highest concentration of organic acids was also found at night time (12:00 a.m.). This increase in organic acids during the night could be related to the accumulation of CO_2 in the form of malic acid in the vacuole [29], and during the day, malic acid is decarboxylated by phosphoenolpyruvate carboxylase (PEPC) [65]. According to Wang et al. [69] and Males and Griffiths [64], stomatal opening in CAM is related to increased relative humidity and decreased transpiration gradient in response to partial pressure of water vapor in the growing environment. Plants with these metabolic traits are more efficient in water use because they open their stomata at night and at cooler times of the day [70–72].

Overall, acclimatization is a complex process that involves the coordination of multiple physiological and biochemical responses to environmental stressors, including oxidative stress [73]. Changes in gene expression and cellular signaling pathways are important mechanisms through which plants acclimatize to environmental stresses, including oxidative stress. These mechanisms enable plants to adapt and maintain their development, even under stressful conditions [74,75].

5. Conclusions

The findings of the study have several potential implications for the cultivation of 'MD-2' pineapple plants. Firstly, the successful cryopreservation of 'MD-2' pineapple shoots using the droplet-vitrification method ensures the long-term conservation of pineapple germplasm, which may be useful for future propagation and breeding programs. Secondly, the morpho-anatomical and physiological characteristics of cryopreserved 'MD-2' pineapple plants were comparable to those of non-cryopreserved plants, indicating that cryopreservation does not negatively impact the growth and development of the plants, thus suggesting that this method may be ready for implementation in pineapple gene bank collections. Finally, we observed that the acclimatization of 'MD-2' pineapple plants in black polyethylene bags produced the best plant development, suggesting that this method could be used to improve plant propagation efficiency in commercial nurseries.

Author Contributions: Conceptualization, A.V.-O. and M.E.M.-M.; experimental design, A.V.-O., J.C.L.-F., N.Q.-B. and M.E.M.-M.; data collection, A.V.-O. and N.Q.-B.; statistical analysis, A.V.-O., J.C.L.-F. and N.Q.-B.; methodology, M.L.-M. preparation of the tables, writing—original paper preparation, A.V.-O.; writing—review and editing, M.L.-M., J.C.B. and M.E.M.-M.; visualization, all authors; project administration, A.V.-O. and M.E.M.-M.; funding acquisition, J.C.B. and M.E.M.-M. All authors have read and agreed to the published version of the manuscript.

Funding: This research received no external funding.

Data Availability Statement: The data presented in this study are available on request from the corresponding author.

Acknowledgments: The authors thank the Centro de Bioplantas and the Faculty of Agricultural Sciences of the Universidad de Ciego de Ávila Máximo Gómez Báez (Cuba), for their support in carrying out the research.

Conflicts of Interest: The authors declare no conflict of interest.

References

1. Jiroutová, P.; Sedlák, J. Cryobiotechnology of plants: A hot topic not only for gene banks. *Appl. Sci.* **2020**, *10*, 4677. [CrossRef]
2. Bettoni, J.C.; Bonnart, R.; Volk, G. Challenges in implementing plant shoot tip cryopreservation technologies. *Plant Cell Tissue Organ Cult.* **2021**, *144*, 21–34. [CrossRef]
3. Bettoni, J.C.; Mathew, L.; Pathirana, R.; Wiedow, C.; Hunter, D.A.; McLachlan, A. Eradication of *Potato Virus S, Potato Virus A*, and *Potato Virus M* from infected *in vitro*-grown potato shoots using in vitro therapies. *Front. Plant Sci.* **2022**, *13*, 1431. [CrossRef]
4. Zhang, A.; Wang, M.-R.; Li, Z.; Panis, B.; Bettoni, J.C.; Vollmer, R. Overcoming challenges for shoot tip cryopreservation of root and tuber crops. *Agronomy* **2023**, *13*, 219. [CrossRef]
5. Martinez-Montero, M.E.; Gonzalez-Arnao, M.T.; Engelmann, F. Cryopreservation of tropical plant germplasm with vegetative propagation-review of sugarcane (*Saccharum* spp.) and pineapple (*Ananas comusus* (L.) Merrill) cases. In *Current Frontiers in Cryopreservation*; Katkov, I., Ed.; IntechOpen: London, UK, 2012; pp. 360–387. [CrossRef]
6. Engelmann, F. Use of biotechnologies for the conservation of plant biodiversity. *In Vitro Cell. Dev. Biol.-Plant* **2011**, *47*, 5–16. [CrossRef]
7. Gonzalez-Arnao, M.T.; Ravelo, M.M.; Urra-Villavicencio, C.; Martinez-Montero, M.M.; Engelmann, F. Cryopreservation of pineapple (*Ananas comosus*) apices. *CryoLetters* **1998**, *19*, 375–382.
8. Gamez-Pastrana, R.; Martinez-Ocampo, Y.; Beristain, C.; Gonzalez-Arnao, M. An improved cryopreservation protocol for pineapple apices using encapsulation-vitrification. *CryoLetters* **2004**, *25*, 405–414.
9. Martinez-Montero, M.E.; Martinez, J.; Engelmann, F.; Gonzalez-Arnao, M. Cryopreservation of pineapple (*Ananas comosus* (L.) Merr) apices and calluses. *Acta Hortic.* **2005**, *666*, 127–131. [CrossRef]
10. Souza, F.V.D.; Kaya, E.; de Jesus Vieira, L.; de Souza, E.H.; de Oliveira Amorim, V.B.; Skogerboe, D. Droplet-vitrification and morphohistological studies of cryopreserved shoot tips of cultivated and wild pineapple genotypes. *Plant Cell Tissue Organ Cult.* **2016**, *124*, 351–360. [CrossRef]
11. Souza, F.V.D.; de Souza, E.H.; Kaya, E.; de Jesus Vieira, L.; da Silva, R.L. Cryopreservation of pineapple shoot tips by the droplet vitrification technique. *Plant Cell Cult. Protoc.* **2018**, *1815*, 269–377. [CrossRef]
12. Villalobos-Olivera, A.; Rodríguez, J.M.; Bernabé, N.Q.; Souza, F.V.D.; Olmedo, J.G.; Martinez-Montero, M.E. Effect of temperature on pre-conditioning pineapple in vitro donor plants for cryopreservation protocol of shoot apices. *Acta Hortic.* **2019**, *1239*, 113–120. [CrossRef]
13. Villalobos-Olivera, A.; García-Brizuela, J.; Olaru, S.; Rodriguez-Ayerbe, P.; Martinez-Montero, M. A dynamical systems approach for pineapple cryopreservation analysis. *IFAC-PapersOnLine* **2019**, *52*, 263–268. [CrossRef]
14. Sakai, A.; Engelmann, F. Vitrification, encapsulation-vitrification and droplet-vitrification: A review. *CryoLetters* **2007**, *28*, 151–172. [PubMed]
15. Panis, B.; Piette, B.; Swennen, R. Droplet vitrification of apical meristems: A cryopreservation protocol applicable to all Musaceae. *Plant Sci.* **2005**, *168*, 45–55. [CrossRef]
16. Thalip, A.A.; Tong, P.S.; Ng, C. The MD2 "Super Sweet" pineapple (*Ananas comosus*). *UTAR Agric. Sci. J.* **2015**, *1*, 14–17.
17. Lakho, M.A.; Jatoi, M.A.; Solangi, N.; Abul-Soad, A.A.; Qazi, M.A.; Abdi, G. Optimizing in vitro nutrient and ex vitro soil mediums-driven responses for multiplication, rooting, and acclimatization of pineapple. *Sci. Rep.* **2023**, *13*, 1275. [CrossRef]
18. Villalobo, A.; González, J.; Santos, R.; Rodríguez, R. Morpho-physiological changes in pineapple plantlets [*Ananas comosus* (L.) Merr.] during acclimatization. *Ciência Agrotecnol.* **2012**, *36*, 624–630. [CrossRef]
19. Harding, K. Genetic integrity of cryopreserved plant cells: A review. *CryoLetters* **2004**, *25*, 3–22.
20. Martinez-Montero, M.E.; Harding, K. Cryobionomics: Evaluating the concept in plant cryopreservation. In *Plant Omics: The Omics of Plant Science*; Barh, D., Khan, M., Davies, E., Eds.; Springer: New Delhi, India, 2015; pp. 655–682. [CrossRef]

21. Wang, B.; Li, J.W.; Zhang, Z.B.; Wang, R.R.; Ma, Y.L.; Blystad, D.R. Three vitrification-based cryopreservation procedures cause different cryo-injuries to potato shoot tips while all maintain genetic integrity in regenerants. *J. Biotechnol.* **2014**, *184*, 47–55. [CrossRef]
22. Czégény, G.; Mátai, A.; Hideg, É. UV-B effects on leaves—Oxidative stress and acclimation in controlled environments. *Plant Sci.* **2016**, *248*, 57–63. [CrossRef]
23. Karim, M.F.; Johnson, G.N. Acclimation of photosynthesis to changes in the environment results in decreases of oxidative stress in *Arabidopsis thaliana*. *Front. Plant Sci.* **2021**, *12*, 683968. [CrossRef]
24. Wang, M.R.; Bi, W.; Shukla, M.R.; Ren, L.; Hamborg, Z.; Blystad, D.R. Epigenetic and genetic integrity, metabolic stability, and field performance of cryopreserved plants. *Plants* **2021**, *10*, 1889. [CrossRef]
25. Ren, L.; Wang, M.-R.; Wang, Q.-C. ROS-induced oxidative stress in plant cryopreservation: Occurrence and alleviation. *Planta* **2021**, *124*, 254. [CrossRef] [PubMed]
26. Villalobos-Olivera, A.; Entensa, Y.; Martínez, J.; Escalante, D.; Quintana, N.; Souza, F.V. Storage of pineapple shoot tips in liquid nitrogen for three years does not modify field performance and fruit quality of recovered plants. *Acta Physiol. Plant.* **2022**, *44*, 65. [CrossRef]
27. Villalobos-Olivera, A.; Ferreira, C.F.; Yanes-Paz, E.; Lorente, G.Y.; Souza, F.V.; Engelmann, F.; Martínez-Montero, M.E.; Lorenzo, J.C. Inter simple sequence repeat (ISSR) markers reveal DNA stability in pineapple plantlets after shoot tip cryopreservation. *Vegetos* **2022**, *35*, 360–366. [CrossRef]
28. Villalobos-Olivera, A.; Nápoles, L.; Mendoza, J.R.; Escalante, D.; Martínez, J.; Concepción, O.; Zevallos, B.E.; Martínez-Montero, M.E.; Cejas, I.; Engelmann, F.; et al. Exposure of pineapple shoot tips to liquid nitrogen and cryostorage do not affect the histological status of regenerated plantlets. *Rom. Biotechnol. Lett.* **2019**, *24*, 1061–1066. [CrossRef]
29. Rodríguez-Escriba, R.C.; Rodríguez, R.; López, D.; Lorente, G.Y.; Pino, Y.; Aragón, C.E. High light intensity increases CAM expression in MD-2 micro-propagated pineapple plants at the end of acclimatization stage. *Am. J. Plant Sci.* **2015**, *6*, 3109–3125. [CrossRef]
30. Aragón, C.; Pascual, P.; González, J.; Escalona, M.; Carvalho, L.; Amancio, S. The physiology of ex vitro pineapple (*Ananas comosus* L. Merr. var MD-2) as CAM or C3 is regulated by the environmental conditions: Proteomic and transcriptomic profiles. *Plant Cell Rep.* **2013**, *32*, 1807–1818. [CrossRef]
31. Murashige, T.; Skoog, F. A revised medium for rapid growth and bio assays with tobacco tissue cultures. *Physiol. Plant.* **1962**, *15*, 473–497. [CrossRef]
32. Nápoles, L.; Cid, M.; Hernández, L.; Alvares, Y.; Zamora, V.; Lorente, G.L.; Rodríguez, R.; Concepción, O. Scale up of in vitro plant production of 'MD2' pineapple free of Pineapple mealybug wilt-associated virus-1, -2 and -3 for introduction to productive scale in Cuba. *Acta Hortic.* **2019**, *1039*, 121–128. [CrossRef]
33. Nishizawa, S.; Sakai, A.; Amano, Y.; Matsuzawa, T. Cryopreservation of asparagus (*Asparagus officinalis* L.) embryogenic suspension cells and subsequent plant regeneration by vitrification. *Plant Sci.* **1993**, *91*, 67–73. [CrossRef]
34. Sakai, A.; Kobayashi, S.; Oiyama, I. Cryopreservation of nucellar cells of navel orange (*Citrus sinensis* Osb. var. *brasiliensis* Tanaka) by vitrification. *Plant Cell Rep.* **1990**, *9*, 30–33. [CrossRef]
35. Pino, Y.; Concepción, O.; Santos, R.; González, J.; Rodríguez, R. Effect of previcur (r) energy fungicide on MD-2 pineapple (*Ananas comosus* var. *comosus*) plantlets during the acclimatization phase. *Pineapple News* **2014**, *21*, 24–26.
36. Ebel, A.I.; Itati Giménez, L.; González, A.M.; Alayón Luaces, P. Evaluación morfoanatómica de hojas "D" de piña (*Ananas comosus* (L.) Merr. var. *comosus*) en respuesta a la implantación de dos sistemas de cultivo en Corrientes, Argentina. *Acta Agronómica* **2016**, *65*, 390–397. [CrossRef]
37. Bettoni, J.C.; Fazio, G.; Carvalho Costa, L.; Hurtado-Gonzales, O.P.; Rwahnih, M.A.; Nedrow, A.; Volk, G.M. Thermotherapy followed by shoot tip cryotherapy eradicates latent viruses and apple hammerhead viroid from in vitro apple rootstocks. *Plants* **2022**, *11*, 582. [CrossRef] [PubMed]
38. Guerra, P.A.; de Souza, E.H.; de Andrade, E.C.; Max, D.D.A.S.; de Oliveira, R.S.; Souza, F.V.D. Comparison of shoot tip culture and cryotherapy for eradication of ampeloviruses associated with Pineapple mealybug wilt in wild varieties. *In Vitro Cell. Dev. Biol.-Plant* **2020**, *56*, 903–910. [CrossRef]
39. Wang, M.R.; Bi, W.L.; Bettoni, J.C.; Zhang, D.; Volk, G.M.; Wang, Q.C. Shoot tip cryotherapy for plant pathogen eradication. *Plant Pathol.* **2022**, *71*, 1241–1254. [CrossRef]
40. Bettoni, J.C.; Marković, Z.; Bi, W.; Volk, G.M.; Matsumoto, T.; Wang, Q.-C. Grapevine shoot tip cryopreservation and cryotherapy: Secure storage of disease-free plants. *Plants* **2021**, *10*, 2190. [CrossRef]
41. Wang, M.R.; Hamborg, Z.; Slimestad, R.; Elameen, A.; Blystad, D.R.; Haugslien, S. Assessments of rooting, vegetative growth, bulb production, genetic integrity and biochemical compounds in cryopreserved plants of shallot. *Plant Cell Tissue Organ Cult.* **2021**, *144*, 123–131. [CrossRef]
42. Bettoni, J.C.; Souza, J.A.; Volk, G.M.; Dalla Costa, M.; da Silva, F.N.; Kretzschmar, A.A. Eradication of latent viruses from apple cultivar 'Monalisa' shoot tips using droplet-vitrification cryotherapy. *Sci. Hortic.* **2019**, *250*, 12–18. [CrossRef]
43. Tavazza, R.; Lucioli, A.; Benelli, C.; Giorgi, D.; D'aloisio, E.; Papacchioli, V. Cryopreservation in artichoke: Towards a phytosanitary qualified germplasm collection. *Ann. Appl. Biol.* **2013**, *163*, 231–241. [CrossRef]
44. Wang, Q.; Mawassi, M.; Li, P.; Gafny, R.; Sela, I.; Tanne, E. Elimination of grapevine virus A (GVA) by cryopreservation of *in vitro*-grown shoot tips of *Vitis vinifera* L. *Plant Sci.* **2003**, *165*, 321–327. [CrossRef]

45. Van Huylenbroeck, J.; Piqueras, A.; Debergh, P. Photosynthesis and carbon metabolism in leaves formed prior and during ex vitro acclimatization of micropropagated plants. *Plant Sci.* **1998**, *134*, 21–30. [CrossRef]
46. González, G.Y.L.; Rodríguez-Escriba, R.C.; Martínez, R.E.I.; Pérez, L.S.; Borroto, Y.L.D.; Sánchez, R.R. Response of 'MD-2' pineapple plantlets (*Ananas comosus* var. *comosus*) to a controlled release fertilizer during the acclimatization stage. *Pineapple News* **2017**, *23*, 37–40.
47. Hu, W.H.; Liu, S.F.; Liaw, S.I. Long-term preconditioning of plantlets: A practical method for enhancing survival of pineapple (*Ananas comosus* (L.) Merr.) shoot tips cryopreserved using vitrification. *CryoLetters* **2015**, *36*, 226–236.
48. Hamid, N.S.; Tan, B.C.; Khalid, N.; Osman, N.; Jalil, M. Adaptive features of *in vitro*-derived plantlets of MD2 pineapple during acclimatization process. *Indian J. Hortic.* **2020**, *77*, 595–602. [CrossRef]
49. Agbidinoukoun, A.; Ouikoun, G.C.; Mikpon, T.; Kamade, G.T.; Badou, B.T.; Aïsso, R.C. Influence of watering solution and phenotype on the growth of in vitro propagated pineapple (Smooth Cayenne Cultivar) plantlets during acclimatization. *Agric. Sci.* **2021**, *12*, 1215–1230. [CrossRef]
50. Souza, J.A.; Bettoni, J.C.; Dalla Costa, M.; Baldissera, T.C.; dos Passos, J.F.M.; Primieri, S. In vitro rooting and acclimatization of 'Marubakaido' apple rootstock using indole-3-acetic acid from rhizobacteria. *Commun. Plant Sci.* **2022**, *12*, 16–23. [CrossRef]
51. Mežaka, I.; Kļaviņa, D.; Kaļāne, L.; Kronberga, A. Large-Scale In Vitro Propagation and Ex Vitro Adaptation of the Endangered Medicinal Plant *Eryngium maritimum* L. *Horticulturae* **2023**, *9*, 271. [CrossRef]
52. Bouzo, C.A.; Favaro, J.C. Container size effect on the plant production and precocity in tomato (*Solanum lycopersicum* L.). *Bulg. J. Agric. Sci.* **2015**, *21*, 325–332.
53. Durand, M.; Mainson, D.; Porcheron, B.; Maurousset, L.; Lemoine, R.; Pourtau, N. Carbon source–sink relationship in Arabidopsis thaliana: The role of sucrose transporters. *Planta* **2018**, *247*, 587–611. [CrossRef]
54. White, A.C.; Rogers, A.; Rees, M.; Osborne, C.P. How can we make plants grow faster? A source–sink perspective on growth rate. *J. Exp. Bot.* **2016**, *67*, 31–45. [CrossRef] [PubMed]
55. Kim, J.K.; Shawon, M.R.A.; An, J.H.; Yun, Y.J.; Park, S.J.; Na, J.K.; Choi, K.Y. Influence of substrate composition and container size on the growth of tissue culture propagated apple rootstock plants. *Agronomy* **2021**, *11*, 2450. [CrossRef]
56. Bettoni, J.C.; Dalla Costa, M.; Gardin, J.P.P.; Kretzchmar, A.A.; Souza, J.A. In vitro propagation of grapevine cultivars with potential for South of Brazil. *Am. J. Plant Sci.* **2015**, *6*, 1806–1815. [CrossRef]
57. Demarco, P.; Gómez Herrera, M.D.; Gonzalez, A.; Alayón Luaces, P. Effects of foliar versus soil water application on ecophysiology, leaf anatomy and growth of pineapple. *Fruits* **2020**, *75*, 44–51. [CrossRef]
58. de Paula Alves, J.; Pinheiro, M.V.M.; Corrêa, T.R.; Alves, G.L.; dos Santos Marinho, T.R.; Batista, D.S. Morphophysiology of *Ananas comosus* L. Merr during in vitro photomixotrophic growth and ex vitro acclimatization. *In Vitro Cell. Dev. Biol.-Plant* **2023**, *59*, 106–120. [CrossRef]
59. Barberis, I.M.; Cárcamo, J.M.; Cárcamo, J.I.; Albertengo, J. Phenotypic plasticity in Bromelia serra Griseb.: Morphological variations due to plant size and habitats with contrasting light availability. *Rev. Bras. Biociências* **2017**, *15*, 3–10.
60. Couto, T.R.; Silva, J.R.; Moraes, C.R.O.; Ribeiro, M.S.; Netto, A.T.; Carvalho, V.S. Photosynthetic metabolism and growth of pineapple (*Ananas comosus* L. Merr.) cultivated ex vitro. *Theor. Exp. Plant Physiol.* **2016**, *3*, 333–339. [CrossRef]
61. Gotoh, E.; Oiwamoto, K.; Inoue, S.I.; Shimazaki, K.-I.; Doi, M. Stomatal response to blue light in crassulacean acid metabolism plants *Kalanchoe pinnata* and *Kalanchoe daigremontiana*. *J. Exp. Bot.* **2019**, *70*, 1367–1374. [CrossRef]
62. Kitajima, K.; Hogan, K.P. Increases of chlorophyll a/b ratios during acclimation of tropical woody seedlings to nitrogen limitation and high light. *Plant Cell Environ.* **2003**, *26*, 857–865. [CrossRef]
63. Mitchell, A.; Arnott, J. Effects of shade on the morphology and physiology of amabilis fir and western hemlock seedlings. *New For.* **1995**, *10*, 79–98. [CrossRef]
64. Males, J.; Griffiths, H. Stomatal biology of CAM plants. *Plant Physiol.* **2017**, *174*, 550–560. [CrossRef] [PubMed]
65. Wang, L.; Han, S.; Wang, S.; Li, W.; Huang, W. Morphological, photosynthetic, and CAM physiological responses of the submerged macrophyte *Ottelia alismoides* to light quality. *Environ. Exp. Bot.* **2022**, *202*, 105002. [CrossRef]
66. Ceusters, N.; Borland, A.M.; Ceusters, J. How to resolve the enigma of diurnal malate remobilisation from the vacuole in plants with crassulacean acid metabolism? *New Phytol.* **2021**, *229*, 3116–3124. [CrossRef]
67. Ceusters, N.; Valcke, R.; Frans, M.; Claes, J.E.; Van den Ende, W.; Ceusters, J. Performance index and PSII connectivity under drought and contrasting light regimes in the CAM orchid *Phalaenopsis*. *Front. Plant Sci.* **2019**, *10*, 1012. [CrossRef]
68. Marques, A.R.; Duarte, A.A.; de Souza, F.A.; de Lemos-Filho, J.P. Does seasonal drought affect C3 and CAM tank-bromeliads from campo rupestre differently? *Flora* **2021**, *282*, 151886. [CrossRef]
69. Wang, H.; Wang, X.Q.; Zeng, Z.L.; Yu, H.; Huang, W. Photosynthesis under fluctuating light in the CAM plant *Vanilla planifolia*. *Plant Sci.* **2022**, *317*, 111207. [CrossRef]
70. Qiu, S.; Xia, K.; Yang, Y.; Wu, Q.; Zhao, Z. Mechanisms Underlying the C3–CAM Photosynthetic Shift in Facultative CAM Plants. *Horticulturae* **2023**, *9*, 398. [CrossRef]
71. Heyduk, K. Evolution of Crassulacean acid metabolism in response to the environment: Past, present, and future. *Plant Physiol.* **2022**, *190*, 19–30. [CrossRef]
72. Hu, R.; Zhang, J.; Jawdy, S.; Sreedasyam, A.; Lipzen, A.; Wang, M.; Ng, V.; Daum, C.; Keymanesh, K.; Liu, D.; et al. Comparative genomics analysis of drought response between obligate CAM and C3 photosynthesis plants. *J. Plant Physiol.* **2022**, *277*, 153791. [CrossRef]

73. Nievola, C.C.; Carvalho, C.P.; Carvalho, V.; Rodrigues, E. Rapid responses of plants to temperature changes. *Temperature* **2017**, *4*, 371–405. [CrossRef] [PubMed]
74. Baťková, P.; Pospíšilová, J.; Synková, H. Production of reactive oxygen species and development of antioxidative systems during in vitro growth and ex vitro transfer. *Biol. Plant.* **2008**, *52*, 413–422. [CrossRef]
75. Chandra, S.; Bandopadhyay, R.; Kumar, V.; Chandra, R. Acclimatization of tissue cultured plantlets: From laboratory to land. *Biotechnol. Lett.* **2010**, *32*, 1199–1205. [CrossRef] [PubMed]

 *horticulturae*

Review

In Vitro Micrografting of Horticultural Plants: Method Development and the Use for Micropropagation

Min-Rui Wang [1,*], Jean Carlos Bettoni [2,*], A-Ling Zhang [1], Xian Lu [1], Dong Zhang [1,*] and Qiao-Chun Wang [1]

[1] State Key Laboratory of Crop Stress Biology for Arid Region, College of Horticulture, Northwest A&F University, Yangling 712100, China; aling.zhang@nwsuaf.edu.cn (A.-L.Z.); luxian@nwafu.edu.cn (X.L.); qiaochunwang@nwafu.edu.cn (Q.-C.W.)

[2] The New Zealand Institute for Plant and Food Research Limited, Batchelar Road, Palmerston North 4410, New Zealand

* Correspondence: wangmr@nwsuaf.edu.cn (M.-R.W.); jean.bettoni@plantandfood.co.nz (J.C.B.); afant@nwsuaf.edu.cn (D.Z.)

Abstract: In vitro micrografting is an important technique supporting the micropropagation of a range of plant species, particularly woody plant species. Over the past several decades, in vitro micrografting has become a strategy to facilitate shoot recovery and acclimatization of in vitro-grown horticultural species. This review focuses on studies on horticultural crops over the past two decades that cover the establishment of in vitro micrografting, discusses factors affecting the success of in vitro micrografting, and provides commentary on the contribution of micrografting applications to the field of micropropagation. Considering the important roles of micrografting in the restoration of vigor and rooting competence, in promotion of shoot recovery following somatic embryogenesis and organogenesis, and in facilitation of shoot regrowth after cryopreservation, the potential use of this technique in facilitation of genetic engineering and safe conservation of horticultural species are specially highlighted.

Keywords: cryopreservation; in vitro grafting; in vitro propagation; somatic embryogenesis; organogenesis

Citation: Wang, M.-R.; Bettoni, J.C.; Zhang, A.-L.; Lu, X.; Zhang, D.; Wang, Q.-C. In Vitro Micrografting of Horticultural Plants: Method Development and the Use for Micropropagation. *Horticulturae* **2022**, *8*, 576. https://doi.org/10.3390/horticulturae8070576

Academic Editor: Hrotkó Károly

Received: 11 June 2022
Accepted: 23 June 2022
Published: 24 June 2022

Publisher's Note: MDPI stays neutral with regard to jurisdictional claims in published maps and institutional affiliations.

1. Overall Developments and Characters of Micrografting

Plant grafting, a common practice for vegetative propagation of crops, refers to the natural or the deliberate connection of two discrete plant segments [1]. Grafting can be used to avoid juvenility of perennial woody species and can confer important agronomic traits to scions such as uniformity of plant architecture and tolerance to biotic and abiotic stresses [1–4]. In addition, the scion–rootstock combination can influence tree vigor, yield and fruit quality, and can extend the harvest season [5,6]. Following the advent of in vitro plant tissue culturing in the early 1900s [7], a grafting system using tissue culture (micrografting) was first demonstrated by Doorenbos [8] in ivy and then Holmes [9] in chrysanthemum in the 1950s, and was later developed and standardized for virus eradication from citrus species by Murashige et al. [10] and Navarro et al. [11]. To date, in vitro micrografting (IVM) has been widely applied (1) in pathogen management to facilitate the eradication, indexing and transmission of pathogens, as well as the assessments of graft incompatibility induced by pathogen infection [11–15]; (2) to facilitate in vitro rooting [16–19], to invigorate regenerating plant tissue cultures during micropropagation [19–22], and for the rapid assessment of graft compatibility [23–26]; and (3) in studies focusing on the molecular mechanism of graft compatibility, as well as the exchange and trafficking of macromolecules between scions and rootstocks [27–30].

IVM is an important technique that facilitates the micropropagation of horticultural crops and forest species because of the following characteristics. Firstly, IVM is often performed on seedlings to obtain rooted plants in species in which in vitro root induction is difficult [16,31]; secondly, it can reduce species-specific responses of scions to the culture medium, as the rootstock mediates the delivery of the hormonal and nutritional requirements necessary for the scion regrowth from the medium [18,32,33]; thirdly, IVM can be performed throughout the year using scions and rootstocks at the same physiological stage, while the success of in vivo grafting is season-dependent [17,21]. IVM is performed in an aseptic controlled environment with a high humidity. This stable in vitro environment and the probable pathogen-free status of micro-scion/rootstock may favor callus formation and rapid establishment of vascular reconnection between scions and rootstocks required for grafting success [34,35]. Micrografting protocols have been developed for many fruit crops including almond [18], apple [36], apricot [23], avocado [37], cacao [38], cashew [39], cherimoya [19], cherry [21], citrus [40,41], guava [4], grape [34], jujube [42], mulberry [43], hazelnut [44], kiwifruit [25], passion fruit [26], olive [45], peach [46], pear [47], pistachio [17], plum [48], walnut [49,50], and watermelon [51].

To highlight how IVM can be used to improve micropropagation in horticulture, this review presents findings from recent studies using IVM, with a focus on factors that affect the micrografting success.

2. Establishment of Micrografting

2.1. Preparation of Scions

The origins and type of scion material are some of the determinants in successful micrografting [34,52–54]. The scions used in micrografting, usually shoots or shoot tips, can be obtained from in vitro or ex vitro grown plants. Scion material has traditionally been sourced from in vitro plants, having the advantage of being free from fungal and bacterial contaminations, the desired size, and being available year-around [35,53,55]. The use of ex vitro material, however, may introduce a seasonality component to the procedure, as the excised plant material may remain in a dormant stage [53]. However, shoot apices newly excised from actively growing trees in the field or in the greenhouse can be used in micrografting procedures [17,34]. Shoot apices sourced from in vivo plants are surface sterilized immediately prior to shoot/shoot tip preparation followed by micrografting. Usually, a short treatment of 70% ethyl alcohol is combined with a longer treatment of sodium hypochlorite or mercuric chloride for the surface sterilization [56].

Tissue browning is a common problem during the establishment of in vitro cultures [41,57]. Likewise, in IVM, wounding in scion/rootstock preparation may also cause browning and oxidation of plant tissues resulting in poor graft success [36,56,58]. The adverse effects of tissue browning may be minimized by presoaking scions in antioxidant solutions [39,59,60]. In cashew (*Anacardium occidentale* L.) and apple (*Malus domestica*), Thimmappaiah et al. [31] and Nunes et al. [61], respectively, reduced phenolic exudation by presoaking the cut edge of the scion with 0.01% ascorbic acid and 0.015% citric acid (1:1) prior to in vitro grafting. Reduced tissue browning in cashew can also be achieved by pre-conditioning in vitro stock shoots in culture media enriched with 0.1% polyvinylpyrrolidone before scion preparation [31]. In developing a protocol for micrografting of native and commercial roses, Davoudi Pahnekolayi et al. [59] showed that silver nitrate, as an antioxidant, played a key role in preventing production of phenolic compounds that could lead to micrografting failure. They found that a quick dip treatment (5–10 min) of wounded explants (scions and rootstocks) with silver nitrate (50 mg L^{-1}) prior to micrografting could prevent tissue browning and consequently increase the survival of micrografts [59]. In contrast, Wu et al. [55] noted that micrografts of *Protea cynaroides* had reduced viability when scions were presoaked in ascorbic acid and citric acid solution; treatment with antioxidant solution induced more browning in scions than in those that were untreated. These results indicated that the fast operation of micrografting was more important in preventing tissue browning than pre-treatment, as suggested by Navarro [62]. Another possible reason could be the insufficient concentration of antioxidants which counterproductively promoted the spread of phenolic oxidation because of the improved wetness of the graft site [55]. Therefore, the response of antioxidants to the reduction/inhibition of phenolic browning may be species-dependent, and their concentrations and/or combinations are other critical points to be addressed.

2.2. Preparation of Rootstocks

In vitro germinated seedlings and segments of in vitro cultured shoots are the two major sources of rootstocks used in micrografting (Table 1). To prepare rootstock seedlings for micrografting, in vitro germination of seeds is the first step, with varied protocols needed to promote germination. Miguelez-Sierra et al. [38] found that cacao seeds did not require any pretreatment for the effective in vitro seed germination; in their protocol, seeds were taken from mature pods, were surface sterilized, inoculated in culture medium, and three-week-old seedlings were used as rootstocks. However, removal of the seed coat is necessary in many cases of in vitro germination. For example, for the preparation of in vitro rootstocks of almond, seeds were firstly removed from their endocarps (hard seed coat) before surface sterilization and in vitro germination [33,63]. Similarly, in pistachio, the mature kernels of seeds were surface sterilized after removing the outer pericarp and shells [17]. Likewise, the mature seeds of cashew [39] and jujube [42] were scarified in concentrated hydrochloric acid and sulfuric acid, respectively, before surface sterilization, to promote seed germination. In some cases, to achieve good germination, embryos have been removed from surface-sterilized seeds and grown in germination medium. In *Protea cynaroides*, a successful shoot tip micrografting technique was developed using 30-day-old in vitro-germinated embryos as rootstock [55]. Following seed germination, the duration of which varies between species, the root is shortened and the seedlings can be either decapitated above the cotyledons, leaving the epicotyls as the site for grafting [33,37], or cut below the cotyledons to make use of the hypocotyls as grafting sites [38,39,42,64].

In species with high levels of adventitious rooting, in vitro shoots can also be used as rootstocks [21,34,61,65]. Following grafting, the grafted shoots are cultured on a medium used for root induction of the rootstock genotype: this requires prior optimization of the rooting medium [65]. In cherry, Bourrain and Charlot [21] obtained a successful grafting rate of 79% when shoots (rootstock) were induced to root prior to grafting. In apple, Obeidy and Smith [66] achieved a graft success up to 45% when rooted in vitro shoots were used as rootstocks and apical 2-cm shoots as scions. When comparing the performance of *Uapaca kirkiana* (Muell. Arg) micrografts arising from in vitro rooted and unrooted rootstocks, Nkanaunena et al. [67] found that the three-month-old rooted rootstocks produced the highest graft success rate (at least 60%), with better development of grafted shoots. The positive response in the success of micrografting from the use of rooted rootstocks may be related to the species studied. Working on grapevine cultivar (cv.) 'Superior' micrografted onto different rootstocks, Sammona et al. [68] found no differences in grafting success between rooted and unrooted rootstocks.

Table 1. Applications of in vitro micrografting for restoration of scion vigor and establishment of plants with roots.

Plant Species (Scion)	Scion Source and Size	Rootstock Source and Age	Grafting Technique	Success Rate (%) and (No. Scions Tested)	Reference
Amygdalus communis (Almond)	Shoots of 1.5–2.0 cm in length	Almond seedlings of 2 weeks old	Top slit	100 (1)	[18]
Anacardium occidentale (Cashew)	In vitro shoot apices	Cashew seedlings, 5–8 cm in height (age not specified)	Apical and side grafting	45–73 (1)	[69]
	Shoots of 3–15 mm in length	Cashew seedlings of 20–25 days old	Top slit and side grafting	80 (top wedge) and 100 (side grafting) (1)	[31]
Annona cherimola (Cherimoya)	Nodal section of 2 cm in length	Cherimoya seedling of 42 days old	Side insertion	31–70 (3)	[19]
Citrus deliciosa (Kinnow mandarin)	Shoot tips less than 1.0 mm in size	*C. jambhiri*, *C. carrizo* and *C. reshnii* seedlings of 15–20 days old	Side reverse T insertion	Up to 66.5 (1)	[70]
Garcinia indica	Apical shoots of 0.5–1.0 cm in length for the initial grafting and 1.0–1.5 cm for the subsequent grafting	*Garcinia indica* seedlings of 2 months old	Top slit	95 (1)	[71]
Malus domestica (Apple)	Field grown shoots (size not specified)	Apple shoots of 3 weeks old	Vertical slit	42–93 (3)	[36]
Olea europea (Olive)	Greenhouse-grown shoots of 1.0–1.5 cm in length	Olive seedlings of 3 weeks old	Top slit	Up to 83 (1)	[45]
Opuntia ficus-indica (Cactus)	Shoots of 0.5 cm in length	Shoots of *O. strepacantha*, *O. robusta*, *O. cochinera*, *O. leucotricha* and *O. ficus-indica*, 1.0 cm in length (age not specified)	Top wedge and horizontal graft	30 (top wedge) to 90 (horizontal) (1)	[54]
Passiflora edulis (Passion fruit)	Nodal segments of 1.5 cm in length	Passion fruit shoots of 2 months old	V-shaped joint with grafting devices	73.3 (1)	[26]
Pelecyphora aselliformis (Cactus)	Apical and subapical segments of 5 and 3 mm, respectively	*O. ficus-indica* shoots, 10 mm in length (age not specified)	Horizontal graft	81 and 97 for the subapical and apical scions, respectively (1)	[72]
Pistacia vera var. Siirt (Pistachio)	Shoot tips of 0.5–10 mm in length	Pistachio seedlings of 10–14 days old	Top slit and top wedge	Up to 80 (1)	[17]
Protea cynaroides (King Protea)	Shoots of 5 mm in length	King Protea seedling of 30 days old	Top slit	80 (1)	[55]
Prunus dulcis (Almond)	Apical shoots of 1.5–2.0 cm in length	Shoots of almond/peach hybrid rootstock of 3–7 weeks old	Top slit	50–70 (2)	[65]
	Shoots (size not specified)	Almond seedlings of 2 weeks old	Top slit	Up to 100 (1)	[63]
	Shoot tips of 4, 8 and 15 mm in length	Almond seedlings of 14 days old	Top slit and top wedge	90–100 (2)	[33]
Prunus avium (Cherry)	Shoot tips of 0.3–1.0 cm in length	*P. avium* × (*P. canescens* × *P. tomentosa*) shoots, 3–4 cm in length (age not specified)	Top slit	Up to 79 (1)	[21]
Pyrus communis ('Old Home' x 'Farmingdale 333') (Pear)	Shoots of 10 mm in length	*P. elaeagrifolia* seedlings of 10–14 days old	Cleft	97.9 (1)	[24]
Rosa hybrida cvs./(Rose)	Shoots 10–15 mm in length	*R. canina* and *R. multiflora* shoots, 20 mm in length (age not specified)	With grafting devices	Up to 100 (2)	[59]
Theobroma cacao (Cacao)	Shoots of 4–6 mm in length	Cacao seedlings of 5–6 weeks old	Not specified	>50 (1)	[73]
	Bud sticks with apical or axillary buds (sourced from potted plants) of 1 cm in length	Cacao seedlings of 3 weeks old	Top slit and side grafting	55–95 (1)	[38]
Vitis vinifera (Grape)	Shoot tips of 0.2–0.5 mm in length	White to slightly coloured hypocotyls from white somatic embryos	Side grafting	18–30 (4)	[74]
	In vitro/in vivo derived shoot tips of 0.3–0.8 mm in length	Shoots of *V. vinifera* × *V. berlandieri*, 1.0 cm in length (age not specified)	Not specified	40–61 (in vitro shoot tips) and 12–17 (in vivo shoot tips) (4)	[34]
Ziziphus mauritiana (Jujube)	Shoots of 5–10 mm in length	Jujube seedlings (7 spp.) of 4 weeks old	Top wedge	28–100 (1)	[42]

2.3. Grafting Techniques

The success of a micrografting procedure depends on the successful union of the rootstock and scion. The skill of the grafter is a key determinant of in vitro graft success. Various grafting techniques have been described, and the choice of which to use may depend on the type and size of the scion propagule and the purpose of the micrografting. The top-slit or top-wedge methods are the most frequently used in vitro grafting techniques and have been tested across a wide range of genera (Table 1). A slit or cleft is made onto the rootstock and wedge-shaped scions inserted into the cleft [17,33,55,75]. When rootstock thinner than scions are being used, micrografting can be performed by a reverse-cleft graft in which a slit is made at the bottom of the scions for the insertion of the rootstocks with a wedged top [76]. When small shoot tips were used as scions, their placement into the slit made at the top or directly over the rootstocks is usually referred as apical micrografting [69]. The insertion of small shoot tips into a slit on one side of the rootstock is termed side grafting (or side insertion) [69,74]. In *Citrus*, side grafting by inserting the shoot apices into inverted T-cuts of rootstocks has been successfully used [70]. Side insertion was also applied when longer shoots or nodal sections were used as scions [19,77,78]. The major methods applied in micrografting are illustrated in Figure 1.

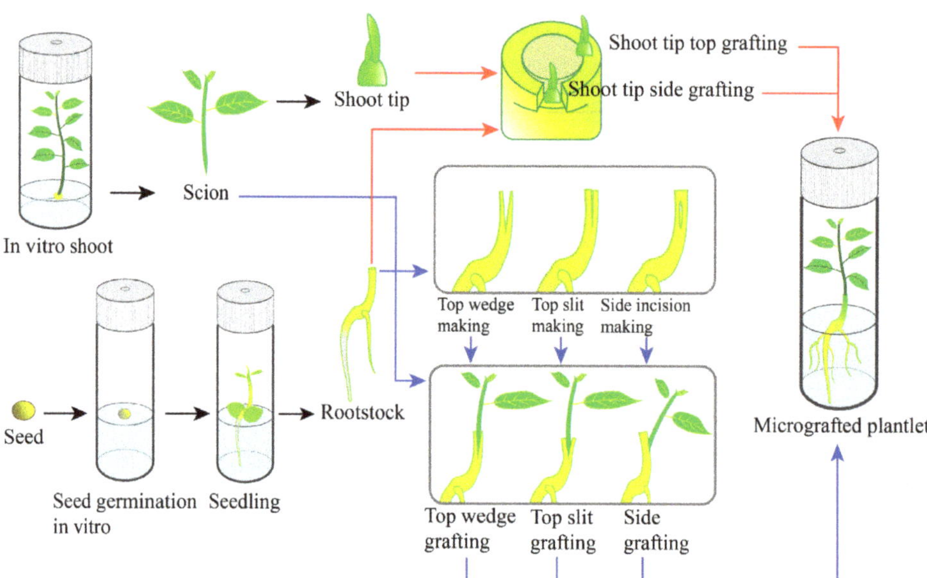

Figure 1. Schematic illustrations of in vitro micrografting methods. An in vitro-germinated seedling decapitated at the epicotyl is demonstrated as the rootstock. Black arrows indicate the preparation of scion and rootstock before the grafting process. Red arrows indicate the micrografting of shoot tips onto the rootstock using the top grafting and side grafting methods. Blue arrows indicate the use of in vitro shoots as scions in micrografting, and the top-wedge grafting, the top-slit grafting and the side grafting are illustrated.

In vitro propagules used in micrografting procedures are highly susceptible to moisture, whereby dehydration of the cut scion or rootstock surface can negatively affect the graft success. Therefore, in order to avoid dehydration, IVM should be performed promptly after the preparation of rootstock and scions to avoid dehydration [21]. In addition, ensuring a firm contact between the rootstock and scion is extremely important in developing a strong graft union [36,55].

Several devices have been used to enable fast and effective union between the rootstock and scion, such as the elastic electric-wire tube [44], aluminum foil [66,76], Parafilm® strip [42], silicon tube [13,79,80], paper bridge [59], silicone chip [30], plastic clamps [50] or alginate gel beads [44,75,81]. These grafting devices are used to support the graft and hold the scion and rootstock together during graft healing, particularly for the top-slit and top-wedge methods.

In addition to grafting devices, the practice of dipping the lower end of the scion in the culture medium before fitting it into the rootstock [12], or applying agar solution on the grafting zone as an adhesive material [36,82] to hold the graft, are also strategies to establish and fix the graft union, particularly when the scion does not fit properly into the rootstock. Pathirana and McKenzie [12] suggested that, in addition to delivering nutrients directly to the graft site, the strategy of dipping the lower end of the scion in the culture medium before grafting keeps the cut surfaces moist until the high relative humidity within the vessel is re-established after closure. This strategy resulted in a high success rate of 75–85% in micrografting of grapevine. Dobránszki et al. [82] described an effective method for IVM of apple using agar–agar solution to stick the scion to the vertical slit of the rootstock. Briefly they placed the V-shape cut scion base in an antioxidant solution (0.15 mg L^{-1}, 0.1 mg L^{-1} ascorbic acid, and 0.1 mg L^{-1} gibberellic acid) to inhibit oxidative browning, followed by treatment with 1% agar–agar solution, and then two drops of agar solution were placed around the graft zone before attaching the scion to the rootstock. With this method, the graft success rate was 95% and all acclimatized plants survived [82]. In contrast, medium-supported grafting resulted in lower micrograft survival rates than the unsupported technique in *Protea cynaroides* [55]. Similarly, working on conifer micrografts, Ponsonby and Mantell [83] and Cortizo et al. [84] found either a reduction in graft union success or no response, respectively, when antioxidant additives or culture medium solution was applied to the micrograft union. Therefore, we suggest that medium-supported grafting be applied as a back-up strategy for IVM.

2.4. Culture Conditions

Various culture conditions have been tested to optimize the regrowth of micrografts with success dependent on the plant species and the source of plant material used. In rootstock seedlings, seeds are usually germinated under continuous darkness for 1 to 6 weeks [40,64,85,86], but successful protocols using seedlings germinated in light conditions have also been reported [27,51,72,87]. Working on grapefruit micrografted onto seedling sour orange, Ali et al. [40] showed that the grafting success was related to the light conditions during seedling development. The frequency of graft success increased from 5 to 50% when rootstock seedlings were obtained from seeds germinated under continuous darkness for two weeks compared with success using seeds germinated under continuous light [40]. Similarly, in *Citrus* cultivars Cadenera Fina and Pera (sweet oranges) micrografted onto different rootstocks, Navarro et al. [11] found a greater graft success rate when rootstock seedlings from seeds germinated and grown in darkness were used (37.5%) than seedlings germinated in light conditions (2.7%). In contrast, working on Tahitian lime and Valencia orange micrografted to seedlings from the mandarin (Cleopatra), Suárez et al. [87] found moderate rates of success that ranged from 14 to 28%, respectively, on rootstocks from seeds germinated under light conditions.

The conditions in which plants are grown following micrografting can influence graft success. Micrografted plants of jujube tree (*Ziziphus mauritiana* 'Gola') were first grown in darkness for ten days and then transferred to light conditions [42]. They found that a period in the dark before and after grafting was important to avoid the photo oxidation at the grafting point as well as to minimize the destruction of the auxins synthesized in the scion [42]. In almond, micrografted plants were cultured on rooting medium and incubated in the dark for 7 days, then transferred to the light of 35–40 μmol m^{-2} s^{-1} for two weeks, and finally to 60 μmol m^{-2} s^{-1} for one week before in vivo acclimatization [65]. Several studies have reported successful micrografts without a dark incubation period following the grafting procedure [13,24,72,82,86].

Different supporting systems have been used with growth media in micrografting procedures. Paper bridges, perlite and vermiculite have been used for supporting micrografted plants in liquid culture medium [63,88,89], as well as the solid and semi-solid culture media [21,25,34,72,86]. For example, liquid medium with perlite or a paper bridge was used as the supporting system in micrografting of almond [63] and lime shoots [90]. Liquid medium has also been used in micrografting cashew [39] and cut rose [59]. An adjustable paper bridge can be made to better support the micrografts cultured with liquid medium [66]. The advantages of using liquid medium lie in the better availability and absorption of nutrients, and the reduced damage to the root system when moving plants [39]. In micrografting of cherry, an agar-solidified medium with vermiculite was successfully used to obtain high-quality grafted plants [21]. In apple, the highest graft success was achieved when micrografted plants were cultured on agar-solidified medium [89].

2.5. Acclimatization of Micrografted Plants

Once micrografted plants are well rooted and showed clear scion regrowth, they can be transferred to potting mix following a gradual transition of light intensity and ventilation to achieve successful acclimatization [17,18,21,77]. Acclimatization is a critical phase within micrografting protocols: significant losses can occur when transplanting micrografted plants to ex vitro conditions [62]. Micrografts are removed from in vitro conditions and rinsed with tap water to remove any remaining medium from the roots, and finally grafted plants are transferred to pots containing substrate [21,22,33]. During the first few days, micrografted plants are maintained in high relative humidity and gradually transferred to ex vitro conditions [43,59,87,91]. Micrografted plants of jujube cultured for one month on growth medium and with the scion having grown to 5–10 cm in length (scions were initially with 5–10 mm) could be successfully acclimatized (83–87%) [42]. Miguelez-Sierra et al. [38] found that micrografted plants of cacao grown in vitro for two weeks could be transferred to ex vitro conditions. They observed that only plants with at least 1 cm of scion elongation and two expanded leaves survived acclimatization [38]. The presence of roots on the rootstock is essential for micrografted plants to survive acclimatization. Hieu et al. [26] found that the micrografted plants of passion fruit without roots did not survive acclimatization; plants with roots formed in vitro achieved the highest survival rate during the acclimatization phase. The survival rate during acclimatization of micrografted plants varies among species. For example, in apple, survival rate of grafted plants reached 100% [82], for almond it was 85–100% [92], it was 82% in cacao [36], and 75% in passionfruit [26], whereas in Tahitian lime and Valencia orange, survival rates ranged from 47 to 50%, respectively [87]. Contrasting results were also reported by Kobayashi et al. [93] in micrografting of sweet orange buds derived from organogenesis. In that instance, the fully developed in vitro micrografts grew slowly in the greenhouse, so the micrografted plants were re-grafted onto three-month-old seedlings of Rangpur lime for rapid acclimatization and normal development of the plants. In general, however, there was a higher acclimatization success rate of micrografted plants than for ungrafted plants when plants were difficult to manage in conventional tissue culture or to establish roots on [65].

3. Factors Affecting the Success of Micrografting

3.1. Scions

In micrografting of cacao, Miguelez-Sierra et al. [38] used two types of bud sticks with either apical or axillary buds as scions. While better graft survival was achieved with the apical shoot apices (85 to 100%), the latter showed a more rapid post-grafting regrowth, resulting in better acclimatization [38]. In micrografts of a cactus (*Pelecyphora aselliformis* Ehrenberg), the use of apical and sub-apical slices as scions produced successful rates of 97% and 81%, respectively [72]. For in vitro grafting of jujube, shoot tips excised from in vitro cultures as scions showed significantly higher survival (70%) and scion growth (3.3 cm) than those collected from mature trees (33% survival and scion growth of 1.3 cm) [42]. Working on micrografts of pistachio, Onay et al. [17] observed that there was a seasonal variation in success rate when micrografting ex vitro material as scions. They found that the month at which shoot tips were harvested from the field growing trees significantly affected graft success: 80% of success at June compared with 10 and 30% for February and December, respectively. In contrast, similar success rates (50 to 80%) were obtained year-round when using in vitro-sourced shoot tips as scions [17].

The size of scions can affect the success of micrografting. Thimmappaiah et al. [39] reported higher success of micrografting (80%) in cashew when scion size was 6–15 mm, while scions of less than 5 mm only had 0.5% of graft success. Similarly, in pistachio, a scion of 4–6 mm resulted in a significantly higher success rate (79%) than 2–4 mm and over 10 mm, which gave 57% and 17% micrografting success, respectively [17]. In micrografting of almond, Yıldırım et al. [33] noted no significant difference in the success rate with scion size ranging from 4 to 15 mm, but significantly improved shoot length and leaf number were observed when larger scion size was used (15 mm). By contrast, in micrografting of cherry, a scion size of 3–5 mm produced higher graft success (42–46%) than the 29% when 10-mm-long scions were used [21].

Channuntapipat et al. [65] compared the effects of various hardness of scions on micrografting survival in almond using wood-stem shoots as rootstocks. Results showed that the highest micrografting survival was found from 'hard scion/hard stem', while no successful graft was found from the combination of 'soft scion/wood stem' [65].

3.2. Rootstocks

Scions can be grafted onto either the hypocotyls or epicotyls of in vitro germinated seedlings. Hypocotyls are usually preferred because epicotyls may contain axillary meristems that compete with the scions and may grow following the grafting, thus adding extra steps of identifying and removing rootstock-derived shoots [78]. In micrografting of cashew, Thimmappaiah et al. [39] noted that side grafting of shoots onto the hypocotyl resulted in higher grafting success (100%) than top-wedge grafting on epicotyls (80%), attributing this to better cambial contact established when side-grafting to hypocotyls. Although comparable micrografting success was achieved when shoot apices of cashew were side micrografted onto the hypocotyl and epicotyl of seedlings, the hypocotyl grafts grew more vigorously than the epicotyl ones [69]. All these studies highlight the advantages of using of hypocotyls as the micrografting sites, when in vitro-germinated seedlings are used as the rootstocks.

Scions grafted onto the same species (homografts) had significantly higher success rates and subsequent growth than heterografts in cactus [54]. In this study, five different combinations of micrografts were tested, with the scion growth of homografts extending from 0.5 to 28.8 cm after 90 days of post-grafting cultures, significantly higher than with the four heterograft combinations [54]. Similarly, graft incompatibility was reported in micrografting in vitro *Ziziphus mauritiana* 'Gola' of west Africa onto an American jujube (*Z. joazeiro*), while achievable results were obtained using those originating from the Old World as rootstocks [42].

The age of rootstock seedlings can also affect the success of micrografting. Working with grapefruit grafted onto sour orange seedlings, Ali et al. [40] found that the highest percentage of successful micrografts was obtained when two- (60%) to three- (40%) week-old seedlings were used, while a further week of seedling growth resulted in a high percentage of unsuccessful grafts (70%). In addition, they observed that most (80%) of the shoot tips grafted onto one-week-old seedlings became quiescent and turned into calluses. Similarly, working on Kinnow mandarin grafted onto Carrizo citrange seedlings, Chand et al. [94] found the highest micrografting success (38%) when 12-day-old seedlings were used, whilst the success was reduced to 24% with 18-day-old rootstock seedlings. They suggested that the reduction in micrografting success with older rootstocks may be due to the harder stem, which made it more difficult to preparing the cut edges for the grafts as well as to insert the scions [94].

3.3. Micrografting Methods

The success of micrografting depends on numerous factors including the grafting method used [87]. Establishing good contact between the microscion and rootstock fosters the reconnection of the cambial tissue and is pivotal to formation of the micrograft union [55,65]. In micrografts of almond, Yıldırım et al. [33] found that the top-slit method resulted in better connection, leading to fusion between scion and rootstock with success rates of 90–100%, while significantly higher numbers of displaced micrografts were detected in top wedge micrografting, which produced only 30–40% of successful grafts. Similar results were produced in micrografting of pistachio by Onay et al. [17] where the top-slit method was compared with the top-wedge method, and 80% success was obtained from the top-slit compared with 60% from the top-wedge.

In Tahitian lime and Valencia orange micrografted onto mandarin seedlings, Suárez et al. [87] only found moderate rates of micrografting success, ranging from 14 to 28%, respectively, when scion shoot tips (<5 mm long) were placed in a slanted position on the decapitated surface of the rootstocks. Shoot tips either positioned at the top or on the side-chip buds of the rootstocks failed completely [87].

Estrada-Luna et al. [54] compared horizontal and wedge grafts for micrografting of *Opuntia* spp. They found the horizontal grafts more successful, owing the reduced scion displacement. In micrografting small *Vitis* shoot apices onto rootstocks of much greater diameter, Torres-vinals [74] observed that side grafting was more successful than top grafting (19% versus 14%). Similarly, in cashew, the side grafting of small shoot apices led to a significantly higher success rate (66%) than the apical grafting method (45%) [69].

3.4. Culture Conditions

The carbon source is a factor that may induce in vitro plant recalcitrance and affect micrografting success [21,95]. In cherry, the highest rates of successful micrografts (79%) were obtained when 30 g L^{-1} of glucose was used as a carbon source compared with sucrose (58%) [21]. In addition, they observed that the incremental increase of 30 to 75 g L^{-1} for both sucrose and glucose did not increase grafting success [21]. In contrast, in micrografts of cut roses, elevating sucrose concentration (50 g L^{-1}) in the culture medium resulted in significantly higher micrografting success than 30 g L^{-1} [59]. Similarly, in Kinnow mandarin and Succari oranges micrografted on 'Rough' lemon seedlings, Naz et al. [96] found that increasing sucrose concentration in the culture medium from 30 to 50 g L^{-1} improved graft success rates from 21 to 33% in both cultivars. In addition, a further increment of graft success (to 38%) was found in Kinnow mandarin grown in culture medium supplemented with 70 g L^{-1} sucrose [96]. Similar improvements in successful micrografting for other *Citrus* species, by increasing sucrose concentration in the culture medium, have been reported by Navarro et al. [11,62], Ali et al. [40], and Singh et al. [97].

Almond micrografts cultured in liquid medium with perlite as a support system showed better survival and scion growth than those supported by paper bridges [63]. Thimmappaiah et al. [39] noted that liquid medium may be better for supporting growth of cashew micrografts as availability and absorption of nutrients was higher and there was less damage to the root system, compared with use of a solid medium. Liquid woody plant medium was successfully used to support *Prunus* micrografts [98]. In using agar-solidified medium to support micrografts of cherry, Bourrain and Charlot [21] found adding vermiculite to the medium increased the grafting success from 12 to 52%. Furthermore, they noticed that the aerated structure of the medium in the presence of vermiculite enhanced the root system and favored the development of secondary roots [21]. Successful micrograft unions in apple, cherry, and *Citrus* were achieved using either agar-solidified medium or vermiculite to hold the grafts during the healing period, with no difference in rate of success [66].

A range of mineral formulations with difference ionic content have been used in micrografting protocols [12,25,42,62,74,98,99]. There is no "best" or "standard" mineral formulation: the selection of a culture medium is influenced by plant species used. In addition, the phytohormones applied in the culture medium can also affect the development of micrografts. Cytokinin and auxin are the most frequently used phytohormones in growth media when micrografts are used [22,26,91]. In pistachio, Onay et al. [17] found that the micrografts cultured in a growth medium with 2.22 μM 6-benzylaminopurine (BAP) were significantly more successful (72%) than those cultured in medium containing 2.46 μM indole-3-butyric acid (IBA) (36%) or in phytohormone-free medium (47%). In contrast, in almond, the medium in which the grafted plants were cultured had no effect on micrografting success [18,33]. It is also worth noting that almond micrografts grown in medium containing 1.0 mg L^{-1} IBA had an increased number and length of roots, while grafts grown in medium with 1.0 mg L^{-1} BAP resulted in higher scion proliferation [18,33]. The different effects of auxin and cytokinin on the survival and growth of Kinnow mandarin micrografts were reported by Kumar et al. [70]. They found that the highest graft survival (66.9%) was observed in graft growth medium supplemented with 3.0 mg L^{-1} 2,4-Dichlorophenoxyacetic acid (2, 4-D), while early bud sprouting and increased scion length were obtained in growth medium supplemented with 1.0 mg L^{-1} BAP [70].

There have been many successful reports of micrografting without the use of phytohormones in the growth medium [21,24,25,34,40,66,72,98]. This would minimize the need to test the genetic stability of regenerants, as high concentrations of plant growth regulators may cause somaclonal variation within in vitro tissue cultures [25,98,100–102].

4. Applications of Micrografting in Micropropagation

4.1. Root Promotion

In vitro rooting is an important stage of micropropagation protocols [56,103]. For some species, the main challenge of micropropagation has been the difficulty in inducing adventitious root formation [18,22,70,104]. IVM is an alternative means to provide roots and overcome rooting difficulties in the vegetative propagation of these species [18,22,55,71], Table 1. For example, IVM of microshoots onto rootstock seedlings was applied to solve the in vitro recalcitrance of *Protea cynaroides*, an important ornamental species endemic to South Africa [55]; *Lens culinaris*, an important pulse crop of Mediterranean area [16]; and some *Prunus* species [21,33,55,66]. The development of suitable micrografting techniques to overcome rooting difficulties of plant species has been highlighted in previous reviews [15,105,106].

Root induction is the limiting step in the in vitro propagation of *Garcinia indica* [71]. To overcome this difficulty, Chabukswar and Deodhar [71] proposed a micrograft protocol where shoot tips were repeatedly grafted on to in vitro juvenile seedlings to restore rooting competence. They conducted micrografting using the 2-month-old in vitro-grown seedlings as rootstocks to reinvigorate in vitro shoots established from 20-year-old trees. The elongated shoots (scions) about 0.5–1.0 cm in length were cut into a V-shape at the bottom and inserted into a vertical incision in the rootstock, after they had been decapitated at the lowest node. After the graft union formed (6–8 weeks), the apical region (1–1.5 cm in length) of the scion was cut, and it was re-grafted onto new in vitro seedling rootstocks. After five successive micrograftings, 75% of the grafts were rooted and successfully acclimatized [71]. In vitro rooting of *Annona cherimola* shoots was also difficult, but it could be achieved after 2–3 consecutive cycles of micrografting onto rootstock seedlings [19]. Similarly, three cycles of in vitro grafting improved the rooting ability in jujube [42]. While numerous studies made use of micrografting in order to promote rooting, in *Juglans rejia* (walnut), the in vitro adult clones did not produce adventitious roots, even after two consecutive cycles of micrografting onto rootstock seedlings [107]. However, an acceptable root induction rate could be obtained after 30 cycles of in vitro subcultures [107]. Further study is therefore still needed to improve the promotion of rooting in walnut. Micrografting for improved rooting often utilizes in vitro germinated seedlings as rootstocks; some examples can be found in Table 1.

4.2. Promotion of Shoot Proliferation

Long-established in vitro plants often demonstrate declined regenerative ability [108,109]. The reduced proliferation could be reversed in vitro following successive grafting onto vigorous rootstocks [19,33,106,110]. In order to improve the micropropagation process of three cherimoya cultivars, Padilla and Encina [19] micrografted its nodal segments onto in vitro germinated seedlings. It was found that proliferation from shoot segments was significantly improved in all micrografted plants compared with conventionally in vitro cultured segments [19]. The restoration of shoot proliferation was also achieved in the almond cultivars Ferragnes and Ferraduel micrografted onto in vitro germinated wild almond seedlings [33].

Farahani et al. [45] repeatedly micrografted mature olive segments (1–1.5 mm in length containing lateral meristem) onto three-week-old germinated seedlings for improved shoot proliferation of the cultivar Zard over a number of culture cycles. The micrografting success, shoot elongation and bud sprouting were improved, particularly after the third successive micrografting [45]. In *Ziziphus mauritiana*, repetitive micrografting ($\geq$2 times) of in vitro shoots onto in vitro germinated seedlings improved the growth of the scions as well as the percentage of rooted micrografts [42]. Improved in vitro rooting and shoot proliferation were also achieved following micrografting in several plant species, such as cherimoya [19], mandarin and sweet orange [106,111]. The improved scion growth was a consequence of reinvigorated rooting either from direct support by the in vitro germinated seedlings [45] or through recovered adventitious rooting [19].

4.3. Embryo Rescue or the Promotion of Organogenesis-Derived Shoot Regrowth

Recovery of plants via de novo organogenesis and somatic embryogenesis can be important for obtaining genetically modified plants, and in vitro mutagenesis [112–115]. However, it can be problematic in some horticultural species owing the difficulties of rooting [116,117] or to inadequate callus maturation and tissue culture [37].

Micrografting has been applied to overcome the inability of many organogenesis-derived regenerants to readily produce roots [77,78,93,113]. The poor rooting ability observed in regenerated shoots of sunflower was resolved by micrografting shoots regenerated from leaves onto in vitro-germinated seedlings using a side insertion method [77]. In this method, best survival (75%) was obtained by inserting the 0.5–1.0-cm-long shoots with a wedge-shaped base into the longitudinal cut at the hypocotyl. The micrografted sunflowers were successfully acclimatized, and they flowered and produced seeds [77]. Kobayashi et al. [93] applied IVM to support the growth of sweet orange (*Citrus sinensis*) regenerated from thin sections of mature stem segments. Using the same regeneration system in four sweet orange cultivars, Almeida et al. [118] tested the use of micrografting with plantlets of Carrizo citrange as rootstocks to support shoot recovery after genetic transformation. IVM was also applied in pepper (*Capsicum annuum*) to obtain rooted transgenic plants regenerated from cotyledon-derived organogenesis [119].

Noticeably, ex vitro microgafting was used in legumes using in vitro-regenerated shoots as scions and ex vitro-germinated seedlings as rootstocks to facilitate grafting and acclimatization simultaneously in field pea (*Pisum sativum* L.) [120] and chickpea (*Cicer arietinum* L) [121]. Similar protocols were also used to obtain rooted plants from cotyledon-derived adventitious pear shoots [122].

IVM was first reported in 1992 to support the recovery of shoot regenerated from somatic embryos (SEs) of cocoa plants [123]. In avocado, Raharjo and Litz [37] proposed an effective micrografting procedure for SE shoot rescue. Briefly, in their study, SE-derived shoots of 5–10 mm in length were grafted (V-shaped cut) onto in vitro rootstock seedlings, and then grafted plants were grown on a phytohormone-free medium. Using this protocol, micrografted plants were established after 3–4 weeks, and 70.5% of the SE-derived shoots were rescued, whereas only 30.4% of nonmicrografted SE shoots survived and normal plantlets were never recovered [37]. In addition, the micrografting protocol was followed by ex vitro grafting, and this has served as a protocol for rescuing transformed avocado materials [37]. Palomo-Ríos et al. [124] also reported successful recovery of transgenic plants by IVM in avocado. In this study, globular somatic embryos established from immature zygotic embryos were transformed using *Agrobacterium*. After selection on kanamycin, the germinated somatic embryos were then elongated to 3–5 mm before micrografting onto in vitro-germinated seedlings to achieve better recovery [124]. Likewise, in seedless sweet orange, micrografting of ovary-derived somatic embryos onto in vitro seedlings was applicable to achieve full recovery or somatic organogenesis [117]. Some examples of applying IVM to promote the shoot recovery from de novo organogenesis and somatic embryogenesis are listed in Table 2.

4.4. Shoot Regrowth after Cryopreservation

Cryopreservation is currently considered an applicable strategy to facilitate long-term, cost-effective maintenance of plant genetic resources [125,126]. Shoot tip cryopreservation of many horticultural species has been established in cryobanks; a high level of post-thaw recovery is a requirement of successful cryopreservation [126,127]. Direct shoot tip recovery could not be obtained in some species, such as *Citrus*; thus, micrografting of cryopreserved shoot tips onto in vitro prepared seedlings was used to overcome this [128–130], Table 2. In successful recovery of citrus shoot tips after cryopreservation, Volk et al. [128] prepared six-week-old in vitro 'Carrizo' citrange seedlings as rootstocks to support the shoot tips cryopreserved by a vitrification protocol. Briefly, in their study, rootstock seedlings with a height of at least 3 cm were decapitated 1 cm above the cotyledonary node with a 2-mm deep incision made into the cut surface, followed by horizontal cut through the seedling to create a "ledge" or "step" at the cut surface. Cryopreserved shoot tips were trimmed (0.2 mm of the basal portion) and placed on this rootstock ledge [128]. This post-thaw protocol resulted in 53% of regrowth on average for eight *Citrus* and *Fortunella* species [128]. Volk et al. [129] applied the same IVM procedure to support the post-thaw recovery of 150 pathogen-free citrus accessions representing 32 taxa after a droplet-vitrification cryopreservation. With this procedure, 24 taxa had mean regrowth levels of over 40% after cryopreservation [129]. There are ongoing efforts to use this successful procedure to recover plants following cryopreservation of a wide range of citrus species [86].

Table 2. Applications of in vitro micrografting in promoting of plants recovery after shoot organogenesis, somatic embryogenesis and shoot tip cryopreservation.

Plant Species (Scion)	Scion Source and Size	Rootstock Source and Age	Grafting Technique	Success Rate (%) and (No. Scions Tested)	Reference
Citrus sinensis (Sweet orange)	Shoots of 1–2 mm in length (sourced from greenhouse plants)	Carrizo citrange seedlings of 2 weeks old	Side insertion	90 (1)	[93]
	Shoots recovered from germinated somatic embryos (size not specified)	Carrizo citrange seedlings (age not specified)	Not specified	90 (1)	[117]
Citrus spp.	Shoot tips (1–1.5 mm in length) cryopreserved by droplet-vitrification	Carrizo citrange seedlings up to 6 weeks old	Side grafting	10–100 (32); average of 56%	[129]
Helianthus annuus (Sunflower)	Shoots (0.5–1 cm in length) from leaf explants	Sunflower seedlings of 7–10 days old	Side insertion	47–85 (7)	[77]
	Shoots (1 cm in length) from cotyledon explants	Sunflower seedlings of 1–2 weeks old	Side insertion	69 (1)	[78]
Lens culinaris (Lentil)	Shoots (1–1.5 cm in length) from cotyledonary nodes	Lentil seedlings of 5–6 days old	Top slit	90–100 (3)	[16]
Persea americana (Avocado)	Shoots regenerated from somatic embryos (SEs)	Avocado seedlings of 7–12 days old	Top slit	70.5 (1)	[37]
	Shoots derived from previously micrografted SE shoots			100 (1)	
	Shoots of 5–10 mm in length			59 (1)	
	Shoots (size not specified) regenerated from genetically transformed SEs	Avocado seedlings of 3 weeks old	Top slit	83.6 (1)	[112]
	Shoots of 3–5 mm in length from genetically transformed SEs	Avocado seedlings of 4 weeks old	Top slit	60–80 (1)	[124]
Solanum lycopersicum F1 hybrids (Tomato)	Shoots (size not specified) regenerated from cotyledon explants	Tomato seedlings of 3 weeks old	Top slit	75–83 (3)	[131]
Ziziphus jujuba (Chinese jujube)	Shoot tips (0.2–1 mm in length) cryopreserved by droplet-vitrification	*Z. spinosa* seedling of 4 weeks old	Side grafting	5–75 (1)	[132]

Micrografting was necessary to support the recovery of cryopreserved Chinese jujube (*Ziziphus jujuba*) shoot tips [132]. Cryopreserved shoot tips cultured on recovery medium, without micrografting, developed only leaves without shoot regrowth. In contrast, a high shoot recovery rate of 75% was obtained when shoot tips were micrografted onto sour jujube (*Ziziphus spinosa*) rootstock seedlings. This procedure was also effective to produce plants free of jujube witches' broom phytoplasmas. Therefore, micrografting provided technical support in cryopreservation and production of phytoplasma-free plants [132].

While micrografting in *Citrus* to support cryopreservation protocols has been moving from research [128] to full-scale implementation [86,129], it has not been well explored in other species. Micrografting has potential to improve the regrowth of cryoprocessed shoot tips of woody plants that are still recalcitrant to cryopreservation. Plants that might benefit from this approach include *Pistacia* species, in which shoot tip cryopreservation resulted in low recovery levels ranging from 5.0 to 17.6% [133]. In avocado, although SEs have been shown to be amenable to cryopreservation, studies focusing on shoot tip cryopreservation are still needed for the safe conservation of elite avocado cultivars [134]. Micrografting could therefore be considered as a tool to support shoot tip regrowth after cryopreservation in the cases of pistachio, avocado, and other recalcitrant plant species.

5. Conclusions and Future Prospects

This review focuses on studies highlighting the use of micrografting in micropropagation of horticultural species in the 21st century. As an important technique supporting basic and applied research, IVM consists of preparation of rootstocks and scions, the micrografting process and post-grafting cultures, steps that have all been developed from in vitro tissue culture techniques.

IVM protocols have been developed for many plant species with differing degrees of success. The successful recovery of plants following micrografting depends on numerous factors, such as the origin and preparation of rootstocks and scions, grafting methods, graft growth conditions, as well acclimatization, with these factors being genotype- and species-specific, just as in in vitro tissue culture. Therefore, further development and optimization of micrografting techniques, especially for recalcitrant species, are needed, to expand the use of micrografting in micropropagation. While successful micrografting protocols have been established in most horticultural species, they rely on technically difficult protocols, performed by well-equipped, skilled and well-trained technicians. Progress is thus still needed to simplify micrografting procedures. In addition, the programs applying IVM for micropropagation always include the transfer of the micrografted plants to ex vitro conditions, and thus, the effects of acclimatization on the survival of micrograft plantlet as well as long-term survival in case of partial graft incompatibility should also be evaluated.

Although progress has been made over the years on in vitro plant tissue cultures, in some plant species, IVM is still applied as a necessary step to provide roots for in vitro grown propagules, thus enabling further acclimatization (Figure 2). In species with problematic in vitro rooting, seedlings produced from in vitro germinated seeds are often used as rootstocks in micrografting procedures. For species that showed decreased rooting and shoot proliferation following prolonged in vitro cultures, consecutive micrografting of shoots onto in vitro germinated seedlings has proved effective in restoration of vigor and rooting competence in some instances (Figure 2).

Moreover, IVM has assisted the recovery of shoots generated from de novo organogenesis and somatic embryogenesis (Figure 2), which have been widely used as sources of explants for genetic transformation. Therefore, IVM has been implemented as a step in supporting the genetic transformation of horticultural species such as citrus and avocado. Likewise, the cryopreservation of citrus shoot tips also relies on IVM to sustain post-thaw recovery and shoot regrowth after cryopreservation (Figure 2). The large-scale cryopreservation of citrus has been established with the assistance of micrografting, ensuring the safe and long-term preservation of its valuable genetic resources. The successes observed in citrus cryopreservation would encourage the use of micrografting in the recovery of cryopreserved species that are still recalcitrant to cryopreservation procedures.

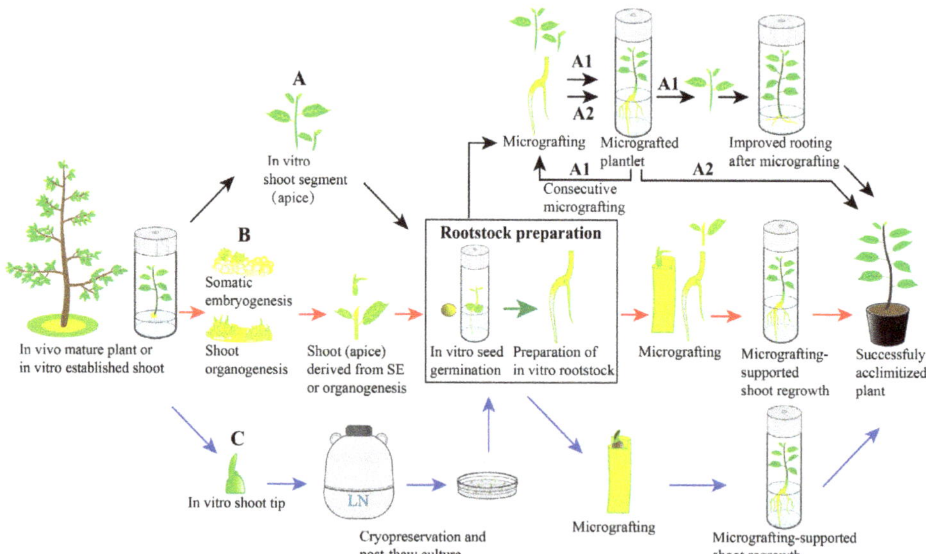

Figure 2. A summary of the applications of in vitro micrografting (IVM) for improved micropropagation. A (black arrows), the use of IVM for in vitro rooting and reinvigoration. A1 indicates the use of consecutive micrografting for reinvigoration of in vitro adventitious rooting and vegetative growth; A2 illustrates the one-step use of micrografting for in vitro rooting. B (red arrows), the use of IVM to assist in the recovery and regrowth of shoots derived from somatic embryogenesis/shoot organogenesis. C (blue arrows), the use of IVM to support the regrowth of cryopreserved shoot tips. The green arrow indicates rootstock preparation.

Author Contributions: M.-R.W. and J.C.B.: data collection, analysis, manuscript writing and revisions; A.-L.Z. and X.L.: data collection and analysis, table and figure preparation; D.Z. and Q.-C.W.: manuscript proposal and manuscript revision. All authors have read and agreed to the published version of the manuscript.

Funding: We acknowledge funding from the National Key R&D Programs of China (project no. 2019YFD1001803) for M.-R.W., X.L., D.Z. and Q.-C.W. and from The New Zealand Institute for Plant and Food Research Limited (project no. 0/601000/01) for J.C.B.

Institutional Review Board Statement: Not applicable.

Informed Consent Statement: Not applicable.

Data Availability Statement: Not applicable.

Conflicts of Interest: The authors declare no conflict of interest.

References

1. Mudge, K.; Janick, J.; Scofield, S.; Goldschmidt, E.E. A History of Grafting. In *Horticultural Reviews*; Janick, J., Ed.; John Wiley & Sons, Inc.: Hoboken, NJ, USA, 2009; Volume 35, pp. 437–493.
2. King, S.R.; Davis, A.R.; Liu, W.; Levi, A. Grafting for Disease Resistance. *Hortscience* **2008**, *43*, 1673–1676. [CrossRef]
3. Goldschmidt, E.E. Plant grafting: New mechanisms, evolutionary implications. *Front. Plant Sci.* **2014**, *5*, 727. [CrossRef] [PubMed]
4. Macan, G.P.F.; Cardoso, J.C. In vitro grafting of *Psidium guajava* in *Psidium cattleianum* for the Management of the *Meloidogyne enterolobii*. *Int. J. Fruit Sci.* **2020**, *20*, 106–116. [CrossRef]
5. Fazio, G. Genetics, breeding, and genomics of apple rootstocks. In *The Apple Genome, Compendium of Plant Genomes*; Korban, S.S., Ed.; Springer: Cham, Switzerland, 2021; pp. 105–130.

6. Martins, V.; Silva, V.; Pereira, S.; Afonso, S.; Oliveira, I.; Santos, M.; Ribeiro, C.; Vilela, A.; Bacelar, E.; Silva, A.P.; et al. Rootstock affects the fruit quality of 'Early Bigi' sweet cherries. *Foods* **2021**, *10*, 2317. [CrossRef] [PubMed]
7. Thorpe, T.A. History of Plant Tissue Culture. In *Methods in Molecular Biology*, 2nd ed.; Loyola-Vargas, V.M., Vázquez-Flota, F., Eds.; Humana Press Inc.: Totowa, NJ, USA, 2006; Volume 318, pp. 9–32.
8. Doorenbos, J. Rejuvenation of *Hedera helix* in graft combinations. Koninklijke Nederlandse Akademie van Wetenschappen: Series C. *Biol. Med. Sci.* **1953**, *57*, 99–102.
9. Holmes, F.O. Elimination of spermivirus from the nightingale chrysanthemum. *Phytopathology* **1956**, *46*, 599–600.
10. Murashige, T.; Bitters, W.P.; Rangan, E.M.; Nauer, E.M.; Roistacher, C.N.; Holliday, P.B. A technique of shoot apex grafting and its utilization towards recovering virus-free citrus clones. *HortScience* **1972**, *7*, 118–119. [CrossRef]
11. Navarro, L.; Roistacher, C.N.; Murashige, T. Improvement of shoot-tip grafting in vitro for virus-free citrus. *J. Am. Soc. Hortic. Sci.* **1975**, *100*, 471–479.
12. Pathirana, R.; McKenzie, M.J. Early detection of grapevine leafroll virus in *Vitis vinifera* using in vitro micrografting. *Plant Cell Tiss. Organ Cult.* **2005**, *81*, 11–18. [CrossRef]
13. Cui, Z.-H.; Agüero, C.B.; Wang, Q.-C.; Walker, M.A. Validation of micrografting to identify incompatible interactions of rootstocks with virus-infected scions of Cabernet Franc. *Aust. J. Grape Wine Res.* **2019**, *25*, 268–275. [CrossRef]
14. Hao, X.-Y.; Jiao, B.-L.; Wang, M.-R.; Wang, Y.-L.; Shang, B.-X.; Wang, J.-Y.; Wang, Q.-C.; Xu, Y. In vitro biological indexing of grapevine leafroll-associated virus 3 in red- and white-berried grapevines (*Vitis vinifera*). *Aust. J. Grape Wine Res.* **2021**, *27*, 483–490. [CrossRef]
15. Chilukamarri, L.; Ashrafzadeh, S.; Leung, D.W.M. In Vitro grafting—Current applications and future prospects. *Sci. Hortic.* **2021**, *280*, 109899. [CrossRef]
16. Gulati, A.; Schryer, P.; McHughen, A. Regeneration and micrografting of lentil shoots. *In Vitro Cell. Dev. Biol.-Plant* **2001**, *37*, 798–802. [CrossRef]
17. Onay, A.; Pirinç, V.; Yıldırım, H.; Basaran, D. In vitro Micrografting of Mature Pistachio (*Pistacia vera* var. Siirt). *Plant Cell Tiss. Organ Cult.* **2004**, *77*, 215–219. [CrossRef]
18. Işıkalan, Ç.; Namli, S.; Akbas, F.; Ak, B.E. Micrografting of almond (*Amygdalus communis*) cultivar 'Nonpareil'. *Aust. J. Crop. Sci.* **2011**, *5*, 61–65.
19. Padilla, I.M.G.; Encina, C.L. The use of consecutive micrografting improves micropropagation of cherimoya (*Annona cherimola* Mill.) cultivars. *Sci. Hortic.* **2011**, *129*, 167–169. [CrossRef]
20. Luo, J.; Gould, J.H. In Vitro shoot-tip grafting improves recovery of cotton plants from culture. *Plant Cell Tiss. Organ Cult.* **1999**, *57*, 211–213. [CrossRef]
21. Bourrain, L.; Charlot, G. In vitro micrografting of cherry (*Prunus avium* L.'Regina') onto 'Piku$^{®}$ 1'rootstock [*P. avium* × (*P. canescens* × *P. tomentosa*)]. *J. Hortic. Sci. Biotechnol.* **2014**, *89*, 47–52. [CrossRef]
22. Koufan, M.; Mazri, M.A.; Essatte, A.; Moussafir, S.; Belkoura, I.; El Rhaffari, L.; Toufik, I. A novel regeneration system through micrografting for *Argania spinosa* (L.) Skeels, and confirmation of successful rootstock-scion union by histological analysis. *Plant Cell Tiss. Organ Cult.* **2020**, *142*, 369–378. [CrossRef]
23. Errea, P.; Garay, L.; Marín, J.A. Early detection of graft incompatibility in apricot (*Prunus armeniaca*) using in vitro techniques. *Physiol. Plant* **2001**, *112*, 135–141. [CrossRef]
24. Dumanoğlu, H.; Çelik, A.; Büyükkartal, H.N.; Dousti, S. Morphological and anatomical investigations on in vitro micrografts of OHxF 333 / *Pyrus elaeagrifolia* interstock / rootstock combination in pears. *J. Agric. Sci.* **2014**, *20*, 269–279.
25. Bao, W.W.; Zhang, X.C.; Zhang, A.L.; Zhao, L.; Wang, Q.C.; Liu, Z.D. Validation of micrografting to analyze compatibility, shoot growth, and root formation in micrografts of kiwifruit (*Actinidia* spp.). *Plant Cell Tiss. Organ Cult.* **2020**, *140*, 209–214. [CrossRef]
26. Hieu, T.; Phong, T.H.; Khai, H.D.; Mai, N.T.N.; Cuong, D.M.; Luan, V.Q.; Tung, H.T.; Nam, N.B.; Nhut, D.T. Efficient production of vigorous passion fruit rootstock for in vitro grafting. *Plant Cell Tiss. Organ Cult.* **2022**, *148*, 635–648. [CrossRef]
27. Fragoso, V.; Goddard, H.; Baldwin, I.T.; Kim, S.-G. A simple and efficient micrografting method for stably transformed *Nicotiana attenuata* plants to examine shoot-root signaling. *Plant Meth.* **2011**, *7*, 1–8. [CrossRef]
28. Paultre, D.S.G.; Gustin, M.P.; Molnar, A.; Oparka, K.J. Lost in transit: Long-distance trafficking and phloem unloading of protein signals in *Arabidopsis* homografts. *Plant Cell* **2016**, *28*, 2016–2025. [CrossRef]
29. Hao, L.; Zhang, Y.; Wang, S.; Zhang, W.; Wang, S.; Xu, C.; Yu, Y.; Li, T.; Jiang, F.; Li, W. A constitutive and drought-responsive mRNA undergoes long-distance transport in pear (*Pyrus betulaefolia*) phloem. *Plant Sci.* **2020**, *293*, 110419. [CrossRef]
30. Tsutsui, H.; Yanagisawa, N.; Kawakatsu, Y.; Ikematsu, S.; Sawai, Y.; Tabata, R.; Arata, H.; Higashiyama, T.; Notaguchi, M. Micrografting device for testing systemic signaling in *Arabidopsis*. *Plant J.* **2020**, *103*, 918–929. [CrossRef]
31. Thimmappaiah; Shirly, R.A.; Sadhana, P.H. In vitro propagation of cashew from young trees. *In Vitro Cell. Dev. Biol. -Plant* **2002**, *38*, 152–156. [CrossRef]
32. Perez-Tornero, O.; Burgos, L. Different media requirements for micropropagation of apricot cultivars. *Plant Cell Tiss. Organ Cult.* **2000**, *63*, 133–141. [CrossRef]
33. Yıldırım, H.; Onay, A.; Süzerer, V.; Tilkat, E.; Ozden-Tokatli, Y.; Akdemir, H. Micrografting of almond (*Prunus dulcis* Mill.) cultivars "Ferragnes" and "Ferraduel". *Sci. Hortic.* **2010**, *125*, 361–367. [CrossRef]
34. Aazami, M.A.; Hassanpouraghdam, M.B. In Vitro Micro-Grafting of Some Iranian Grapevine Cultivars. *Rom. Biotech. Lett.* **2010**, *15*, 5576–5580.

35. Ashrafzadeh, S. In Vitro grafting–twenty-first century's technique for fruit tree propagation. *Acta Agric. Scand. B Soil Plant Sci.* **2020**, *74*, 584–587. [CrossRef]
36. Dobránszki, J.; Jámbor-Benczúr, E.; Hudák, I.; Magyar-Tábori, K. Model experiments for establishment of in vitro culture by micrografting in apple. *Int. J. Hortic. Sci.* **2005**, *11*, 47–49. [CrossRef]
37. Raharjo, S.H.T.; Litz, R.E. Micrografting and ex vitro grafting for somatic embryo rescue and plant recovery in avocado (*Persea americana*). *Plant Cell Tiss. Organ Cult.* **2005**, *82*, 1–9. [CrossRef]
38. Miguelez-Sierra, Y.; Hernández-Rodríguez, A.; Acebo-Guerrero, Y.; Baucher, M.; El Jaziri, M. In vitro micrografting of apical and axillary buds of cacao. *J. Hort. Sci. Biotech.* **2017**, *92*, 25–30. [CrossRef]
39. Thimmappaiah; Puthra, G.T.; Anil, S.R. In vitro grafting of cashew (*Anacardium occidentale*). *Sci. Hortic.* **2002**, *92*, 177–182. [CrossRef]
40. Ali, H.M.A.; Elamin, O.M.; Ali, M.A. Propagation of grapefruit (*Citrus paradisi* macf) by shoot tip micrografting. *Gezira J. Agric.* **2004**, *2*, 37–49.
41. Singh, P.; Patel, R.M. Factors affecting in vitro degree of browning and culture establishment of pomegranate. *Afr. J. Plant Sci.* **2016**, *10*, 43–49. [CrossRef]
42. Danthu, P.; Touré, M.; Soloviev, P.; Sagna, P. Vegetative propagation of *Ziziphus mauritiana* var. Gola by micrografting and its potential for dissemination in the Sahelian Zone. *Agrofor. Syst.* **2004**, *60*, 247–253. [CrossRef]
43. Ma, F.; Guo, C.; Liu, Y.; Li, M.; Ma, T.; Mei, L.; Hsiao, A.I. In vitro shoot-apex grafting of mulberry (*Morus alba* L.). *Hortscience* **1996**, *31*, 460–462. [CrossRef]
44. Nas, M.N.; Read, P.E. Simultaneous micrografting, rooting and acclimatization of micropropagated American chestnut, grapevine and hybrid hazelnut. *Eur. J. Hortic. Sci.* **2003**, *68*, 234–237.
45. Farahani, F.; Razeghi, S.; Peyvandi, M.; Attaii, S.; Mazinani, M.H. Micrografting and micropropagation of olive (*Olea europea* L.) Iranian cultivar: Zard. *Afri. J. Plant Sci.* **2011**, *5*, 671–675.
46. Edriss, M.H.; Baghdady, G.A.; Abdrabboh, G.A.; Abd El-Razik, A.M.; Abdel Aziz, H.F. Micro-grafting of Florida prince peach cultivar. *Nat. Sci.* **2015**, *13*, 54–60.
47. Hassanen, S.A. In vitro grafting of pear (*Pyrus* spp.). *World Appl. Sci. J.* **2013**, *24*, 705–709. [CrossRef]
48. Vozárová, Z.; Nagyová, A.; Nováková, S. In vitro micrografting of different *Prunus* species by cherry-adapted *Plum pox virus* isolate. *Acta Virol.* **2018**, *62*, 109–111. [CrossRef]
49. Wang, G.; Li, X.; Chen, Q.; Tian, J. Studies on factors affecting the microshoot grafting survival of walnut. *Acta Hortic.* **2010**, *861*, 327–332. [CrossRef]
50. Ribeiro, H.; Ribeiro, A.; Pires, R.; Cruz, J.; Cardoso, H.; Barrroso, J.M.; Peixe, A. Ex vitro rooting and simultaneous micrografting of the walnut hybrid rootstock 'Paradox' (*Juglans hindsi* × *Juglans regia*) cl. 'Vlach'. *Agronomy* **2022**, *12*, 595. [CrossRef]
51. Zhang, N.; Zeng, H.X.; Shi, X.F.; Yang, Y.X.; Cheng, W.S.; Ren, J.; Li, Y.H.; Li, A.C.; Tang, M.; Sun, Y.H.; et al. An efficient in vitro micrografting technology of watermelon. *Acta Hortic.* **2015**, *1086*, 65–70. [CrossRef]
52. Edriss, M.H.; Burger, D.W. Micro-grafting shoot-tip culture of citrus on three trifoliolate rootstocks. *Sci. Hort.* **1984**, *23*, 255–259. [CrossRef]
53. Deogratias, J.M.; Castellani, V.; Dosba, F.; Juarez, J.; Arregui, J.M.; Ortega, C.; Ortega, V.; Llacer, G.; Navarro, L. Study of growth parameters on apricot shoot-tip grafting in vitro (STG). *Acta Hortic.* **1991**, *293*, 363–372. [CrossRef]
54. Estrada-Luna, A.A.; López-Peralta, C.; Cárdenas-Soriano, E. In vitro micrografting and the histology of graft union formation of selected species of prickly pear cactus (*Opuntia* spp.). *Sci. Hortic.* **2002**, *92*, 317–327. [CrossRef]
55. Wu, H.C.; du Toit, E.S.; Reinhardt, C.F. Micrografting of *Protea cynaroides*. *Plant Cell Tiss. Organ Cult.* **2007**, *89*, 23–28. [CrossRef]
56. Teixeira da Silva, J.A.; Gulyás, A.; Magyar-Tábori, K.; Wang, M.-R.; Wang, Q.-C.; Dobránszki, J. In vitro tissue culture of apple and other *Malus* species: Recent advances and applications. *Planta* **2019**, *249*, 975–1006. [CrossRef] [PubMed]
57. Jones, A.M.; Saxena, P.K. Inhibition of phenylpropanoid biosynthesis in *Artemisia annua* L.: A novel approach to reduce oxidative browning in plant tissue culture. *PLoS ONE* **2013**, *8*, e76802. [CrossRef]
58. Jonard, R.; Hugard, J.; Macheix, J.J.; Martinez, J.; Mosella-Chancel, L.; Poessel, J.L.; Villemur, P. In Vitro micrografting and its applications to fruit science. *Sci. Hortic-Amst.* **1983**, *20*, 147–159. [CrossRef]
59. Davoudi Pahnekolayi, M.; Tehranifar, A.; Samiei, L.; Shoor, M. Optimizing culture medium ingredients and micrografting devices can promote in vitro micrografting of cut roses on different rootstocks. *Plant Cell Tiss. Organ Cult.* **2019**, *137*, 265–274. [CrossRef]
60. Naddaf, M.E.; Rabiei, G.; Moghadam, E.G.; Mohammadkhani, A. In vitro production of PPV-free Sweet cherry (*Prunus avium* cv. Siahe-Mashhad) by meristem culture and micro-grafting. *J. Plant Bioinform. Biotechnol.* **2021**, *1*, 51–59. [CrossRef]
61. Nunes, J.C.O.; Abreu, M.F.; Dantas, A.C.M.; Pereira, A.J.; Pedrotti, E.L. Caracterização morfológica de microenxertia em macieira. *Rev. Bras. Frutic.* **2005**, *27*, 80–83. [CrossRef]
62. Navarro, L. Application of shoot-tip grafting in vitro to woody species. *Acta Hortic.* **1988**, *227*, 43–55. [CrossRef]
63. Asadi Zargh Abad, M.; Shekafandeh, A. In vitro grafting of 'Sahand' cultivar on two wild almond rootstocks and evaluation of its some physiological and biochemical traits vis-a-vis different rootstocks. *Plant Cell Tiss. Organ Cult.* **2021**, *145*, 507–516. [CrossRef]
64. Juárez, J.; Aleza, P.; Navarro, L. Applications of citrus shoot-tip grafting in vitro. *Acta Hort.* **2015**, *1065*, 635–642. [CrossRef]
65. Channuntapipat, C.; Sedgley, M.; Collins, G. Micropropagation of almond cultivars Nonpareil and Ne Plus Ultra and the hybrid rootstock Titan x Nemaguard. *Sci. Hortic.* **2003**, *98*, 473–484. [CrossRef]
66. Obeidy, A.A.; Smith, M.A.L. A versatile new tactic for fruit tree micrografting. *Hortic. Sci.* **1991**, *26*, 776. [CrossRef]

67. Nkanaunena, G.A.; Kwapata, M.B.; Bokosi, J.M.; Maliro, M.F.A. The effect of age and type of rootstock, method of scion placement and light intensity levels on the performance of *Uapaca kirkiana* (Muell. Arg) micrografts cultured in vitro. *UNISWA J. Agric.* **2001**, *10*, 22–29. [CrossRef]
68. Sammona, O.S.; Abde Elhamid, N.A.; Samaan, M.S.F. Effects of some factors on the micropropagation and micrografting of some grape rootstocks in vitro. *J. Agric. Sci.* **2018**, *26*, 133–146. [CrossRef]
69. Mneney, E.E.; Mantell, S.H. In Vitro micrografting of cashew. *Plant Cell Tiss. Organ Cult.* **2001**, *66*, 49–58. [CrossRef]
70. Kumar, R.A.J.; Kaul, M.K.; Saxena, S.N.; Singh, A.K.; Khadda, B.S. Standardization of protocol for *in Vitro* shoot tip grafting in Kinnow mandarin (*Citrus deliciosa*). *Indian J. Agric. Sci.* **2014**, *84*, 1376–1381.
71. Chabukswar, M.M.; Deodhar, M.A. Restoration of rooting competence in a mature plant of *Garcinia indica* through serial shoot tip grafting in vitro. *Sci. Hortic.* **2006**, *108*, 194–199. [CrossRef]
72. Badalamenti, O.; Carra, A.; Oddo, E.; Carimi, F.; Sajeva, M. Is in vitro micrografting a possible valid alternative to traditional micropropagation in Cactaceae? *Pelecyphora aselliformis* as a case study. *SpringerPlus* **2016**, *5*, 1–4. [CrossRef]
73. Jenderek, M.M.; Holman, G.E.; Irish, B.M.; Souza Vidigal, F. In Vitro Seed Germination and Rootstock Establishing for Micrografting of *Theobroma cacao* L. In Proceedings of the ASHS 2015 Annual Conference, New Orleans, LA, USA, 4–7 August 2015.
74. Torres-Viñals, M.; Sabaté-Casaseca, S.; Aktouche, N.; Grenan, S.; Lopez, G.; Porta-Falguera, M.; Torregrosa, L. Large-scale production of somatic embryos as a source of hypocotyl explants for *Vitis vinifera* micrografting. *Vitis* **2004**, *43*, 163–168. [CrossRef]
75. Mhatre, M.; Bapat, V.A. Micrografting in grapevine (*Vitis* spp.). In *Protocols for Micropropagation of Woody Trees and Fruits*; Jain, S.M., Häggman, H., Eds.; Springer: Dordrecht, The Netherlands, 2007; pp. 249–258.
76. Liu, X.; Liu, M.; Ning, Q.; Liu, G. Reverse-cleft in vitro micrografting of *Ziziphus jujuba* Mill. Infected with jujube witches' broom (JWB). *Plant Cell Tiss. Organ Cult.* **2012**, *108*, 339–344. [CrossRef]
77. Zhang, Z.; Finer, J.J. Sunflower (*Helianthus annuus* L.) organogenesis from primary leaves of young seedlings preconditioned by cytokinin. *Plant Cell Tiss. Organ Cult.* **2015**, *123*, 645–655. [CrossRef]
78. Zhang, Z.; Finer, J.J. Use of cytokinin pulse treatments and micrografting to improve sunflower (*Helianthus annuus* L.) plant recovery from cotyledonary tissues of mature seeds. *In Vitro Cell. Dev. Biol. Plant* **2016**, *52*, 391–399. [CrossRef]
79. Gebhardt, K.; Goldbach, H. Establishment, graft union characteristics and growth of *Prunus* micrografts. *Physiol. Plant.* **1988**, *72*, 153–159. [CrossRef]
80. Hao, X.-Y.; BI, W.-L.; Cui, Z.-H.; Pan, C.; Xu, Y.; Wang, Q.-C. Development, histological observations and *Grapevine leafroll-associated virus-3* localisation in in vitro grapevine micrografts. *Ann. Appl. Biol.* **2017**, *170*, 379–390. [CrossRef]
81. Saleh, S.S.; AbdAllah, M.N.; El Botaty, E.S.M. Using new micro grafting techniques in grape. *J. Plant Prod.* **2019**, *10*, 1015–1022. [CrossRef]
82. Dobránszky, J.; Magyar-Tábori, K.; Jámbor-Benczúr, E.; Lazányi, J. New in vitro micrografting method for apple by sticking. *Int. J. Hort. Sci.* **2000**, *6*, 79–83. [CrossRef]
83. Ponsonby, D.J.; Mantell, S.H. In vitro establishment of *Picea pungens* f. *glauca* and *P. sitchensis* seedling rootstocks with an assessment of their suitabilities for micrografting with scions of various *Picea* species. *J. Hortic. Sci.* **1993**, *68*, 463–475. [CrossRef]
84. Cortizo, M.; Afonso, P.; Fernández, B.; Rodríguez, A.; Centeno, M.; Ordás, R. Micrografting of mature stone pine (*Pinus pinea* L.) trees. *Ann. For. Sci.* **2004**, *61*, 843–845. [CrossRef]
85. Sertkaya, G. Effects of different rootstocks in micrografting on growing of Washington Navel orange plants obtained by shoot tip grafting. *Biotechnol. Biotechnol. Equip.* **2004**, *18*, 82–88. [CrossRef]
86. Volk, G.M.; Jenderek, M.M.; Walters, C.; Bonnart, R.; Shepherd, A.; Skogerboe, D.; Hall, B.D.; Moreland, B.; Krueger, R.; Polek, M. Implementation of *Citrus* shoot tip cryopreservation in the USDA-ARS National Plant Germplasm System. *Acta Hortic.* **2019**, *1234*, 329–334. [CrossRef]
87. Suárez, I.E.; Álvarez, C.; Díaz, C.L. Micrografting of Valencia orange and Tahiti lime. *Temas Agrar.* **2021**, *26*, 26–35. [CrossRef]
88. Lahoty, P.; Singh, J.; Bhatnagar, P.; Raipurohit, D.; Singh, B. In vitro multiplication of Nagpur mandarin (*Citrus reticulata* Blanco) through STG. *Vegetos* **2013**, *26*, 318–324. [CrossRef]
89. Bhatt, K.M.; Banday, F.A.; Mir, M.A.; Rather, Z.A.; Hussain, G. In vitro grafting in apple (*Malus domestica* Borkh) cv. Lal Ambri. *Karnataka J. Agric. Sci.* **2013**, *26*, 399–402.
90. Prasaei, Z.; Hedayat, M.; Rastgoo, S.; Bayat, F. Evaluation rootstock, sucrose concentration and culture medium supporting systems in the micrografting and acclimatization of lime (*Citrus aurantifulia*). *Agric. Biotechnol. J.* **2018**, *10*, 1–17.
91. Zhu, B.; Cao, H.-N.; Zong, C.-W.; Piao, R.-Z.; Chen, L.; Zhou, L. Micrografting technology in grapevine (*Vitis vinifera* L.). *Fruit Veg. Cereal Sci. Biotechnol.* **2007**, *1*, 60–63.
92. Yıldırım, H.; Akdemir, H.; Süzerer, V.; Ozden, Y.; Onay, A. In vitro micrografting of the almond cultivars "Texas", "Ferrastar" and "Nonpareil". *Biotechnol. Biotechnol. Equip.* **2013**, *27*, 3493–3501. [CrossRef]
93. Kobayashi, A.K.; Bespalhok, J.C.; Pereira, L.F.P.; Vieira, L.G.E. Plant regeneration of sweet orange (*Citrus sinensis*) from thin section of mature stem segments. *Plant Cell Tiss. Organ Cult.* **2003**, *74*, 99–102. [CrossRef]
94. Chand, L.; Sharma, S.; Kajla, S. Effect of rootstock and age of seedling on success of in vitro shoot tip grafting in Suárez. *Indian J. Hort.* **2016**, *73*, 8–12. [CrossRef]
95. Benson, E.E. In vitro plant recalcitrance in vitro plant recalcitrance: An introduction. *In Vitro Cell. Dev. Biol.-Plant* **2000**, *36*, 141–148. [CrossRef]

96. Naz, A.A.; Jaskani, M.J.; Abbas, H.; Qasim, M. In vitro studies on micrografting technique in two cultivars of citrus to produce virus free plants. *Pak. J. Bot.* **2007**, *39*, 1773–1778.

97. Singh, A.D.; Meetei, N.T.; Kundu, S.; Salma, U.; Mandal, N. In vitro micrografting using three diverse indigenous rootstocks for the production of *Citrus tristeza virus*-free plants of Khasi mandarin. *In Vitro Cell. Dev. Biol.-Plant* **2019**, *55*, 180–189. [CrossRef]

98. Conejero, A.; Romero, C.; Cunill, M.; Mestre, M.A.; Martínez-Calvo, J.; Badenes, M.L.; Llácer, G. In vitro shoot-tip grafting for safe *Prunus* budwood exchange. *Sci. Hortic.* **2013**, *150*, 265–370. [CrossRef]

99. Revilla, M.A.; Pacheco, J.; Casares, A.; Rodríguez, R. In vitro reinvigoration of mature olive trees (*Olea europaea* L.) through micrografting. *In Vitro Cell. Dev. Biol. Plant* **1996**, *32*, 257–261. [CrossRef]

100. Gao, X.; Yang, D.; Cao, D.; Ao, M.; Sui, X.; Wang, Q.; Kimatu, J.N.; Wang, L. In vitro micropropagation of *Freesia hybrid* and the assessment of genetic and epigenetic stability in regenerated plantlets. *J. Plant Growth Regul.* **2010**, *29*, 257–267. [CrossRef]

101. Sales, E.K.; Butardo, N.G. Molecular analysis of somaclonal variation in tissue culture derived bananas using MSAP and SSR markers. *Int. J. Biol. Vet. Agric. Food Eng.* **2014**, *8*, 572–579. [CrossRef]

102. Krishna, H.; Alizadeh, M.; Singh, D.; Singh, U.; Chauhan, N.; Eftekhari, M.; Sadah, R.K. Somaclonal variation and their applications in horticultural crops improvement. *3 Biotech* **2016**, *6*, 54. [CrossRef]

103. Souza, J.A.; Bettoni, J.C.; Dalla Costa, M.; Baldissera, T.C.; dos Passos, J.M.; Primieri, S. In vitro rooting and acclimatization of 'Marubakaido' apple rootstock using indole-3-acetic acid from rhizobacteria. *Commun. Plant Sci.* **2022**, *12*, 16–23. [CrossRef]

104. Amghar, I.; Ibriz, M.; Ibrahimi, M.; Boudra, A.; Gaboun, F.; Meziani, R.; Iraqi, D.; Mazri, M.A.; Diria, G.; Abdelwahd, R. In vitro root induction from Argan (*Argania spinosa* (L.) Skeels) adventitious shoots: Influence of ammonium nitrate, auxins, silver nitrate and putrescine, and evaluation of plantlet acclimatization. *Plants* **2021**, *10*, 1062. [CrossRef]

105. Hussain, G.; Wani, M.S.; Mir, M.A.; Rather, Z.A.; Bhat, K.M. Micrografting for fruit crop improvement. *Afr. J. Biotechnol.* **2014**, *13*, 2474–2483. [CrossRef]

106. Vidoy-Mercado, I.; Narváez, I.; Palomo-Ríos, E.; Litz, R.E.; Barceló-Muñoz, A.; Pliego-Alfaro, F. Reinvigoration/rejuvenation induced through micrografting of tree species: Signaling through graft union. *Plants* **2021**, *10*, 1197. [CrossRef] [PubMed]

107. Dolcet-Sanjuan, R.; Claveria, E.; Gruselle, R.; Meier-Dinkel, A.; Jay-Allemand, C.; Gaspar, T. Practical factors controlling in vitro adventitious root formation from walnut shoot microcuttings. *J. Am. Soc. Hort. Sci.* **2004**, *129*, 198–203. [CrossRef]

108. Hackett, W.P. Juvenility, maturation, and rejuvenation in woody plants. *Hortic. Rev.* **1985**, *7*, 109–155. [CrossRef]

109. Chakraborty, T.; Chaitanya, K.V.; Akhtar, N. Analysis of regeneration protocols for micropropagation of *Pterocarpus santalinus*. *Plant Biotechnol. Rep.* **2022**, *16*, 1–15. [CrossRef]

110. Abousalim, A.; Mantell, S.H. Micrografting of pistachio (*Pistacia vera* L. cv. Mateur). *Plant Cell Tiss. Organ Cult.* **1992**, *29*, 231–234. [CrossRef]

111. Huang, L.-C.; Hsiao, C.-K.; Lee, S.-H.; Huang, B.-L.; Murashige, T. Restoration of vigor and rooting competence in stem tissues of mature citrus by repeated grafting of their shoot apices onto freshly germinated seedlings in vitro. *In Vitro Cell. Dev. Biol. Plant* **1992**, *28*, 30–32. [CrossRef]

112. Raharjo, S.H.T.; Witjaksono, N.F.N.; Gomez-Lim, M.A.; Padilla, G.; Litz, R.E. Recovery of avocado (*Persea americana* Mill.) plants transformed with the antifungal plant defensin gene PDF1.2. *In Vitro Cell. Dev. Biol.-Plant* **2008**, *44*, 254–262. [CrossRef]

113. Duclercq, J.; Sangwan-Norreel, B.; Catterou, M.; Sangwan, R.S. De novo shoot organogenesis: From art to science. *Trends Plant Sci.* **2011**, *16*, 597–606. [CrossRef]

114. Tiwari, V.; Chaturvedi, A.K.; Mishra, A.; Jha, B. An efficient method of *Agrobacterium*-mediated genetic transformation and regeneration in local Indian cultivar of groundnut (*Arachis hypogaea*) using grafting. *Appl. Biochem. Biotechnol.* **2015**, *175*, 436–453. [CrossRef]

115. Liu, Y.; Yang, X.; Zhao, Y.; Yang, Y.; Liu, Z. An effective method for *Agrobacterium tumefaciens*-mediated transformation of *Jatropha curcas* L. using cotyledon explants. *Bioengineered* **2020**, *11*, 1146–1158. [CrossRef]

116. De Pasquale, F.; Giuffrida, S.; Carimi, F. Minigrafting of shoots, roots, inverted roots, and somatic embryos for rescue of in vitro *Citrus* regenerants. *J. Am. Soc. Hort. Sci.* **1999**, *124*, 152–157. [CrossRef]

117. Cardoso, J.C.; Curtolo, M.; Latado, R.R.; Martinelli, A.P. Somatic embryogenesis of a seedless sweet orange (*Citrus sinensis* (L.) Osbeck). *In Vitro Cell. Dev. Biol.-Plant* **2017**, *53*, 619–623. [CrossRef]

118. Almeida, W.A.B.; Mourão Filho, F.A.A.; Pino, L.E.; Boscariol, R.L.; Rodriguez, A.P.M.; Mendes, B.M.J. Genetic transformation and plant recovery from mature tissues of *Citrus sinensis* L. Osbeck. *Plant Sci.* **2003**, *164*, 203–211. [CrossRef]

119. Balázs, E.; Bukovinszki, Á.; Csányi, M.; Csilléry, G.; Divéki, Z.; Nagy, I.; Mitykó, J.; Salánki, K.; Mihálka, V. Evaluation of a wide range of pepper genotypes for regeneration and transformation with an *Agrobacterium tumefaciens* shooter strain. *S. Afr. J. Bot.* **2008**, *74*, 720–725. [CrossRef]

120. Das, A.; Kumar, S.; Nandeesha, P.; Yadav, I.S.; Saini, J.; Chaturvedi, S.K.; Datta, S. An efficient in vitro regeneration system of fieldpea (*Pisum sativum* L.) via shoot organogenesis. *J. Plant Biochem. Biotechnol.* **2014**, *23*, 184–189. [CrossRef]

121. Roy, P.K.; Lodha, M.L.; Mehta, S.L. In vitro regeneration from internodal explants and somaclonal variation in chickpea (*Cicer arietinum* L). *J. Plant Biochem. Biotechnol.* **2001**, *10*, 107–112. [CrossRef]

122. Nakajima, I.; Ito, A.; Moriya, S.; Saito, T.; Moriguchi, T.; Yamamoto, T. Adventitious shoot regeneration in cotyledons from Japanese pear and related species. *In Vitro Cell. Dev. Biol.-Plant* **2012**, *48*, 396–402. [CrossRef]

123. Aguilar, M.E.; Villalobos, V.M.; Vasquez, N. Production of cocoa plants (*Theobroma cacao* L.) via micrografting of somatic embryos. *In Vitro Cell. Dev. Biol.-Plant* **1992**, *28*, 15–19. [CrossRef]

124. Palomo-Ríos, E.; Cerezo, S.; Mercado, J.A.; Pliego-Alfaro, F. Agrobacterium-mediated transformation of avocado (*Persea americana* Mill.) somatic embryos with fluorescent marker genes and optimization of transgenic plant recovery. *Plant Cell Tissue Organ Cult.* **2017**, *128*, 447–455. [CrossRef]
125. Bettoni, J.C.; Bonnart, R.; Volk, G.M. Challenges in implementing plant shoot tip cryopreservation technologies. *Plant Cell Tissue Organ Cult.* **2021**, *144*, 21–34. [CrossRef]
126. Wang, M.-R.; Lambardi, M.; Engelmann, F.; Pathirana, R.; Panis, B.; Volk, G.M.; Wang, Q.-C. Advances in cryopreservation of in vitro-derived propagules: Technologies and explant sources. *Plant Cell Tissue Organ Cult.* **2021**, *144*, 7–20. [CrossRef]
127. Panis, B. Sixty years of plant cryopreservation: From freezing hardy mulberry twigs to establishing reference crop collections for future generations. *Acta Hortic.* **2019**, *1234*, 1–7. [CrossRef]
128. Volk, G.M.; Bonnart, R.; Krueger, R.; Lee, R. Cryopreservation of citrus shoot tips using micrografting for recovery. *Cryoletters* **2012**, *33*, 418–426. [PubMed]
129. Volk, G.M.; Bonnart, R.; Shepherd, A.; Yin, Z.; Lee, R.; Polek, M.; Krueger, R. Citrus cryopreservation: Viability of diverse taxa and histological observations. *Plant Cell Tissue Organ Cult.* **2017**, *128*, 327–334. [CrossRef]
130. Yi, J.-Y.; Balaraju, K.; Baek, H.-J.; Yoon, M.-S.; Kim, H.-H.; Lee, Y.-Y. Cryopreservation of Citrus limon (L.) Burm. F Shoot Tips. Using a Droplet-vitrification Method. *Korean J. Plant Res.* **2018**, *31*, 684–694.
131. Grigoriadis, I.; Nianiou-Obeidat, I.; Tsaftaris, A.S. Shoot regeneration and micrografting of micropropagated hybrid tomatoes. *J. Hortic. Sci. Biotechnol.* **2005**, *80*, 183–186. [CrossRef]
132. Wang, R.R.; Mou, H.Q.; Gao, X.X.; Chen, L.; Li, C.; Li, M.F.; Wang, Q.C. Cryopreservation for eradication of Jujube witches' broom phytoplasma from Chinese jujube (*Ziziphus jujuba*). *Ann. Appl. Biol.* **2015**, *166*, 218–228. [CrossRef]
133. Akdemir, H.; Onay, A. Biotechnological Approaches for Conservation of the Genus *Pistacia*. In *Sustainable Development and Biodiversity-Biodiversity and Conservation of Woody Plants*; Ahuja, M.R., Mohan Jain, S., Eds.; Springer Nature: Cham, Switzerland, 2017; Volume 17, pp. 221–244.
134. O'Brien, C.; Hiti-Bandaralage, J.; Folgado, R.; Hayward, A.; Lahmeyer, S.; Folsom, J.; Mitter, N. Cryopreservation of woody crops: The avocado case. *Plants* **2021**, *10*, 934. [CrossRef]

MDPI

St. Alban-Anlage 66

4052 Basel

Switzerland

www.mdpi.com

Horticulturae Editorial Office

E-mail: horticulturae@mdpi.com

www.mdpi.com/journal/horticulturae